Growing Roses in Cold Climates

Also Published by the University of Minnesota Press

Growing Perennials in Cold Climates
  Mike Heger, Debbie Lonnee, and John Whitman

Growing Shrubs and Small Trees in Cold Climates
  Debbie Lonnee, Nancy Rose, Don Selinger, and John Whitman

Trees and Shrubs of Minnesota
  Welby R. Smith

Gardening with Prairie Plants: How to Create Beautiful Native Landscapes
  Sally Wasowski; photography by Andy Wasowski

Growing Roses in Cold Climates

Richard Hass,
Jerry Olson,
and John Whitman

Revised and Updated Edition

University of Minnesota Press
Minneapolis
London

All photographs by John Whitman with the exception of the following: 'Nastarana' by Esther Filson (page 69); 'Buff Beauty' and 'Prosperity' by Heirloom Old Garden Roses (pages 31, 36); 'Robin Hood' by Roses of Yesterday and Today, Inc. (page 36); 'Vanity' by Roses Unlimited (page 36); 'Conundrum' and 'Leading Lady' by Elena J. Williams (page 57); Oso Happy™ Candy Oh! by Proven Winners ColorChoice (page 85); 'Sigrid' by Bailey Nurseries, Inc. (page 85). Photographs of winter protection methods by Rich Hass (page 198). Photographs of John Whitman throughout the book by Donna Whitman.

Originally published in 1998 by Contemporary Books, a division of NTC/Contemporary Publishing Group, Inc.

First University of Minnesota Press edition, 2012

Published by the University of Minnesota Press
111 Third Avenue South, Suite 290
Minneapolis, MN 55401-2520
http://www.upress.umn.edu

Library of Congress Cataloging-in-Publication Data

Hass, Richard, author.
    Growing roses in cold climates / Richard Hass, Jerry Olson, and John Whitman.—Revised and updated edition.
    Includes bibliographical references and index.
    ISBN 978-0-8166-7593-7 (pb : alk. paper)
    1. Rose culture—Snowbelt states.   2. Rose culture—Canada.   3. Roses—Snowbelt states.   4. Roses—Canada.   I. Olson, Jerry, author.
    II. Whitman, John, author.   III. Title.
    SB411.H245 2012
    635.9'33734—dc23

                                                                                        2011042435

Printed in China on acid-free paper

The University of Minnesota is an equal-opportunity educator and employer.

19 18 17 16 15 14 13 12   10 9 8 7 6 5 4 3 2 1

CONTENTS

PREFACE

The three books in the Cold Climate series represent three decades of work. *Growing Roses in Cold Climates* is the third book in the series and has been written with the hope that it encourages gardeners to add roses to their gardening palette. Millions of gardeners in the United States and Canada grow them. However, in recent years many negative comments have been made about roses. Here are the most common:

1) Roses are impossible or extremely difficult to grow in cold climates.
2) You cannot grow roses without using an arsenal of chemicals.
3) Roses are expensive.
4) Taking care of roses is just too tough and time-consuming.
5) Roses have prickly stems and are hard to work with.

Some roses are difficult to grow. Others are just as easy as many perennials and shrubs. Some roses do require spraying while others flourish with benign neglect. Specific roses can be expensive while others are less costly than many perennials and most shrubs and small trees. All plants require care and a reasonable amount of effort and time to flourish. Roses are no exception. Yes, most roses have prickly stems. A pair of goatskin leather gloves and a long-sleeved shirt work wonders for the following rewards:

1) Flowers that come in an almost infinite variety of colors, shapes, sizes, and scents.
2) Continuous or repeat bloom throughout the entire growing season (on most roses).
3) Gorgeous cut flowers for the home and as gifts for others.
4) Colorful and edible hips attractive to wildlife, lovely in the fall landscape, and ideal for wreaths.
5) Plants for beds, borders, containers, hedges, naturalized plantings, pillars, and trellises.

By reading this book you will learn exactly which roses to choose to match your personality and landscape. If you are growing roses for the first time, it will help you overcome a fear of failure. Many home gardeners have expressed this fear to me over the years. The only way to overcome this is by growing roses and having success doing it. As with anything, you need reliable advice on what and what not to do. This growing guide will build your confidence and lead to success with these gorgeous plants. Success breeds passion. Rose growers are among the most passionate gardeners. This includes many organic gardeners.

Combine *Growing Roses in Cold Climates* with use of the Internet. The latter is particularly helpful in providing you with photos of the hundreds of roses mentioned in this growing guide. Neither can take the place of the other, but combined they provide a powerful source of information on the world's most popular flower. To paraphrase Vincent Van Gogh's comment on painting, if you hear a voice inside telling you not to grow roses, then by all means grow them . . . and that voice will be silenced.

—John Whitman

ACKNOWLEDGMENTS

Thanks to these wonderful people: Rob Amell, Jodi Molnau Anderson, Julia Anderson, Kim Bartko, Monica Baziuk, Jim Beardsley, Jim Blume, Norma Booty, Lloyd Brace, Lyle Brandt, Tom Carruth, Nancy Caskey, Louise Clements, George Cleveland, Rachelle Cordova, Cy DeCosse, Brian Donahue, Dana Draxten, Lloyd Duerr, Jack Falker, Ron Ferguson, Esther Filson, Debbie Frey, Kristen Gilbertson, Diane Hass, Lori Horman, Bill Ihle, Bill Jones, Pamela Juárez, Anne Knudsen, Betsy Kulak, Patty Leasure, Debbie Lonnee, Clair Martin, Marcia and Joe Massee, Ryan McGrath, Diana McLaughlin, Michelle Meyer, Bob Mugass, Albert I. Nelson, Brad Nichols, John Nolan, Daniel Ochsner, Deborah Orenstein, Todd Orjala, Bob Osborne, Ron Paschina, Bill Patterson, Ted Pew, Ken Pierskalla, Craig Reggelbruigge, Marcia Richards, Paulette Rickard, Laurie Robinson, Clay and Helen Rohrer, George Ross, Don Selinger, Joe Shaven, Leon Snyder, Norton Stillman, Allen Summers, Rob Taylor, Kristian Tvedten, Regina Wells, Donna Whitman, Patricia Wiley, Kathy Wilhoite, Elena Williams, Rob Wilson, David Zlesak, and Kathy Zuzek.

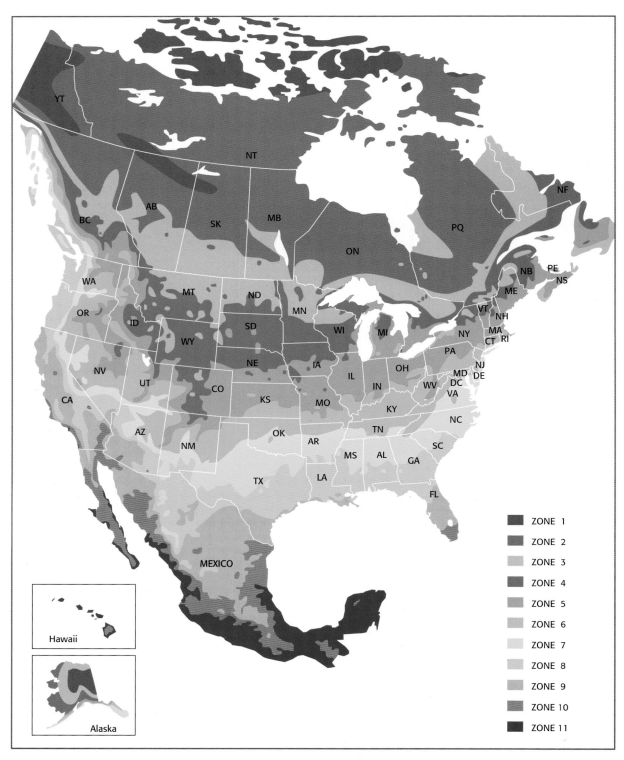

ZONE 1
ZONE 2
ZONE 3
ZONE 4
ZONE 5
ZONE 6
ZONE 7
ZONE 8
ZONE 9
ZONE 10
ZONE 11

Hawaii

Alaska

Growing Roses in Cold Climates is aimed primarily at gardeners living in Zones 1–5. It will also be helpful to gardeners in portions of Zone 6. A cold climate is characterized by temperatures that can dip below −20°F (−29°C), as well as by snow during the winter.

INTRODUCTION

If you live in a region in which winter temperatures can dip below −20°F (−29°C), then—no matter how hot your summers are—you live in what rose growers consider to be a cold climate. Living in a colder climate means that the roses you choose require special attention to make them grow well. Some roses are known to be hardy; they stand up to wintry chills and drying winds without protection. Others, known as tender roses, need winter protection in order to survive and thrive. *Growing Roses in Cold Climates* tells you exactly how to grow both types of roses.

The information in *Growing Roses in Cold Climates* is the result of more than seventy years of hands-on, down-to-earth experience in rose growing. Along the way we have been helped, encouraged, and trained by some of the finest rose growers, and we owe a great deal to these mentors. The result of this professional guidance and personal experience is a system for growing beautiful roses under even the most adverse conditions. The system is a blueprint for getting roses to grow prolifically year after year. It shows you how to create just the right environment in which roses will flourish.

Our goal in this updated and revised edition is to provide detailed growing information on the most popular roses proven to thrive in colder climates. The book covers specific rose groups and a number of roses within each of these groups that have proven to do well in cold climates. We have removed many of the plants from the first edition to make room for the finest new introductions. We have also expanded the number of plants from 700 to 875 roses. The original five-star rating system is still in place. We awarded the top rating of five stars to fewer than 50 roses, to identify those with which we have experienced the best success. The enormous variety of roses listed gives every rose grower, from the novice to the most skilled, an incredible range of colors, forms, textures, and scents from which to work.

As you look through *Growing Roses in Cold Climates,* you will quickly discover that the information on each rose is unlike the quick sketches so typical of other rose books. You will find all 875 roses listed in the index. This makes it easy to find which group a rose belongs in and how to take care of it. Once you turn to the section on the rose you choose, you will find the essential growing information for that rose concentrated in one place. The individual sections are deliberately repetitive to save you time. Unlike other rose books, there is far less need to flip back and forth to find succinct, plant-specific information on wide-ranging topics applicable to that rose group.

You will also find as you read that this book is written in simple, everyday language so that anyone can understand even the more sophisticated aspects of growing roses. Some technical terms are included, generally in parenthesis, but our goal is to help you grow great roses, no matter what your level of knowledge or expertise. The book is as honest as we could make it. We do not gloss over problems, nor do we exaggerate them. In Part I, the lists of varieties in each rose group are laid out in clear, concise, and easy-to-use charts. Part II, "The Basics of Growing Roses," provides everything you need to know to buy, grow, and nurture these lovely, woody plants. In it, you will discover a number of growing tips and secrets. The newest tips are particularly helpful coming from recent experimentation, personal observation, and studies. We continue to emphasize the system on which the original edition was based because it works so well. We hope that even expert rose growers will find some fresh ideas here.

Growing Roses in Cold Climates is the book we would have liked to have had at the beginning of our gardening careers. It would have saved us so much wasted money, time, and energy—not to mention disappointment—as we sadly watched lovely roses we had carefully selected and nurtured die off before their prime. In its depth and presentation of hard-to-find information, we believe this book was and continues to be one of a kind. It is our hope that it helps you grow beautiful roses under the most adverse conditions.

We have worked hard to make this the very best book possible on cold-climate rose growing. This updated and completely revised edition includes many new and exciting rose plants. It also provides updated source lists to help you find all of the roses listed. However, we realize that there is always something to be learned. If you have any tips or gardening secrets you would like to share with us, or if you would like to criticize, correct, or add information, please write to us at P.O. Box 212, Long Lake, MN 55356. Please include a self-addressed, stamped envelope if you would like a reply.

—Richard Hass, Jerry Olson, and John Whitman

Nicole®/'Koricole'

PART I

The Roses

The following sections give you detailed information on how to grow different types of roses. In each listing you'll find detailed instructions on exactly how to grow that type of rose. Every section follows an identical format, so that you will get into a rhythm in using the information. Read only what's important to you. Each section describes everything from the most simple to more technical aspects of rose growing. Match the information to your needs and preferred method of growing. While the listings follow the same format, you'll encounter subtle differences under various headings that can make a big difference in your success with that particular type of rose.

CHAPTER I
INDIVIDUAL LISTINGS

Whether you are just beginning to grow roses or have been doing so for years, each section in this chapter will guide you in simple, clear, concise language to your dream: stunning plants having bountiful bloom, with a minimal amount of wasted effort, money, and time. For most of us, the ultimate purpose of rose growing is a sense of peace, serenity, and joy. Rose growing is a process, and to many, a passion. Having good information is just one step in the right direction in this wonderful journey.

Varieties

You will find more than 875 varieties of roses listed in Part I. We have rated each rose using a star system, with five stars being the highest rating possible. This is different from rating systems used by rose societies, but we believe it is more accurate for cold climates, since other ratings are based on results from test gardens throughout the United States. Still, we strongly urge you to visit local rose gardens. Seeing actual plants gives you the most accurate picture of what a plant will look like under ideal growing conditions, particularly plants in your area. We also hope you will join a rose society. Members often invite each other

to their homes. These private gardens can be as useful as, or even more useful than, public ones in selecting roses. Seeing a rosebush also gives you a chance to sniff the blossoms for a better idea of the plant's true fragrance, which no one can accurately describe in any growing guide. We try to do our best, but the task is difficult. The tables use both a strength range—None, Very slight, Slight, Moderate, Strong, Very Strong—and descriptions of scent.

Sources

Each section also includes an alphabetized list of mail-order sources for bare root plants. Keep the following tips in mind regarding these sources:

- Most mail-order companies now have Web sites that list the roses they sell. Just enter the company's name to get to the site. Many companies sell plants directly to you from these sites.
- Readers often post comments on rose sources online. These are worth looking at, especially if they show a pattern over a period of time.
- If you do not have access to the internet or prefer to deal directly with a person, then you can mail a letter or call the company to speak to someone. You can also ask whether the company prints a catalog.

- Catalogs are less common as costs of printing have risen dramatically. If catalogs are available, there may be a charge. The cost is often applied toward your first purchase.
- Each source listed has a mailing address and telephone number. When contacting a company through the mail, always include a stamped, self-addressed envelope for a reply. If you are requesting a catalog, the company will let you know if there is a charge or will simply send you one for free.
- If you order plants by phone, get the name of the person you talk to and keep it on file.
- If you order in writing, note the date you mail the letter. If plants do not arrive as requested, you may have to follow up with a call. Many catalogs suggest accepting "substitutions" for plants which are sold out. Whether you are willing to do this is a matter of personal choice. You can avoid this problem by ordering very early.
- Since you are growing plants in a cold climate, specify the date you would like to receive your shipment. Stress this. Plants mailed in extreme cold often die; they also require extra care to keep alive until planting time.
- Roses are sold growing on their own roots or budded or grafted onto rootstock. Roses growing on their own roots are preferred because they are generally hardier in cold climates and produce only desirable canes. They do tend to be smaller plants, sometimes less vigorous, and more expensive. Many roses are budded or grafted, either because they are difficult to root from cuttings or grow poorly on their own roots. These tend to be larger and less expensive, but they can produce undesirable canes from the rootstock. When ordering roses through the mail, make sure that the rootstock is hardy. Do not buy plants with *Rosa fortuniana* rootstock.
- Also, ask about the age and size of the plant. Budded (not own root) plants are artificially graded according to size and number of canes: 1 (larger, more canes), 1½ (not so large, fewer canes), 2 (lowest rating other than a cull—the equivalent to a discard). In grading plants, good companies follow the recommendations of the American Association of Nurserymen. Some classes of roses, such as Miniatures, aren't graded at all.
- Ask what kind of guarantee the company has for its plants.
- When you receive plants in the mail, check to see that you got what you ordered.
- Check the plants immediately to make sure that they are not dried out, damaged, or dead. A dead cane has a gray look and will crack if bent; also, if you cut into it, it will be brown rather than light white or green.
- Plant all roses as soon as possible, following the exact planting directions provided.
- Keep records of what you've ordered, and label all plants. If a plant turns out to be an "imposter," let the company know and get your money back. Few companies deliberately mail the wrong plants, but it does happen that plants get mixed up by mistake.
- In a few instances, there are legal restrictions that prohibit the sale of plants between certain states or foreign countries. This is to stop the spread of diseases and insect infestations. Also, when ordering plants from a foreign country, there may be customs inspections that delay shipment so severely as to cause the death of plants. Do not blame the commercial growers listed in this book for adhering strictly to the law or for shipments held up for reasons beyond their control. The inconvenience and cost associated with these problems may cause some companies to refuse to sell plants outside of a specific geographical region. Most mail-order sources are clear about this in their catalogs and online sites.
- A number of companies sell plants directly to rose societies. By joining a local chapter, you will often have access to roses that would be unavailable otherwise.
- Finally, the following publication lists thousands of roses with exact sources for the ardent rose grower: *Combined Rose List,* by Beverly R. Dobson & Peter Schneider, P.O. Box 677, Mantua, OH 44255. Sources for roses are also listed at www.plantinfo .umn.edu. You can see photos of specific roses at www.HelpMeFind.com/Roses. The American Rose

Society has a website for members only with photos and descriptions of specific roses. Get information online at ars.org, by mail at P.O. Box 30,000, Shreveport, LA 71130, by phone (800) 637-6534, or by email at ars@ars-hq.org.

Using the Growing Guide

Please read the chapters in Part II if you are not an experienced rose grower before delving into individual sections of Part I. The chapters in Part II contain specific tips and procedures not found in the rest of the book.

While we have tried to keep the language simple and easy to understand throughout this book, you may run into unfamiliar words. Read the glossary in advance to understand their meaning. The glossary also contains information related to these terms that will help you be successful in growing roses in cold climates.

The Five-Star System

We have awarded stars to each rose. The highest possible rating is five stars (*****). The process of awarding these stars involved many factors, from vigor to number of blooms per season. Hardiness was also a consideration in awarding stars, especially for roses left standing in the winter. While this star system does not match those of rose societies, it does take into account the difference between performance in cold versus warm climates. It is subjective, but based on decades of growing and judging roses. A rose with a low rating can still be a lovely plant and well worth buying if grown properly using the system outlined in this book. With the tens of thousands of roses available on the market, we will have invariably left out some fine roses. We are constantly testing new ones and will add the best of these in future editions. If you believe we have missed a superstar tested for three or more years in your garden, we would appreciate hearing from you. Write to Best Books, Inc., P.O. Box 212, Long Lake, MN 55356 and include a self-addressed, stamped envelope if you would like a reply.

'John Cabot'

CLIMBING ROSES

Climbing roses are those with long canes that are ideal for vertical display. Stems continue to grow longer each year if properly protected. Climbers have stiffer stems and larger flowers than Ramblers, which are not hardy in cold climates but grow profusely in the South. Climbing roses are among the most beautiful and quick-growing. Blooms are prolific and often repeat if spent blossoms are removed; in some cases, bloom is constant. Climbers can be long-lived. Unfortunately, only a few are hardy north of the Mason-Dixon line, and only those with good winter protection. For that reason we suggest using specific Shrub roses that form long canes in their place. The Shrub roses we prefer as replacements for Climbers are listed under the heading "Varieties" later in this section. Actual descriptions of each Shrub rose are in the section titled "Shrub Roses" (pp. 92–120).

How Climbing Roses Grow

Climbers are generally grown on budded stock, meaning that buds or scions from one rose are budded or grafted onto rootstock of a different rose. Climbers grow as much as 7 to 8 feet (more than 2 meters) in a season if given support. Climbers have no tendrils, so they must be tied to the support. They are ideal for arbors, fences, trellises, and pergolas. They bloom best if canes run horizontally, which causes more side branches or laterals to appear. These side branches are loaded with blossoms. Normally, expect good bloom only after two full seasons of growth. It takes time for these plants to form an extensive root system and really establish themselves.

The Shrub roses, suggested as replacements for true Climbers, tend to be more vigorous and are often grown on their own roots. Otherwise, they are quite similar.

Where to Plant

Site and Light Climbers thrive in full sun. They need support to grow properly. Place them on the sunniest side of the support. Keep the plant completely on one side of the support. Do not interweave the cane with the support as you would with many climbing perennials. Use appropriate ties as outlined under "Staking" later in the section.

Soil and Moisture Climbing roses need rich soil that retains moisture but drains freely. Replace clay or rock with loam purchased from a garden center or in bags as potting soil. Add lots of organic matter, such as compost, leaf mold, peat moss, or rotted manures. These keep the soil moist and cool during dry periods. They also help the soil drain freely, encourage the growth of beneficial soil microorganisms, and attract worms which keep the soil aerated and fertilized. Soil should be light, loose, and airy. This allows quick root growth and good drainage so that plenty of oxygen can get to the tender feeder roots. A handful of bonemeal or superphosphate mixed into the soil provides additional nutrients for the young plant. Proper soil preparation takes a little time, but it is critical to good growth.

Spacing Climbers need good air circulation. Keep them at least 24 inches (60 cm) away from the foundation of a house. Poor air circulation and high heat near a foundation can cause spider mite problems. Furthermore, cement foundations may leach harmful chemicals into the soil, may make soil more alkaline (raise the pH), and may draw out helpful moisture. If you're planting a Climber near a support such as a post or split-rail fence, the stem should be approximately 12 inches (30 cm) from the post itself. This makes burying the plant for winter protection using the Minnesota Tip Method (explained in Part II) easier.

Planting

Bare Root Plant bare root roses as soon as the ground can be worked in spring. Follow the explicit directions outlined in Part II.

The bud union should be placed about one-half below and one-half above the soil.

Potted Plants You'll often find several Climbing roses in local garden centers or nurseries. Pick out plants with three or more healthy canes and luxuriant foliage. Check plants carefully for any signs of disease or insect infestations to avoid buying infected plants. Avoid leaving the plant in the car while you do other errands, since the heat buildup can damage it. Also, if the plant will be exposed to wind, protect it well by wrapping it in thick plastic or paper. As soon as you get home, remove the protective covering, and water the soil if it's at all dry. Plant the rose immediately as outlined in Part II. Do not assume that budded roses have been properly planted in a pot. Make sure that the bud union is placed at the correct depth in the planting hole.

How to Care for Climbing Roses

Water Keep the soil evenly moist throughout the growing season. Deep watering is much more effective than frequent sprinkling. Soak the ground thoroughly once, then soak it again so that the water is absorbed to a depth of 18 inches (45 cm). The simplest way to do this is to let a hose run at the base of the plant for 10 minutes or longer. During heat waves, watering with a sprinkler is highly recommended to discourage spider mites. You may have to leave a sprinkler running for several hours to saturate the soil sufficiently.

Mulch After the soil warms up to at least 60°F (15.6°C), apply mulch around the base of the plant. The mulch should come close to but not touch the canes. The most commonly used mulch is shredded, not whole, leaves. Place at least a 3-inch (7.5-cm) layer over the soil. If you have enough leaves to make a thicker mulch, so much the better. Leaves keep the soil moist and cool which encourages rapid root growth. The mulch feeds microorganisms and worms which enrich and aerate the soil. The mulch also inhibits weed growth and makes it much easier to pull any weeds that do sprout. Other good mulches include pine needles and grass clippings. If the latter are fresh, use only 2 inches (5 cm) to avoid heat buildup and odor. These mulches are all inexpensive and effective. If you use chipped wood or shredded bark, apply additional fertilizer to the soil, since these rob the soil of nitrogen as they decompose. Since mulch is eaten by soil microorganisms and worms, it needs to be replaced regularly throughout the growing season. Remove and compost all mulch in the fall.

Fertilizing If combining the use of inorganic and organic fertilizers, be careful during the first year. Use inorganic fertilizer only after the plant has leafed out and is growing vigorously.

In subsequent years, as early as possible in spring, sprinkle 10-10-10 (or 5-10-5) granular fertilizer around the base of each plant. Use common sense in judging the amount, more for bigger plants, less for smaller. Hand sprinkle, never placing fertilizer against the stem of the plant. Water immediately to dissolve the granules and carry nutrients to the roots.

In both the first and subsequent years do the following: One to two weeks after the first feeding with 10-10-10 granules soak the base of each plant with a liquid containing 20-20-20 water soluble fertilizer following directions on the label. Use common sense, giving larger plants more, smaller plants less. This feeding will stimulate sensational bloom as the plant matures.

One week later soak the base of each plant with a liquid containing fish emulsion. Follow the directions on the label. Feed according to the size of the plant.

One week later give all plants the 20-20-20 treatment again. One week later, give all plants the fish emulsion soaking. Keep this rotation going until the middle of August. At this time, stop all use of fertilizers containing nitrogen, since nitrogen stimulates new growth which dies off in most winters. Use 0-10-10 fertilizer after mid-August if desired.

Additional fertilizing with Epsom salts (magne-sium sulfate) is highly recommended. Give ⅓ cup (about 75 g) for larger plants, less for smaller, two times a year: in late May and in early July. Magnesium sulfate neutralizes the soil and makes it easier for the plant to take in nutrients. This stimulates the growth of new canes, or basal breaks, from the base of the plant.

Good fertilizers for organic growers are alfalfa meal (rabbit pellets), blood meal, bonemeal, compost, cow manure, fish emulsion, fish meal, rotted horse manure, and Milorganite. Add bonemeal to the soil before planting. The others are effective added to the soil at planting time and as additional feedings on the surface of the soil throughout the season.

Weeding Rose gardens should be weed free. If you have a large rose garden, an herbicide such as Roundup® is most commonly recommended for bed preparation. It is absorbed by the weed to kill it completely. These systemic herbicides are often the only practical way to kill perennial weeds. Ideally, prepare the bed or planting hole in late summer or early fall. This way, all herbicide residue will have broken down completely by planting time the following spring. Use mulch to keep annual weeds in check. Pull up by hand any that sprout through the mulch. They'll pop up easily through this moist material.

Staking Climbing roses do not have any way of attaching themselves to support. You have to do the work for them.

Attach the canes to the support with ties in a loose, figure-eight knot. Green ties are less noticeable than other colors. Check the ties every few weeks to loosen them if necessary as the cane expands.

Keep all cane on one side of the support. Never intertwine it with the support as you would some annual and perennial vines. This will make it much easier to take canes down for winter protection without crimping or breaking them.

Climbers bloom more profusely if canes run horizontally along a support.

If you lay one cane along the ground, shoots will often spring up at intervals.

Pegging To increase bloom on repeat-blooming Climbers, bend long, arching canes over until the tip touches the ground. Do this without crimping the base of the cane. Attach the cane to the ground with metal hooks or pieces of wood. This technique induces the plant to send out many branches (laterals) and side branches (sublaterals) from the bent cane. In this way, the plant is tricked into producing much more abundant bloom than if it were left to grow naturally on its own.

Disbudding Some growers remove flower buds in the first year. The idea is to direct all energy to the growth of canes and roots. Remove the buds as soon as they appear, until the end of the season.

Do not remove flower buds on mature Climbers, since the object is to get as much bloom as possible.

Deadheading Remove faded blossoms to encourage additional bloom throughout the season. The proper way to do this is covered under "Pruning."

Pruning During the first year, pruning may cause Climbers to revert to bush form. The only recommended pruning is the removal of dead, diseased, or broken canes.

In subsequent years, prune after the last frost when buds begin to swell but before any leaves are on the plant. Cut off any cane that dies back after the winter. Remove as little growth as possible.

How you prune is important to prevent decay at the end of the cane. Make all cuts at a 45-degree angle, about ¼ inch (6 mm) above a bud on healthy canes. The angle stops water from collecting on the end of the cane, which can induce decay. If you cut too close to the bud, you may kill it. If you cut too far away, the entire tip of the cane may die back. Generally, make your cut just above an outward-facing bud for better plant shape.

Use sharp pruning shears. Place the sharpest part of the blade next to the part of the plant you want to save. This gives you a clean cut, which prevents the tip of the cane from dying back.

Second pruning: After the plant has finished its first bloom, cut all side branches or laterals back so that each has only two or three eyes (buds). Cut just above a dormant bud, the place where a leaf connects to the cane. Each lateral will now produce a second round of bloom from canes emerging from the dormant buds.

After each bloom, follow the same procedure. Four cycles of bloom are possible in a good season, though two or three are more likely.

After 3 to 4 years, canes may produce little or no bloom during a season. Mark these. Then cut them back to the base of the plant in the spring. Some growers remove old cane every two years, but this is a matter of choice. Cutting back or removing old cane often stimulates new cane to develop from the base of the plant (basal shoots), and this new cane will produce abundant flowers.

Never remove cane that is still producing flowers.

On a Climber that blooms only once you can cut canes back either immediately after the blooming period or the following spring just as new growth begins.

Never prune in late summer or early fall. This causes new growth, which is likely to die back during the winter. The dead cane often gets infected with disease or invites insect attack.

Winter Protection Winter protection varies with each variety, from next to none to the Minnesota Tip Method. The descriptions of recommended varieties in the table at the end of the section tell you what method to use. Using the Minnesota Tip Method is hard work but often necessary. The method is covered in detail in Part II.

Following are some special tips that apply to Climbers:

Do not remove ties holding up a Climber until the trench for burying the plant has been dug and the plant thoroughly loosened at the base with a spading fork.

If you followed the directions under "Staking," the plant will be growing on only one side of the support. Untie all but the top ties first. When you untie the top one, the entire cane will be loose and begin to flop over.

Once all the canes have been untied, tie them together in a number of places with polyester twine. Pull them together gently, trying not to break any canes. If you do break a cane, snip off the broken portion with pruning shears.

When pushing the plant over by using the spading fork as a wedge underneath the roots, *avoid crimping the cane, especially at the base of the plant.* Make sure to loosen the soil around the base of the plant so that the canes are nearly horizontal to the ground. This is the only way to avoid undue stress at the base.

When you raise the plants in spring, tie them immediately to their support as previously outlined in this section. Always keep cane on one side of the support.

Problems

Insects Climbers are vulnerable to insects. Consider a preventive spraying program. Spider mites can be a problem if a Climber is planted too close to a house where temperatures rise and where the plant rarely gets sprinkled by rain. Discourage mites by correct planting and misting the plants frequently during dry weather. If any mites appear, kill them immediately with a miticide.

Disease Climbers are more prone to disease than many other types of roses. Use a preventive spray program. The most common problems are black spot and powdery mildew. Prevention is always easier than control.

Propagation

One of the best ways to propagate Climbing roses is by air or ground layering. Refer to Chapter 7 in Part II for detailed information regarding these two techniques.

Special Uses

Cut Flowers Some of the sprays on Climbers are quite beautiful as cut flowers. Remove them from mature plants only.

Dried Flowers All roses can be used for dried flowers, but Climbers are not the ones most sought after for this purpose.

Sources

Angel Gardens, P.O. Box 1106, Alachua, FL 32616, (352) 359-1133

Antique Rose Emporium, 9300 Lueckmeyer Rd., Brenham, TX 77833, (800) 441-0002

Brushwood Nursery, 431 Hale Lane, Athens, GA 30607, (706) 548-1710

Burlington Rose Nursery, 24865 Rd. 164, Visalia, CA 93292, (559) 747-3624

Chamblee's Rose Nursery, 10926 US Hwy 69 N, Tyler, TX 75706, (800) 256-7673

Countryside Roses, 5016 Menge Ave., Pass Christian, MS 39571, (228) 452-2697

David Austin Roses Ltd., 15059 Hwy 64 W, Tyler, TX 75704, (800) 328-8893

Edmunds' Roses, 335 S High St., Randolph, WI 53956, (888) 481-7673

Fritz Creek Gardens, P.O. Box 15226, Homer, AK 99603, (907) 235-4969

Garden Valley Ranch, 498 Pepper Rd., Petaluma, CA 94952, (707) 795-0919

Greenmantle Nursery, 3010 Ettersburg Rd., Garberville, CA 95542, (707) 986-7504

Heirloom Roses, 24062 Riverside Dr. NE, Saint Paul, OR 97137, (503) 538-1576

High Country Roses, P.O. Box 148, Jensen, UT 84035, (800) 552-2082

Hortico, Inc., 723 Robson Rd., RR# 1, Waterdown, ON L0R 2H1 Canada, (905) 689-6984

Inter-State Nurseries, 1800 E Hamilton Rd., Bloomington, IL 61704, (309) 663-6797

Jackson & Perkins, 2 Floral Ave., Hodges, SC 29653, (800) 872-7673

Mary's Plant Farm & Landscaping, 2410 Lanes Mill Rd., Hamilton, OH 45013, (513) 894-0022

McKay Nursery Co., P.O. Box 185, Waterloo, WI 53594, (920) 478-2121

North Creek Farm, 24 Sebasco Rd., Phippsburg, ME 04562, (207) 389-1341

Northland Rosarium, 9405 S Williams Lane, Spokane, WA 99224, (509) 448-4968

Palatine Fruit & Roses, 2108 Four Mile Creek Rd., RR #3, Niagara-on-the-Lake, ON L0S 1J0 Canada, (905) 468-8627

Pickering Nurseries, 3043 County Rd. 2, RR #1, Port Hope, ON L1A 3V5 Canada, (905) 753-2155

Raintree Nursery, 391 Butts Rd., Morton, WA 98356, (800) 391-8892

Regan Nursery, 4268 Decoto Rd., Fremont, CA 94555, (800) 249-4680

Rogue Valley Roses, P.O. Box 116, Phoenix, OR 97535, (541) 535-1307

Rose Fire, Ltd., 09394 State Rte. 34, Edon, OH 43518, (419) 272-2787

Rosemania, 4920 Trail Ridge Dr., Franklin, TN 37067, (888) 600-9665

Roses of Yesterday and Today, 803 Brown's Valley Rd., Watsonville, CA 95076, (831) 728-1901

Roses Unlimited, 363 N Deerwood Dr., Laurens, SC 29360, (864) 682-7673

S & W Greenhouse, Inc., P.O. Box 30, 533 Tyree Springs Rd., White House, TN 37188, (615) 672-0599

Spring Valley Roses, P.O. Box 7, Spring Valley, WI 54767, (715) 778-4481

Two Sisters Roses, 1409 N Redbud Lane, Newcastle, OK 73065, (no phone by request)

Vintage Gardens (custom propagation), 4130 Gravenstein Hwy N, Sebastopol, CA 95472, (707) 829-2035

White Flower Farm, P.O. Box 50, Litchfield, CT 06759, (800) 503-9624

Witherspoon Rose Culture, 3312 Watkins Rd., Durham, NC 27707, (800) 643-0315

VARIETIES

Choose varieties by how much work you're willing to do to protect them during the winter. The hardiness ratings in the table are for the *unprotected* cane of these roses. The other qualities of individual roses will naturally be a part of your decision-making process.

In colder climates a number of growers use Shrub roses with long canes to replace true Climbers. Good Shrubs for this purpose are Bonica® (hardy to 10°F/−12.2°C), 'Captain Samuel Holland' (hardy to −15°F/−26°C), Dortmund® (hardy to −5°F/−20°C), 'Henry Kelsey' (hardy to −15°F/−26°C), 'John Cabot' (hardy to −20°F/−29°C), 'John Davis' (hardy to −20°F/−29°C), 'Louis Jolliet' (hardy to −20°F/−29°C), 'Poltsjärnen' (hardy to −20°F/−29°C), 'Quadra' aka 'J. F. Quadra' (hardy to −20°F/−29°C), Sea Foam® (hardy to +20°F/−7°C), and 'William Baffin' (hardy to −40°F/−40°C). These are all described in the Varieties table of the Shrub rose section. Note that the crowns of these plants are far hardier than the temperatures listed here which are for *unprotected* cane only.

VARIETIES	COLOR	PETAL COUNT	FRAGRANCE	BLOOM	HARDINESS
Altissimo®****	Medium red	5	None to slight	Repeat	+10−20°F

Very vigorous. Beautiful, 4-to 5-inch (10- to 12.5 cm) flowers with bright yellow stamens. Flowers appear cupped to flat and have richest color in cool weather. Faint aroma of cloves. Dark, dense foliage with purplish tones when emerging. May form orange red hips. Use the Minnesota Tip Method for winter protection.

| America™**** | Orange pink | 40−45 | Strong spicy | Repeat | +20°F |

Moderately vigorous. Stunning coral buds open into orange salmon pink flowers up to 3 to 4 inches (7.5 to 10 cm) across. Nice, spicy scent. Mid to deep green, semi-glossy foliage. Use the Minnesota Tip Method for winter protection.

'William Baffin'

Bonica®

'Henry Kelsey'

'Captain Samuel Holland'

VARIETIES	COLOR	PETAL COUNT	FRAGRANCE	BLOOM	HARDINESS
'Awakening'****	Light pink	26–40	Moderate to strong	Repeat	+0°F

A 1935 sport of 'New Dawn' in what was formerly Czechoslovakia. Lovely 3-inch (7.5-cm) flowers with a quartered appearance. Light green, glossy foliage. Requires some winter protection. Lay it down on ground and cover with cardboard and leaves, or make the effort to use the Minnesota Tip Method.

'Blaze'***	Medium to dark red	9–16	Slight	Intermittent	+20°F

Vigorous. Forms clusters of 3-inch (7.5 cm) flowers on leathery, mid to dark green foliage. Use the Minnesota Tip Method for winter protection.

'Don Juan'***	Dark red	35	Very strong	Repeat	+20°F

Velvety blooms up to 5 inches (12.5 cm) across. Dark green, glossy foliage—almost leathery. Prickly cane. Requires the Minnesota Tip Method for winter protection. Worth lots of extra attention because of its remarkable fragrance!

Joseph's Coat®***	Red orange yellow	23–28	Light fruity	Repeat	+20°F

Marginally vigorous in cold climates. Clusters of flowers just under 3 inches (7.5 cm) across. Interesting combination of yellow and red coloration. Dark, semi-glossy foliage. Expect no more than two good flushes of bloom. Use the Minnesota Tip Method for winter protection.

'New Dawn'*****	Light pink	35–40	Strong	Repeat	+0°F

Very vigorous. Lovely light pink blossoms with deeper colored centers just under 3 inches (7.5 cm) across. Fruity or tea-like scent. Glossy, dark green foliage. Detach the cane from its support in late fall, lay on the ground, and cover with cardboard and leaves. This is all the protection this climber usually needs. One of the best true Climbers for cold-climate growers.

'Piñata'****	Yellow orange red	28	Slight fruity	Repeat	+20°F

A true semi-climber. Stunning yellow and red buds open into flowers up to 3 inches (7.5 cm) across and tinged either orange or red (or both). Nice, glossy, green foliage. Grows slowly. Use the Minnesota Tip Method for winter protection.

Ramblin' Red™***	Medium red	26–40	Slight	Repeat	−15°F

Seedling of 'Henry Kelsey.' A climbing rose with 4-inch (10-cm) bright red flowers with yellow centers. Foliage is dark green turning to purple at the end of the season. May form orange red hips. Susceptible to black spot. Take cane off its support, lay it on the ground, and cover it with soil and a thick layer of leaves, just to be safe.

Sky's The Limit™***	Light to medium yellow	20	Slight fruity	Repeat	+15°F

A climbing rose producing nice clusters of ruffled clear yellow flowers in the first year of growth. Foliage is medium green. Reliable repeat bloom. Take cane off its support, lay it on the ground, and cover it with soil and a thick layer of leaves.

'White Dawn'****	Pure white	30–35	Moderate	Repeat	+20°F

A true climber, a sport of 'New Dawn.' Forms lovely clusters of gardenia-like flowers on glossy dark green foliage. Use the Minnesota Tip Method for winter protection.

'Hannah Gordon'/'Tabris'

FLORIBUNDA ROSES

Floribundas are the second-largest class of roses (Hybrid Teas are the largest). They are excellent landscape and bedding plants noted for long-lasting color and nearly continuous bloom. Floribundas form bouquet-like clusters of flowers, which individually are smaller, less beautiful, and less fragrant than those of Hybrid Teas. Most plants are quite bushy and relatively compact. They are often used as low hedges or in mass plantings for a brilliant display of color, since they bloom heavily from June until frost. Most with a few exceptions are susceptible to disease and will require winter protection to survive.

How Floribundas Grow

Most Floribundas are budded or grafted onto rootstock of a different plant. The cane produced from the rootstock produces inferior flowers and should be removed whenever it appears. The budded canes produce many-flowered stems, ideal for cut flowers. As flowers are removed, the plant will often send out new bloom. The plants are not truly everblooming but do offer good repeat or intermittent bloom. The rose regenerates by sending up new cane from the base of the plant.

Where to Plant

Site and Light Floribundas thrive in full sun. They need at least 6 hours of direct light per day to bloom properly. Lots of sun dries off foliage and inhibits disease. It also induces lush growth, resulting in more stems with many side branches. These in turn create magnificent, long-lasting bloom. Avoid planting Floribundas under trees, which not only shade the plants but also compete for nutrients and water.

Soil and Moisture Floribundas need rich soil that retains moisture but drains freely. Replace clay or rock with loam purchased from a garden center or in bags as potting soil. Add lots of organic matter, such as compost, leaf mold, peat moss, or rotted manures. These keep the soil moist and cool during dry periods. They help the soil drain freely, encourage the growth of beneficial soil microorganisms, and attract worms which keep the soil aerated and fertilized. Soil should be light, loose, and airy. This allows quick root growth and good drainage so that plenty of oxygen can get to the tender feeder roots. A handful of bonemeal or superphosphate mixed into the soil also provides nutrients for the young plant. Proper soil preparation takes a little time, but it is critical to good growth.

Spacing Space according to the potential size of the plant. Sizes are given in the Varieties table at the end of this section. Plants need good air circulation to prevent disease. Floribundas look best when planted in large groupings, certainly in groups of no fewer than three to five plants of the same variety.

Planting

Bare Root Plant bare root Floribundas as soon as the ground can be worked in spring. A good bare root plant will have three good canes about ½ inch (12 mm) in diameter. The planting steps are outlined in detail in Part II. The bud union should be planted one-half below and one-half above the soil.

Potted Plants Most nurseries and garden centers stock the most popular Floribundas. Look for plants with as many healthy canes as possible and luxuriant foliage. A good plant will have several fine canes and be growing vigorously. Check the plant for any signs of disease or insect infestations to avoid buying a diseased plant. Take the plant directly home. Don't let it sit in the car while you run errands. If you have to expose the plant to wind, protect it well with plastic or paper. Once home, remove the protective covering, and water the soil immediately if it's at all dry.

The actual planting method is covered in detail in Part II.

How to Care for Floribunda Roses

Water Keep the soil evenly moist throughout the growing season. Soak it to a depth of 18 inches (45 cm) at each watering. Water carries nutrients to the feeder roots and encourages lush growth and long-lasting bloom. It is almost impossible to overwater Floribundas if the soil drains freely.

Mulch After the soil warms up to at least 60°F (15.6°C), apply a mulch around the base of the plant. The mulch should come close to but not touch the cane. The most commonly used mulch is shredded, not whole, leaves. Place at least a 3-inch (7.5-cm) layer over the soil. If you have enough leaves to make a thicker mulch, so much the better. Leaves keep the soil moist and cool which encourages rapid root growth. The mulch feeds microorganisms and worms which enrich and aerate the soil. The mulch also inhibits weed growth and makes pulling any weeds that do sprout much easier. Other good mulches include pine needles and grass clippings. The latter should be only 2 inches (5 cm) deep, or they will heat the soil and may smell. These mulches are all inexpensive and effective. If you use chipped wood or shredded bark, apply additional fertilizer to the soil, since these rob the soil of nitrogen as they decompose. Because mulch is eaten by soil microorganisms and worms, it needs to be replenished regularly throughout the growing season. Remove and compost all mulch in the fall.

Fertilizing If combining the use of inorganic and organic fertilizers, be careful during the first year. Use inorganic fertilizer only after the plant has leafed out and is growing vigorously.

In subsequent years, as early as possible in spring, sprinkle 10-10-10 (or 5-10-5) granular fertilizer around the base of each plant. Use common sense in judging the amount, more for bigger plants, less for smaller. Hand sprinkle, never placing fertilizer against the stem of the plant. Water immediately to dissolve the granules and carry nutrients to the roots.

In both the first and subsequent years do the following: One to two weeks after the first feeding with 10-10-10 granules soak the base of each plant with a liquid containing 20-20-20 water soluble fertilizer following directions on the label. Use common sense, giving larger plants more, smaller plants less. This feeding will stimulate sensational bloom as the plant matures.

One week later soak the base of each plant with a liquid containing fish emulsion. Follow the directions on the label. Feed according to the size of the plant.

One week later give all plants the 20-20-20 treatment again. One week later, give all plants the fish emulsion soaking. Keep this rotation going until the middle of August. At this time, stop all use of fertilizers containing nitrogen, since nitrogen stimulates new growth which dies off in most winters. Use 0-10-10 fertilizer after mid-August if desired.

Additional fertilizing with Epsom salts (magnesium sulfate) is highly recommended. Give ⅓ cup (about 75 g) for larger plants, less for smaller, two times a year: in late May and in early July. Magnesium sulfate neutralizes the soil and makes it easier for the plant to take in nutrients. This stimulates the growth of new canes (basal breaks) from the base of the plant.

Good fertilizers for organic growers are alfalfa meal (rabbit pellets), blood meal, bonemeal, compost, cow manure, fish emulsion, rotted horse manure, and Milorganite. Add bonemeal to the soil before planting. The others are effective added to the soil at planting time and as additional feedings on the surface of the soil throughout the season.

Weeding Weeds compete with roses for water and nutrients. They also act as hosts for insects which carry disease. Before planting, kill all perennial weeds with an herbicide such as Roundup®. Use a mulch to inhibit the growth of annual weeds. Pull up by hand any weeds that sprout through the mulch. Avoid all but shallow hoeing, since hoeing could damage the root system.

Staking Floribundas tend to be compact and low growers. Some of the larger Floribundas occasionally require staking. You can avoid the need for staking by planting larger plants in a protected area where wind gusts are uncommon. If support is necessary, place stakes into the ground early in the year and attach the plant loosely with ties made of soft material. Use a figure-eight knot when tying the stem to the support.

Disbudding Some growers remove flower buds in the first year. This directs all energy to the canes and roots. Remove buds just as they appear, for best results.

You can let mature Floribundas bloom without disbudding. This creates a cluster effect of vibrant color. The center bud opens first, then the ones on the outside burst into bloom as the central bud fades.

Some growers prefer a different look and remove the larger, central bud. Snip this out with pruners just as it begins to form. The side buds will then be larger. Disbudding in this manner is a question of personal taste and nothing more. It has no effect on the health of the plant.

Deadheading Snip off spent blossoms regularly to encourage additional bloom. Many plants will repeat bloom, but none are truly everblooming. When removing the last flower in a cluster, snip the stem back with a slanting cut ¼ inch (6 mm) above a five-leaflet leaf well below the original cluster. You want to cut the stem back far enough to remove the entire upper portion of the original cluster.

Pruning Let the plant grow freely in the first year, removing only diseased or damaged cane. Like some gardeners, you may prefer to remove all flower buds during the first season until late summer. This creates a healthier, more vigorous bush. In late summer, let the plant finally bloom. The late flowering will help the plant end its cycle and begin to harden it off for the winter season.

The following spring, remove dead or diseased canes. Also remove any canes that are crisscrossing in the center of the plant. Ideally, you want canes growing outward. However, you should remove as little cane as possible.

In subsequent years, follow the same procedure. Each year, leave as much live cane alone as possible. This advice contradicts that given in most other rose books, which suggest cutting cane back by one-third the first year, by two-thirds the second year, and all the way back the third. In cold climates you need all the good cane you can get.

Do not remove the smaller canes emerging from the base of the plant. These may look spindly, but they will mature into free-flowering, healthy canes in time.

Winter Protection   Floribundas need winter protection. Without it they will often die out completely. Use the Minnesota Tip Method outlined in Part II. Most Floribundas will survive unprotected to about 26°F/−3.3°C (the mid-20s). 'Chuckles' and 'Nearly Wild' are exceptions and are hardy to at least −30°F (−34.4°C), although 'Chuckles' needs a little protection to ensure hardiness to that temperature. Simply mound soil around its base and as far up the canes as possible.

Problems

Insects   Floribundas are prone to the same insect problems as other roses. Prevent problems with a routine spraying of insecticide every 7 to 10 days throughout the season. If spider mites appear, use a miticide immediately to stop them from forming large colonies that can destroy your plants. Mites are most common in hot, dry weather.

Disease   Floribundas are quite susceptible to disease. Prevent these with a routine spraying of fungicide every 7 to 10 days. Prevention is much easier than control. Varieties with yellow to orange coloration are most susceptible to black spot. If you are an organic grower, choose other colorations.

Propagation

It is possible to propagate Floribundas by budding and grafting, but these techniques are so difficult for the average gardener that you're better off buying plants. Some specialized growers have been able to get a few varieties growing on their own roots through cuttings, but this is extremely difficult even for commercial growers. Buy your Floribundas.

Special Uses

Boutonnieres   Bella Rosa® is truly outstanding as a boutonniere. Unfortunately, it is unlikely to be in bloom in cold climates for spring proms. However, it makes a lively corsage for summer weddings.

Cut Flowers   Floribunda flower clusters are like brilliant bouquets and are ideal for cutting. Avoid taking cuttings longer than 6 inches (15 cm) on younger plants. On mature plants cut ¼ inch (6 mm) above an outward-facing bud just above a five-leaflet leaf. Keep cut stems under water and in shade while you work. Varieties vary in how long they will keep, but average 4 days. If they keep for a week, that's exceptional. One of the longest-lasting varieties is Iceberg®. When taking cuttings, leave as much foliage on the plant as possible. This is critical for the plant's health.

Dried Flowers   It is possible to dry flowers of Floribundas, but few people do. Generally speaking, the individual blooms are not as stunning or scented as those of other roses. But, that shouldn't stop you from trying if a certain rose appeals to you.

Sources

Angel Gardens, P.O. Box 1106, Alachua, FL 32616, (352) 359-1133
Antique Rose Emporium, 9300 Lueckmeyer Rd., Brenham, TX 77833, (800) 441-0002
Burlington Rose Nursery, 24865 Rd. 164, Visalia, CA 93292, (559) 747-3624
Chamblee's Rose Nursery, 10926 US Hwy 69 N, Tyler, TX 75706, (800) 256-7673

Countryside Roses, 5016 Menge Ave., Pass Christian, MS 39571, (228) 452-2697

David Austin Roses Ltd., 15059 Hwy 64 W, Tyler, TX 75704, (800) 328-8893

Edmunds' Roses, 335 S High St., Randolph, WI 53956, (888) 481-7673

Garden Valley Ranch, 498 Pepper Rd., Petaluma, CA 94952, (707) 795-0919

Goodness Grows, Inc., P.O. Box 311, 332 Elberton Rd., Lexington, GA 30648, (706) 743-5055

Greenmantle Nursery, 3010 Ettersburg Rd., Garberville, CA 95542, (707) 986-7504

Heirloom Roses, 24062 Riverside Dr. NE, Saint Paul, OR 97137, (503) 538-1576

High Country Roses, P.O. Box 148, Jensen, UT 84035, (800) 552-2082

Hortico, Inc., 723 Robson Rd., RR# 1, Waterdown, ON L0R 2H1 Canada, (905) 689-6984

Inter-State Nurseries, 1800 E Hamilton Rd., Bloomington, IL 61704, (309) 663-6797

Jackson & Perkins, 2 Floral Ave., Hodges, SC 29653, (800) 872-7673

Mary's Plant Farm & Landscaping, 2410 Lanes Mill Rd., Hamilton, OH 45013, (513) 894-0022

McKay Nursery Co., P.O. Box 185, Waterloo, WI 53594, (920) 478-2121

Northland Rosarium, 9405 S Williams Lane, Spokane, WA 99224, (509) 448-4968

Palatine Fruit & Roses, 2108 Four Mile Creek Rd., RR #3, Niagara-on-the-Lake, ON L0S 1J0 Canada, (905) 468-8627

Pickering Nurseries, 3043 County Rd. 2, RR #1, Port Hope, ON L1A 3V5 Canada, (905) 753-2155

Regan Nursery, 4268 Decoto Rd., Fremont, CA 94555, (800) 249-4680

Rogue Valley Roses, P.O. Box 116, Phoenix, OR 97535, (541) 535-1307

Rosemania, 4920 Trail Ridge Dr., Franklin, TN 37067, (888) 600-9665

Roses of Yesterday and Today, 803 Brown's Valley Rd., Watsonville, CA 95076, (831) 728-1901

Roses Unlimited, 363 N Deerwood Dr., Laurens, SC 29360, (864) 682-7673

S & W Greenhouse, Inc., P.O. Box 30, 533 Tyree Springs Rd., White House, TN 37188, (615) 672-0599

Spring Valley Roses, P.O. Box 7, Spring Valley, WI 54767, (715) 778-4481

Two Sisters Roses, 1409 N Redbud Lane, Newcastle, OK 73065, (no phone by request)

Vintage Gardens (custom propagation), 4130 Gravenstein Hwy N, Sebastopol, CA 95472, (707) 829-2035

White Flower Farm, P.O. Box 50, Litchfield, CT 06759, (800) 503-9624

Wisconsin Roses, 7939 31st Ave., Kenosha, WI, 53142, (262) 358-1298

Witherspoon Rose Culture, 3312 Watkins Rd., Durham, NC 27707, (800) 643-0315

VARIETIES

Here is a wonderful assortment of Floribundas that thrive in cold climates. Choose ones for your garden by flower color, flower form, scent, and height. Four of the hardiest are 'Chuckles,' Easy Going™, 'Eutin,' and 'Nearly Wild.' For winter protection, mound soil around their base and leaves over their crowns. Some floribundas are offered as plants grown on their own roots to improve winter hardiness. 'Chuckles,' Honey Perfume™, Mardi Gras®, and 'Tuscan Sun' are a few examples. 'Day Breaker' is rarely bothered by Japanese beetles. When colors are separated by a forward slash (/), the first color describes the upper surface of the petal and the color or colors after the (/) describe the undersides (reverse) of the petals. The slash may also be followed by information on the center or eye of the flower, which may contrast nicely to the color of the petals.

VARIETIES	COLOR	PETAL COUNT	FRAGRANCE	HEIGHT
Amber Queen®****	Apricot yellow pink	25–40	Moderate spicy	36″
'Angel Face'****	Bluish purple	25–30	Strong lemony	36″
'Apricot Nectar'****	Apricot yellow pink	40	Strong fruity	48″
Bella Rosa®****	Deep to pale pink	34	Slight	36″
Betty Boop™****	Yellowish white edged coral	6–12	Slight	48″
'Betty Prior'***	Medium pink	5	Moderate	36″
Bolivar™***	Deep pink/cream	55+	None	36″
'Bordure Rose' ('Strawberry Ice')***	Deep to light pink	22–25	Slight	30″
'Brass Band'****	Apricot orange	30–35	Slight	36″
'Brilliant Pink Iceberg'****	Deep pink/cream white	18–25	Slight	36″
Burgundy Iceberg™***	Light purple/cream	20–30	Moderate	36″
Caramel Antike® (Antique Carmel®)***	Caramel yellow	100–120	Slight	48″
'Cathedral' ('Coventry Cathedral')***	Pink apricot yellow blush	18–24	Slight	36″
'Charisma'***	Yellow orange red	40–50	Slight	30″
Charles Aznavour® (Matilda®)***	White edged pink	17–25	None	30″
'Cherish'****	Orange pink	28	Slight	36″
Chihuly®****	Red yellow	25+	Slight	24″
'Chuckles'****	Deep pink	5	Moderate fruity	24″
Cinco de Mayo™****	Rusty red orange	20–25	Slight	30″
'Circus'****	Orange yellow edged red	48–58	Light to moderate	30″
City of Belfast®***	Orange red	17–25	Slight	36″
Day Breaker™ ('Daybreaker')*****	Light peach yellow apricot	24–35	Slight	42″
'Deep Purple'***	Purplish red mauve	30–45	Moderate	36″
Dicky® (Anisley Dickson®)***	Orange pink coral	35	Slight	36″
Disneyland® Rose****	Coppery orange pink	30–35	Slight	24″
Easy Does It™****	Peachy orange apricot	25–30	Moderate	36″
Easy Going™***	Peach golden yellow	25–30	Moderate	36″

VARIETIES	COLOR	PETAL COUNT	FRAGRANCE	HEIGHT
Ebb Tide™***	Plum purple	35+	Strong	42″
'Else Poulsen'***	Medium pink	10	Slight	36″
'English Miss'***	Light pink	60	Strong	36″
Escapade®***	Pinkish mauve/white eye	12	Slight	30″
Eureka™****	Yellow apricot	25–30	Slight	30″
Europeana®*****	Dark red	25–30	Slight	48″
'Eutin'***	Medium red	17–25	Slight	36″
Evening Star®****	Creamy white	17–25	Slight	40″
'Eye Paint' (Eyepaint®)***	Scarlet/white gold eye	5	Slight	36″
'Fashion'***	White yellow edged pink	23	Moderate	36″
'Fire King'****	Orange red	48	Moderate	36″
'First Edition'***	Orange pink	28	Slight	30″
'Floradora'***	Orange red	25	Slight	36″
'French Lace'****	White pink blush	30–35	Slight	36″
'Gay Princess'***	Light pink	17–25	Moderate	36″
'Gene Boerner'****	Medium pink	35	Slight	30″
George Burns™***	Yellow red cream stripes	30–35	Moderate	24″
'Ginger'***	Orange red	28	Moderate	24″
'Gingersnap'****	Bright orange yellow	30–35	Slight	24″
'Golden Slippers'***	Orange pink/golden eye	23	Moderate	30″
'Gruss an Aachen'****	Light pink/apricot pink	40–45	Slight	30″
Gypsy Carnival™***	Deep red/yellow	50	Strong	40″
'Hannah Gordon' ('Tabris')****	White edged light red	25–30	Slight	36″
H.C. Anderson®***	Dark red	9–16+	Slight	36″
Honey Perfume™***	Yellow apricot	30	Spicy	30″
Hot Cocoa™***	Russet red	25–30	Moderate	40″
Iceberg®****	White	20–25	Strong	30″
'Ice White' (Vision Blanc®)****	White cream	25	Slight	36″
Impatient®***	Orange red	20–25	Slight	30″
Intrigue®***	Light purple	20–30	Moderate	30″
'Ivory Fashion'****	Cream white/yellow eye	17	Light to moderate	30″
'Judy Garland'***	Yellow edged orange red	35	Moderate to strong	36″
Julia Child™****	Golden yellow	26–40+	Strong	30″
Lavaglut® ('Lava Flow')***	Dark red	25–30	Slight	36″
'Lavender Pinocchio'****	Pinkish mauve/gold eye	28	Moderate	36″
'Little Darling'*****	Light pink apricot	27	Spicy	30″
Livin' Easy™***	Apricot orange	25–30	Slight	36″
Lovestruck®***	Salmon pink yellow	17–26	Slight	24″
Mardi Gras®****	Orange yellow/yellow	17–25	Slight	36″
Margaret Merril®***	White/light yellow eye	28	Strong	48″
Marina®***	Pinkish orange	17–25	Moderate	36″

'Apricot Nectar'

'Bordure Rose'/'Strawberry Ice'

Dicky®/Anisley Dickson®

Europeana®

'First Edition'

'French Lace'

'Gingersnap'

'Sea Pearl'

VARIETIES	COLOR	PETAL COUNT	FRAGRANCE	HEIGHT
Marmalade Skies™	Orange pink	22	None to slight	36″
Moondance™*****	Creamy white	25—40	Moderate	48″
Mountbatten®***	Medium yellow	45	Moderate	36″
'Nearly Wild'****	Medium pink	5	Moderate	24″
Nicole® ('Koricole')***	White edged pink	30—35	Slight	36″
Orangeade®***	Orange red/yellow eye	10	Slight	30″
Our Lady of				
Guadalupe™****	Silvery pink	25	Slight	36″
'Permanent Wave***	Deep pink red	9—16	None to slight	36″
Playboy®***	Red orange yellow	7—10	Slight	36″
'Pleasure'***	Medium pink	33	Slight	36″
Pretty Lady®****	Creamy white pink	25	Slight	30″
Rainbow Sorbet®***	Golden yellow edged pink	15—25	Moderate	40″
'Raspberry Ice' (see 'Hannah Gordon')				
Red Abundance®*****	Dark red	28—36	Slight	24″
'Red Pinocchio'***	Dark red	28	Moderate	30″
'Redgold'***	Red golden yellow	25	Slight	36″
Regensberg®***	Deep pink/white eye	21	Moderate	30″
'Royal Occasion'				
(Montana®)****	Orange red	20	Slight	36″
Sarabande®****	Orange red/golden eye	13	Slight	30″
'Saratoga'***	White/gold eye	33	Strong	30″
Scentimental™***	Red white swirls	25—35	Spicy	40″
'Sea Pearl'****	Pink shaded cream	24	Slight	36″
Sexy Rexy®****	Medium pink	40—50	Slight	30″
'Sheila's Perfume'****	Yellow tinted red	20—25	Strong	30″
'Showbiz'****	Bright medium red	20—30	Slight	24″
Spanish Sunset™***	Orange pink	10—12	Slight	30″
'Summer Fashion'***	Yellow white edged pink	20—30	Moderate	24″
Sun Flare®***	Medium yellow	20—30	Slight	24″
'Sunsprite'****	Deep yellow	20—30	Strong	24″
'Sweet Dream'***	Peachy apricot	17—25	Moderate	24″
'Sweet Vivien'***	Cream edged pink/gold eye	17	Slight	30″
Tamango®***	Dark red	35	Slight	30″
Träumerei®***	Orange red	17—25	Strong	36″
Trumpeter®****	Red orange	35—45	Slight	30″
'Tuscan Sun'****	Coppery apricot orange pink	25	Slight	36″
'Vera Dalton'***	Medium pink	24	Moderate	24″
'Vogue'***	Deep pink/cream	25	Moderate	30″
Walking on Sunshine™***	Bright to light yellow	25—30	Slight	30″
White Licorice™***	White with lemon tones	35—40	Strong	36″
'Woburn Abbey'***	Orange/yellow center	25	Moderate	30″

'Camelot'

GRANDIFLORA ROSES
(FLORIBUNDA HYBRID TEA TYPE ROSES)

The Grandiflora is a cross between a Hybrid Tea and a Floribunda. The flowers look like those of a Hybrid Tea but often form clusters of buds. Their stems are not quite as long as Hybrid Teas, but still ideal for cutting. Plants are generally quite tall, stately, and vigorous. This firm, upright growth makes them suitable for informal hedges. Grandifloras have good repeat bloom if spent flowers are removed. On some plants, bloom is almost continuous.

How Grandiflora Roses Grow

Almost all Grandifloras are budded or grafted plants. A bud or slip from a named variety is budded or grafted onto another plant's rootstock. The plants may form suckers from the rootstock which must be removed. Grandifloras will form numerous stems if properly watered and cared for. These in turn branch, forming a shrublike plant with prolific bloom.

Where to Plant

Site and Light Grandifloras thrive in full sun. For good bloom they need at least 6 hours of direct sun a day. Sun is important for vigorous growth and the reduction of foliar diseases that could weaken or kill the plant. Avoid placing Grandifloras under eaves or drip lines where they may be damaged by rainfall. Also avoid planting them under trees which reduce the amount of light significantly. Plant them as hedges or specimen plants, or in the back of borders, always giving them plenty of room for good air circulation and lots of light.

Soil and Moisture Grandifloras need rich soil that retains moisture but drains freely. Replace clay or rock with loam purchased from a garden center or in bags as potting soil. Add lots of organic matter, such as compost, leaf mold, peat moss, or rotted manures. These keep the soil moist and cool during dry periods. They also help the soil drain freely, encourage the growth of beneficial soil microorganisms, and attract worms which keep the soil aerated and fertilized. Soil should be light, loose, and airy. This allows quick root growth and good drainage so that plenty of oxygen can get to the tender feeder roots. A handful of bonemeal or superphosphate mixed into the soil provides additional nutrients for the young plant. Proper soil preparation takes a little time, but it is critical to good growth.

Spacing Some of the Grandifloras will grow 6 feet (180 cm) tall. Give all plants enough space so that there is good air circulation between plants to prevent disease. Proper spacing and pruning allows adequate light to the stems and leaves. This promotes vigorous growth which results in heavier bloom.

Planting

Bare Root Plant bare root Grandifloras as soon as the ground can be worked in spring. A good bare root plant will have three good canes about ½ inch (12 mm) in diameter. The bud union should be planted about one-half below and one-half above the soil. The planting steps are outlined in detail in Part II.

Potted Plants You'll often find a few of the more popular Grandifloras in garden centers or nurseries. Pick out plants with healthy canes and luxuriant foliage. Be sure to check plants carefully for any signs of disease or insect infestations to avoid buying infected plants. Avoid leaving the plant in the car while you do other errands, since the heat buildup can damage it. Also, if the plant will be exposed to wind, protect it well by wrapping it in thick plastic or paper. As soon as you get home, remove the plastic, and water the soil if it's at all dry. For best results, follow the detailed steps for planting potted roses given in Part II.

How to Care for Grandiflora Roses

Water Grandifloras need lots of water. Keep the soil evenly moist at all times. Saturate the soil each time you water. Proper watering carries nutrients to the feeder roots, making plants vigorous. Healthy plants resist disease and insect infestations more easily than weak ones. They also bloom more profusely, more often, and with better color.

Mulch After the soil warms up to 60°F (15.6°C), apply a mulch around the base of the plant. The mulch should come close to but not touch the canes. The most commonly used mulch is shredded, not whole, leaves. Place at least a 3-inch (7.5-cm) layer over the soil. If you have enough leaves to make a thicker mulch, so much the better. Leaves keep the soil moist and cool, which encourages rapid root growth. The mulch feeds microorganisms and worms which enrich and aerate the soil. The mulch also inhibits weed growth and makes pulling any weeds that do sprout much easier. Other good mulches include pine needles and grass clippings. The latter should be only 2 inches (5 cm) deep, or they will heat the soil and may smell. If you use chipped wood or shredded bark, apply additional fertilizer to the soil, since these rob the soil of nitrogen as they decompose. Because mulch is eaten by soil microorganisms and worms, it needs to be replenished regularly throughout the growing season. Remove and compost all mulch in the fall.

Fertilizing If combining the use of inorganic and organic fertilizers, be careful during the first year. Use inorganic fertilizer only after the plant has leafed out and is growing vigorously.

In subsequent years, as early as possible in spring, sprinkle 10-10-10 granular fertilizer around the base of each plant. Use common sense in judging the amount, more for bigger plants, less for smaller. Hand sprinkle, never placing fertilizer against the stem of the plant. Water immediately to dissolve the granules and carry nutrients to the roots.

In both the first and subsequent years do the

following: One to two weeks after the first feeding with 10-10-10 granules soak the base of each plant with a liquid containing 20-20-20 water soluble fertilizer following directions on the label. Use common sense, giving larger plants more, smaller plants less. This feeding will stimulate sensational bloom as the plant matures.

One week later soak the base of each plant with a liquid containing fish emulsion. Follow the directions on the label. Feed according to the size of the plant.

One week later give all plants the 20-20-20 treatment again. One week later, give all plants the fish emulsion soaking. Keep this rotation going until the middle of August. At this time, stop all use of fertilizers containing nitrogen, since nitrogen stimulates new growth which dies off in most winters. Use 0-10-10 fertilizer after mid-August if desired.

Additional fertilizing with Epsom salts (magnesium sulfate) is highly recommended. Give ⅓ cup (about 75 g) for larger plants, less for smaller, two times a year: in late May and in early July. Magnesium sulfate neutralizes the soil and makes it easier for the plant to take in nutrients. This stimulates the growth of new canes (basal breaks) from the base of the plant.

Good fertilizers for organic growers are alfalfa meal (rabbit pellets), blood meal, bonemeal, compost, cow manure, fish emulsion, rotted horse manure, and Milorganite. Add bonemeal to the soil before planting. The others are effective added to the soil at planting time and as additional feedings on the surface of the soil throughout the season.

Weeding Destroy all perennial weeds before planting, using an herbicide such as Roundup®. Inhibit annual weed growth by using a thick mulch as outlined earlier. If any weeds emerge, remove them immediately, since they compete with the plant for water and nutrients. They also act as hosts for destructive insects. Pull up all annual weeds by hand to prevent damaging the plant's root system.

Staking Some of the larger Grandifloras can get quite heavy with bloom. Canes should not sway forcefully in heavy winds. It is wise to support larger canes with stakes. Always use a soft material tied loosely in a figure-eight pattern to anchor the cane to the stake.

Disbudding Some growers remove flower buds in the first year. This directs all of the plant's energy into forming canes and roots. Remove buds just as they appear for best results.

Removing buds from mature Grandifloras makes little sense. The desired effect is to have a cluster of lovely, Hybrid Tea–like flowers close together on the end of the cane.

Deadheading As blossoms fade, snip them off. When all of the blooms in a cluster have faded, snip the stem back to a point ¼ inch (6 mm) above a five-leaflet leaf below the original flower cluster. This will encourage further bloom.

Pruning During the first year, do not prune, other than to remove dead, broken, or diseased cane. Snip off all buds before they bloom, but do not remove any foliage. This channels the food to the root system and stems for a healthier plant. Let the plant bloom in late summer to help the plant prepare for winter dormancy.

During the second year, leave basal breaks alone, but remove all other spindly growth. Ideally, you want three or more healthy stems, each ½ to ¾ inch (12 to 18 mm) in diameter. Cut out crossing canes. In this way you shape the plant so that it has an open center. This is good for air circulation to prevent disease and allows more light to strike the plant for good bloom.

During the third year, remove any cane that did not produce well in the previous season. Mark these each year for removal with a tag. If an older cane is producing strong side branches (laterals), do not remove it. The laterals often produce magnificent bloom.

Winter Protection Grandifloras need good winter protection to survive in cold climates. The best is the Minnesota Tip Method explained in Part II.

Grandifloras will survive unprotected to about 26°F/−3.3°C (the mid-20s). 'Earth Song' is listed here and under Shrub Roses, since it's one of the hardiest of the group.

Problems

Insects Grandifloras are susceptible to insect infestations. A preventive spraying program is recommended. Spray every 7 to 10 days with an insecticide to keep bugs at bay. If weather gets hot and dry, use a miticide to prevent spider mites from appearing.

Disease Grandifloras are susceptible to a number of diseases. A regular preventive spraying program is recommended. Spray every 7 to 10 days with a fungicide to keep fungal infections from forming. Alternate the type of fungicide used for best protection.

Propagation

Although it is possible to propagate Grandifloras through budding and grafting, few people take the time to do it. Some companies produce a few Grandifloras on their own roots, taking stem cuttings, but this is successful only with a few varieties and so complex that the home gardener is best to avoid it.

Special Uses

Cut Flowers Cut stems no longer than 6 inches (15 cm) from young plants. As plants mature, take longer stems. Make cuts ¼ inch (6 mm) above an outward-facing bud. Flowers are often firm and long. So are the stems as the plant matures. When cutting flowers, leave as much foliage on the plant as possible. This is important for the plant's health.

Dried Flowers Although it's possible to dry Grandiflora flowers, other types of roses produce more beautiful blossoms for arrangements or more strongly scented petals for potpourris.

Sources

Chamblee's Rose Nursery, 10926 US Hwy 69 N, Tyler, TX 75706, (800) 256-7673

Edmunds' Roses, 335 S High St., Randolph, WI 53956, (888) 481-7673

Garden Valley Ranch, 498 Pepper Rd., Petaluma, CA 94952, (707) 795-0919

Goodness Grows, Inc., P.O. Box 311, 332 Elberton Rd., Lexington, GA 30648, (706) 743-5055

Heirloom Roses, 24062 Riverside Dr. NE, Saint Paul, OR 97137, (503) 538-1576

Hortico, Inc., 723 Robson Rd., RR# 1, Waterdown, ON L0R 2H1 Canada, (905) 689-6984

Mary's Plant Farm & Landscaping, 2410 Lanes Mill Rd., Hamilton, OH 45013, (513) 894-0022

McKay Nursery Co., P.O. Box 185, Waterloo, WI 53594, (920) 478-2121

Northland Rosarium, 9405 S Williams Lane, Spokane, WA 99224, (509) 448-4968

Palatine Fruit & Roses, 2108 Four Mile Creek Rd., RR #3, Niagara-on-the-Lake, ON L0S 1J0 Canada, (905) 468-8627

Pickering Nurseries, 3043 County Rd. 2, RR #1, Port Hope, ON L1A 3V5 Canada, (905) 753-2155

Regan Nursery, 4268 Decoto Rd., Fremont, CA 94555, (800) 249-4680

Roses Unlimited, 363 N Deerwood Dr., Laurens, SC 29360, (864) 682-7673

S & W Greenhouse, Inc., P.O. Box 30, 533 Tyree Springs Rd., White House, TN 37188, (615) 672-0599

Two Sisters Roses, 1409 N Redbud Lane, Newcastle, OK 73065, (no phone by request)

Witherspoon Rose Culture, 3312 Watkins Rd., Durham, NC 27707, (800) 643-0315

'Earth Song'

'Love'

New Year®

'Pink Parfait'

VARIETIES

Along with generally producing more flowers of hybrid form, Maria Shriver™ and Rock 'n' Roll™ are very fragrant. Two varieties with good disease resistance are excellent choices for organic gardeners: About Face™ and 'Earth Song.' Although 'Love' and Queen Elizabeth® have been around for years, they are still among the very finest roses you can buy because they are vigorous and long lived with proper care. A number of growers are now producing Grandifloras on their own roots to improve winter hardiness. A few examples are 'Centennial,' 'Kiss Me,' and 'Sweetness.'

VARIETIES	COLOR	PETAL COUNT	FRAGRANCE	HEIGHT
About Face™***	Bronzy red gold orange	35	Slight	48"
'Aquarius'****	Medium to light pink	30—35	Slight	60"
'Arizona'***	Pinkish orange	25—40	Strong	48"
'Camelot'****	Salmon pink	40—55	Spicy	48"
Candelabra™***	Rusty orange salmon	25—30	Slight	36"
Caribbean™***	Apricot orange	30—35	Moderate	40"
'Centennial'***	Apricot to yellow cream	22	Slight	30"
'Ch-Ching!'***	Golden yellow edged pink	30	Strong	40"
Cherry Parfait™****	Red white	35	Slight	36"
Crimson Bouquet™***	Velvety dark red	20—25	Slight	48"
'Dick Clark'***	Medium red/yellowish cream	25—30	Slight	48"
'Earth Song'***	Deep pink	25—30	Moderate	60"
'El Catalá'***	Red silvery white	24—40	Slight	36"
Fame!™****	Deep pink red/gold center	30—35	Slight	36"
'Fragrant Plum'***	Lavender mauve	20—25	Fruity	40"
Gold Medal®*****	Medium yellow	30—35	Slight	48"
'Kiss Me'****	Clear deep pink	20—25	Strong	30"
Lagerfeld™***	Lilac purple	30—35	Strong	60"
'Love'****	Red white reverse	40—45	Slight	48"
'Lucky Lady'***	Light pink	26—40	Slight	60"
Maria Shriver™***	White	40—	Strong	48"
'Mélody Parfumée'***	Lavender purple plum	26—40	Strong	40"
'Montezuma'****	Orange pink	30—35	Slight	48"
'Mount Shasta'****	White pink blush	17—25	Moderate	48"
New Year®***	Medium yellow/orange	17—25	Slight	40"
Octoberfest™*****	Orange yellow	35	Slight	54"
'Olé'****	Orange red	40—50	Moderate	40"
'Pearlie Mae'***	Yellow pink	30—36	Moderate	48"
'Pink Parfait'***	Pink white shading	17—25	Slight	48"
Prominent®***	Orange red	30—35	Slight	40"
Queen Elizabeth®*****	Medium pink	38	Moderate	60"
Rock & Roll™***	Red white	41+	Strong	36"
'Scarlet Knight'***	Medium red	17—25	Slight	36"
Shreveport®***	Orange yellow	44—51	Slight	36"
'Sonia'****	Coral pink	30	Fruity	40"
Strike It Rich™****	Gold yellow edged pink	30	Strong	40"
'Sweet Fragrance'***	Coral orange to apricot	25—40	Moderate	40"
'Tournament of Roses'****	Medium pink	25—30	None	36"
Waiheke™***	Orange pink	30	Spicy	36"
White Lightnin'™***	White	30—35	Strong	36"
Wild Blue Yonder™***	Bluish wine purple	25—30	Strong	48"

'Buff Beauty'

HYBRID MUSK ROSES

The Hybrid Musks are barely hardy, dying back to snow or ground level each year. Many die out completely. This is not a group recommended for casual cold-climate gardeners. Almost all form lovely clusters of flowers which bloom heavily in June and will repeat bloom until fall. These flowers, for the most part, are very fragrant—one of the reasons rose growers take a chance on them. The plants are also quite disease resistant and don't require as much spraying as other roses. Furthermore, they can grow in partial light to deep shade.

How Hybrid
Musk Roses Grow

Hybrid Musk roses are grown on budded stock. Place the bud union 2 to 3 inches (5 to 7.5 cm) below ground. If the rootstock forms suckers, remove them. The latter will produce bloom different from those on the budded cane. Deeper planting encourages root formation from the budded portion of the plant. This increases your chances of getting the plant through the winter. Hybrid Musks often suffer severe dieback in extreme cold. However, plants are vigorous growers and bloom extremely well on new cane. So, despite dieback, you can get excellent and profuse bloom with an exquisite fragrance. Foliage is often extremely attractive as well, varying in coloration by variety and growing conditions.

Where to Plant

Site and Light Unlike most other roses, Hybrid Musks not only tolerate, but do well in partial light to deep shade. This gives you great leeway in where you plant them.

Soil and Moisture Hybrid Musks need rich soil that retains moisture but drains freely. Replace clay or rock with loam purchased from a garden center or in bags as potting soil. Add lots of organic matter, such as compost, leaf mold, peat moss, or rotted manures. These keep the soil moist and cool during dry periods. They also help the soil drain freely, encourage the growth of beneficial soil microorganisms, and attract worms which keep the soil aerated and fertilized. Soil should be light, loose, and airy. This allows quick root growth and good drainage so that plenty of oxygen can get to the tender feeder roots. A handful of bonemeal or superphosphate mixed into the soil provides additional nutrients for the young plant. Proper soil preparation takes a little time, but it is critical to good growth.

Spacing Give Hybrid Musks plenty of space for good air circulation. Some of them will grow wider than they are tall, so place them at least as far apart as their potential height, if not more.

Planting

Virtually all Hybrid Musks are sold as bare root plants. Occasionally, you'll find potted plants at special arboretum sales or in local nurseries, but this is quite rare.

Bare Root Plant bare root Hybrid Musks directly in the garden as soon as the ground can be worked in spring. Each plant should have at least three canes. Some plants may have more than three, and that's fine, as long as they are evenly spaced. If the canes are too close together, it's best to cut out one or more, but try to end up with no fewer than three healthy canes. Place the bud union 2 to 3 inches (5 to 7.5 cm) below the soil. This may stimulate root growth from the upper budded portion of cane, which makes the plant hardier. The exact steps on bare root planting are covered in Part II.

Potted Plants Occasionally, you'll find potted Hybrid Musk roses, but, again, this is rare. Plants should have at least three healthy canes with lush foliage. Check the leaves carefully for any sign of insects or disease. Buy only healthy plants. Get any plant you buy home quickly. Don't let a plant sit in a car while you run errands. If it must be exposed to

wind, wrap the entire exposed stem and cane in plastic or paper. As soon as you get home, remove the protective covering, and water the plant immediately. Plant the potted rose exactly as outlined in Part II.

How to Care for Hybrid Musk Roses

Water Keep the soil evenly moist throughout the growing season. Deep watering is much more effective than frequent sprinkling. Soak the ground thoroughly once, then soak it again so that the water is absorbed to a depth of 18 inches (45 cm). The simplest way to do this is to let a hose run at the base of the plant.

Mulch After the soil warms up to 60°F (15.6°C) apply a mulch around the base of the plant. The mulch should come close to but not touch the canes. The most commonly used mulch is shredded (not whole) leaves. Place at least a 3-inch (7.5-cm) layer over the soil. If you have enough leaves to make a thicker mulch, so much the better. Leaves keep the soil moist and cool which encourages rapid root growth. The mulch feeds microorganisms and worms which enrich and aerate the soil. The mulch also inhibits weed growth and makes pulling any weeds that do sprout much easier. Other good mulches include pine needles and grass clippings. The latter should be only 2 inches (5 cm) deep, or they will heat the soil and may smell. These mulches are all inexpensive and effective. If you use chipped wood or shredded bark, apply additional fertilizer to the soil, since these rob the soil of nitrogen as they decompose. Because mulch is eaten by soil microorganisms and worms, it needs to be replenished regularly throughout the growing season. Remove and compost all mulch in the fall.

Fertilizing If combining the use of inorganic and organic fertilizers, be careful during the first year. Use inorganic fertilizer only after the plant has leafed out and is growing vigorously.

In subsequent years, as early as possible in spring, sprinkle 10-10-10 (or 5-10-5) granular fertilizer around the base of each plant. Use common sense in judging the amount, more for bigger plants, less for smaller.

Hand sprinkle, never placing fertilizer against the stem of the plant. Water immediately to dissolve the granules and carry nutrients to the roots.

In both the first and subsequent years do the following: One to two weeks after the first feeding with 10-10-10 granules soak the base of each plant with a liquid containing 20-20-20 water soluble fertilizer following directions on the label. Use common sense, giving larger plants more, smaller plants less. This feeding will stimulate sensational bloom as the plant matures.

One week later soak the base of each plant with a liquid containing fish emulsion. Follow the directions on the label. Feed according to the size of the plant.

One week later give all plants the 20-20-20 treatment again. One week later, give all plants the fish emulsion soaking. Keep this rotation going until the middle of August. At this time, stop all use of fertilizers containing nitrogen, since nitrogen stimulates new growth which dies off in most winters. Use 0-10-10 fertilizer after mid-August if desired.

Additional fertilizing with Epsom salts (magnesium sulfate) is highly recommended. Give ⅓ cup (about 75 g) for larger plants, less for smaller, two times a year: in late May and in early July. Magnesium sulfate neutralizes the soil and makes it easier for the plant to take in nutrients. This stimulates the growth of new canes (basal breaks) from the base of the plant.

Good fertilizers for organic growers are alfalfa meal (rabbit pellets), blood meal, bonemeal, compost, cow manure, fish emulsion, rotted horse manure, and Milorganite. Add bonemeal to the soil before planting. The others are effective added to the soil at planting time and as additional feedings on the surface of the soil throughout the season.

Weeding Kill all perennial weeds before planting. An herbicide such as Roundup® is recommended. Use mulch to keep annual weeds in check. Hand weed the few that sprout through the mulch. Never cultivate deeply because the roots run quite close to the surface and can be damaged easily.

Staking Hybrid Musks rarely grow tall enough in cold climates to be used as Pillar or Climbing roses.

However, they do occasionally send out a tall cane that looks best if supported by a stake, fence, or trellis. Attach such canes to the support with a soft material tied in a figure-eight knot (or any manner recommended in Part II). You want to keep the cane upright but not have the tie so tight it damages the cane as it grows.

Disbudding Some growers remove flower buds in the first year, directing all energy to the plant itself. Remove buds just as they appear, for best results.

On mature plants, no one removes buds. Doing so would just stop the bush from reaching its natural beauty which shows up as a shower of blossoms.

Deadheading Hybrid Musks often die back each season. Blooming often takes place late in summer. Removing spent blossoms is not necessary.

Pruning Each spring, cut back canes to live wood. Dead cane looks dark, is brittle, and will snap if you bend it. Cut canes all the way back to ground level if necessary. On some canes, you'll find live wood, which is pale green or whitish when cut with pruners. Cut the cane ¼ inch (6 mm) above an outward-facing bud. Plants often will produce new canes each season. Don't remove any cane, since new growth often produces excellent bloom. In short, prune as little as possible.

Winter protection Ideally, it would be best to use the Minnesota Tip Method for winter protection. However, most people don't have the stamina to do this with these larger plants. Instead, after a hard frost, mound up soil around the base of each plant. Do this as far up as possible. Hope for lots of snow to protect additional cane. You're gambling but saving lots of time and effort. For complete protection, do use the Minnesota Tip Method outlined in detail in Part II. With the more minimal protection, much of the plant will survive to about −20°F (−29°C).

Ardent lovers of Hybrid Musks grow them in containers. In late fall they slide the rootball of the plants into black plastic bags and bury them in the garden as they would other tender roses.

Problems

Insects Hybrid Musks can be invaded by any of the common rose pests, but they are vigorous plants and fight invasions quite well. Use insecticides or miticides only in extreme cases.

Disease These plants are quite disease resistant. Most death is caused by severe winter weather and inadequate winter protection. However, if fungal infections begin to occur, use spray as necessary to keep them in check.

Propagation

Cuttings If you're willing to go to the effort of growing Hybrid Musks in cold climates, you might be interested in propagating them from either hardwood or softwood cuttings.

For hardwood cuttings, choose a healthy cane from the current year's growth. The cane should be no smaller in diameter than a pencil. Push against the thorns. If they snap off easily, the cane is at the right maturity for cutting. All cuttings should have three sets of leaves.

For softwood cuttings, we suggest you take cuttings at two different times to increase your odds of success: Take some just as growth buds form in the spring; take others later on after the plant has leafed out. Each cutting should have three sets of leaves.

In addition to these special tips for Hybrid Musks, you can refer to Chapter 7 in Part II for exact steps in taking both hardwood and softwood cuttings.

Special Uses

Cut Flowers While these roses do not produce flowers in the classic shape desired by many people, they are fragrant and look stunning in informal arrangements. In short, they do make excellent cut flowers.

Dried Flowers The best use for dried Hybrid Musk flowers is in a potpourri. Pick petals at the peak of their fragrance and before they begin to wilt. (See Chapter 8 for instructions on making and storing potpourris.)

Sources

Angel Gardens, P.O. Box 1106, Alachua, FL 32616, (352) 359-1133

Antique Rose Emporium, 9300 Lueckmeyer Rd., Brenham, TX 77833, (800) 441-0002

Burlington Rose Nursery, 24865 Rd. 164, Visalia, CA 93292, (559) 747-3624

Countryside Roses, 5016 Menge Ave., Pass Christian, MS 39571, (228) 452-2697

David Austin Roses Ltd., 15059 Hwy 64 W, Tyler, TX 75704, (800) 328-8893

ForestFarm, 990 Tetherow Rd., Williams, OR 97544, (541) 846-7269

Greenmantle Nursery, 3010 Ettersburg Rd., Garberville, CA 95542, (707) 986-7504

Heirloom Roses, 24062 Riverside Dr. NE, Saint Paul, OR 97137, (503) 538-1576

High Country Roses, P.O. Box 148, Jensen, UT 84035, (800) 552-2082

Hortico, Inc., 723 Robson Rd., RR# 1, Waterdown, ON L0R 2H1 Canada, (905) 689-6984

Jackson & Perkins, 2 Floral Ave., Hodges, SC 29653, (800) 872-7673

Linda's Antique Roses, 405 Oak Ridge Dr., San Marcos, TX 78666, (512) 353-3220

Mary's Plant Farm & Landscaping, 2410 Lanes Mill Rd., Hamilton, OH 45013, (513) 894-0022

Northland Rosarium, 9405 S Williams Lane, Spokane, WA 99224, (509) 448-4968

Palatine Fruit & Roses, 2108 Four Mile Creek Rd., RR #3, Niagara-on-the-Lake, ON L0S 1J0 Canada, (905) 468-8627

Pickering Nurseries, 3043 County Rd. 2, RR #1, Port Hope, ON L1A 3V5 Canada, (905) 753-2155

Regan Nursery, 4268 Decoto Rd., Fremont, CA 94555, (800) 249-4680

Rogue Valley Roses, P.O. Box 116, Phoenix, OR 97535, (541) 535-1307

Roses of Yesterday and Today, 803 Brown's Valley Rd., Watsonville, CA 95076, (831) 728-1901

Roses Unlimited, 363 N Deerwood Dr., Laurens, SC 29360, (864) 682-7673

S & W Greenhouse, Inc., P.O. Box 30, 533 Tyree Springs Rd., White House, TN 37188, (615) 672-0599

Two Sisters Roses, 1409 N Redbud Lane, Newcastle, OK 73065, (no phone by request)

Vintage Gardens (custom propagation), 4130 Gravenstein Hwy N, Sebastapol, CA 95472, (707) 829-2035

Wayside Gardens, 1 Garden Lane, Hodges, SC 29695, (800) 213-0379

Witherspoon Rose Culture, 3312 Watkins Rd., Durham, NC 27707, (800) 643-0315

VARIETIES

These truly beautiful roses are hardy only to 15°F (−9°C) (some would say a few degrees lower) without winter protection. With minimal protection they often are crown hardy to about −20°F (−29°C). Cover the crowns of these plants with potting soil and a thick layer of whole leaves in late fall for optimal winter protection. Or use the Hass Method of placing 3 bags of whole leaves against the sides of the plant for additional protection from cold and desiccating winds (see p. 198). Deep snow is your best ally. Or consider growing them in pots that can be buried in late fall. That way hardiness is not really an issue. In cold climates these plants may not reach the heights listed. However, they bloom well on new growth if you can get them through the winter. Many of them are quite fragrant (musky as their name implies), especially in humid weather. They are so lovely, they are worth the extra attention they demand.

VARIETIES	COLOR	PETAL COUNT	FRAGRANCE	HEIGHT
'Ballerina'****	Medium pink/white	5	Slight	36"

Upright, mounded to arching. Canes spread out. Dainty 1-inch (2.5 cm) flowers with white dot in the center and yellow stamens. Showers of these flowers cover the entire bush. Often referred to as "hydrangea-like." Almost continuous bloom. Light to mid green, semi-glossy foliage. Cane with few thorns. Tiny red hips. A good choice for organic gardeners (zone 4 to 5).

'Belinda'****	Medium pink	9–16	Variable	48"

Spreading. Bright red buds open into large clusters of lovely cupped pink flowers 1 inch (2.5 cm) across with a white dot and bright yellow stamens in the center of each. Excellent repeat bloom. Attractive, dark green, glossy foliage. Cane with few thorns. Orange hips (zone 6).

'Bishop Darlington'****	Peach pink yellow	17	Moderate	48"

Upright vase. Vigorous, informal. Coral pink buds burst into small clusters of creamy pink flowers 3 inches (7.5 cm) or larger across. Good repeat bloom. Dark green to bronze semi-glossy foliage on slender canes. Reported to form hips—but none to our knowledge (zone 5 to 6).

'Bloomfield Dainty'***	Medium yellow	5–8	Moderate	60"

Arching. Moderately vigorous. Deep orange buds open into canary yellow buttercup-like flowers more than 2 inches (5 cm) across and canary yellow in color with lovely golden centers. Flowers fade to pink. Good repeat bloom. Medium to dark green, glossy foliage. Slender canes are prickly (zone 6).

'Buff Beauty'*****	Apricot cream	50	Moderate	48"

Spreading. Bush often gets wider than it is tall. Will act as a Climber in more southerly locations. Apricot yellow buds open into clusters of translucent flowers up to 3 inches (7.5 cm) wide. These are often yellowish or creamy with a tea-like fragrance. Excellent repeat bloom. Dark green foliage. New cane a lovely reddish hue. No hips. A gem! Consider growing in a container to be buried in late fall for winter protection (zone 5 to 6).

'Cornelia'****	Deep pink yellow	41+	Moderate	60"

Mounded, spreading. Coral red buds produce masses of 2-inch (5 cm) blossoms tinged yellow aging to pink. Some growers would call the color coppery or almost apricot. Good repeat bloom. Dark green, glossy foliage often tinged bronze. No hips. Consider growing this in a container to be buried in winter. A good choice for organic gardeners (zone 6)

'Felicia'****	Pink ivory salmon	41+	Strong	48"

Upright. Lax. Clusters of flowers up to 3 inches (7.5 cm) across, silvery pink with yellow at the petals' base. Often cupped or ruffled. Excellent repeat bloom. Medium green foliage and cane with few but large thorns. Wonderful cut flower. No hips. Consider growing this plant in a container to be buried in late fall. A good choice for organic gardeners (zone 6).

'Prosperity'

'Robin Hood'

Sally Holmes®

'Vanity'

VARIETIES	COLOR	PETAL COUNT	FRAGRANCE	HEIGHT
'Francesca'****	Apricot yellow	41+	Moderate	60"

Upright, arching. Deep yellow orange buds burst into clusters of loose flowers typically more than 3 inches (7.5 cm) across. Musky, fruity, or honeysuckle scent. Good repeat bloom. Leathery, dark green foliage. No hips (zone 6).

VARIETIES	COLOR	PETAL COUNT	FRAGRANCE	HEIGHT
'Kathleen'***	Light pink	5	Moderate	40"

Upright vase. Coral pink buds open into clusters of flowers 2 inches (5 cm) across or less with yellow to golden centers. Flowers often white tinged pink, like apple blossoms. Good repeat bloom. Medium to dark green glossy foliage on slender stems. May form small red hips (zone 5 to 6).

VARIETIES	COLOR	PETAL COUNT	FRAGRANCE	HEIGHT
'Pax'***	White	9–16+	Moderate to strong	40"

Spreading, open, to arching. Lovely long creamy pink buds open into large clusters of flowers up to 4 inches (10 cm) wide, creamy colored with conspicuous red gold centers. Excellent repeat bloom. Leathery, dark green glossy foliage tinged red when immature. No hips. (zone 6).

VARIETIES	COLOR	PETAL COUNT	FRAGRANCE	HEIGHT
'Penelope'****	White tinged pink	17–25	Moderate	36"

Upright spreading vase. Pinkish buds open into clusters of salmon to creamy pink 3-inch (7.5-cm) flowers with yellow centers. Excellent repeat bloom. Dense bush with glossy, deep green foliage. Excellent for cutting. Reddish pink hips often tinged orange. Grow in a pot and bury in late fall for optimal winter protection—worth the effort (zone 6).

VARIETIES	COLOR	PETAL COUNT	FRAGRANCE	HEIGHT
'Prosperity'**	White tinged pink	9–16	Moderate	36"

Compact bushy, mounded. Creamy pink buds open into loose clusters of small 1½- inch (3.75-) flowers, ivory white flushed pink with lemony yellow to golden centers. Good repeat bloom. Glossy, medium to dark green foliage. Said to form hips but have not seen them (zone 5 to 6).

VARIETIES	COLOR	PETAL COUNT	FRAGRANCE	HEIGHT
'Robin Hood' ('Robin des Bois')***	Medium red	5–8	None to slight	48"

Upright. Large clusters of 1½ inch (3.75 cm) cherry red blossoms with white eye and golden center bloom on a vigorous plant with nearly constant bloom. Attractive glossy bronzy, green foliage. Forms small, red hips. A good choice for organic gardeners (zone 6).

VARIETIES	COLOR	PETAL COUNT	FRAGRANCE	HEIGHT
Sally Holmes®****	White tinged pink	5	None to slight	60"

Upright in cold climates to climbing in warmer areas. Abundant lovely light apricot buds burst into clusters of white 3-inch (7.5-cm) flowers with prominent yellow stamens. Medium green, semi-glossy foliage. No hips. A good choice for organic gardeners. Sources vary in listing this plant as a Hybrid Musk or Shrub Rose (zone 5).

VARIETIES	COLOR	PETAL COUNT	FRAGRANCE	HEIGHT
'Skyrocket' ('Wilhelm')***	Dark red	24–30	Moderate	60"

Upright and open. Small clusters of flowers roughly 3 inches (7.5 cm) across with pale yellow stamens. Honey-like fragrance. Good repeat bloom. Glossy, medium to dark green foliage. Cane has very few thorns. Bright orange hips (zone 6).

VARIETIES	COLOR	PETAL COUNT	FRAGRANCE	HEIGHT
'Vanity'****	Deep pink to red	8–15	Moderate	60"

Arching, lax growth. Clusters of deep pink to red wavy flowers 2 inches (5 cm) or more across with yellow center. Good repeat bloom. Glossy, mid to dark green foliage. Excellent for cut flowers. Possible orange hips (zone 6).

VARIETIES	COLOR	PETAL COUNT	FRAGRANCE	HEIGHT
'Will Scarlet'****	Medium red to pink	15–20	Slight	60"

Upright, open, arching growth in colder climates, more of a Climber to the south. Sport of 'Wilhelm.' Deep red buds open into clusters of scarlet red flowers up to 3 inches (7.5 cm) across with yellow stamens. Medium to dark green foliage. Glossy orange red hips (zone 5 to 6).

VARIETIES	COLOR	PETAL COUNT	FRAGRANCE	HEIGHT
'Wind Chimes'****	Medium pink	5	Moderate	48"

Arching growth. Rosy flowers 1 inch (2.5 cm) or more across, with white center and prominent gold stamens. Mid to deep green, semi glossy foliage. Excellent repeat bloom with long stems that are ideal for cutting either when in bloom or when covered with a delightful display of orange to orange red hips in fall (zone 6).

Pristine®

HYBRID TEA ROSES

Hybrid Teas are the most popular roses, because they produce beautifully pointed buds which open into swirls of petals. These truly elegant blooms are often large and fragrant. They have an almost conelike shape and a varying number of petals. Usually, there is one flower per stem, making them ideal for cutting. However, some varieties produce a central flower surrounded by several others. Hybrid Teas bloom off and on throughout the season in flushes, rather than producing blooms continuously. They are the most prized of bedding plants, often mixed in with perennial flowers, or used as accent or specimen plants on their own. They are, however, more finicky than many other roses and generally require routine spraying for disease and insects. They also must be tipped each year to protect them from winter cold and drying winds. For many growers, their beauty makes this worthwhile.

How Hybrid Tea Roses Grow

Some Hybrid Teas are grown on their own roots, but most are budded or grafted. A bud or slip (scion) from a named variety is budded or grafted to a different rose plant's rootstock. The rootstock may send out suckers which must be removed as soon as they appear. Plants vary in form from elegantly tall to spreading or bushy. They are always aristocratic. Some Hybrid Teas are not hardy in colder climates (even with winter protection), so choose from field-tested varieties listed in the table.

Where to Plant

Site and Light Hybrid Teas thrive in full sun. They need at least 6 hours of direct sun a day to grow properly. Sun encourages vigorous growth which results in more stems, more branches, and less disease. Avoid planting them under trees which steal valuable water and nutrients from them. Also avoid planting under eaves, unless you are willing to water constantly throughout the growing season.

Soil and Moisture Hybrid Teas need rich soil that retains moisture but drains freely. Replace clay or rock with loam purchased from a garden center or in bags as potting soil. Add lots of organic matter, such as compost, leaf mold, peat moss, or rotted manures. These keep the soil moist and cool during dry periods. They also help the soil drain freely, encourage the growth of beneficial soil microorganisms, and attract worms which keep the soil aerated and fertilized. Soil should be light, loose, and airy. This allows quick root growth and good drainage so that plenty of oxygen can get to the tender feeder roots. A handful of bonemeal or superphosphate mixed into the soil provides additional nutrients for the young plant. Proper soil preparation takes a little time, but it is critical to good growth.

Spacing Some of the Hybrid Teas are tall. Give all plants enough space so that there is good air circulation between them to prevent disease. Proper spacing also allows lots of light to all parts of the plant, which increases vigorous growth and prolific bloom.

Planting

Bare Root Bare root Hybrid Tea roses are widely available through mail-order catalogs, in garden centers and nurseries, and in most retail chain stores. Always buy them as early in the season as possible, since they deteriorate rapidly in the packaging. Plant them as soon as the ground can be worked in spring.

Each plant should have at least three canes. Some plants may have more than three, and that's fine, as long as they are evenly spaced. If the canes are too close together, it's best to cut out one or more, but end up with no fewer than three healthy canes.

When planting Hybrid Tea roses, the bud union should be one-half below, and one-half above the soil.

The planting steps are outlined in detail in Part II.

Potted Plants You'll often find a wide assortment of Hybrid Teas in local garden centers or nurseries. Pick out plants with healthy canes and luxuriant foliage. Check plants carefully for any signs of disease or insect infestations to avoid buying infected plants. Avoid leaving the plant in the car while you do other errands, since the heat buildup can damage it. Also, if the plant will be exposed to wind, protect it well by wrapping it in paper or plastic. As soon as you get home, remove the covering, and water the soil if it's at all dry. Pot exactly as outlined in Part II.

How to Care for Hybrid Tea Roses

Water Keep the soil evenly moist at all times. Roots are often quite shallow and prone to damage from drought. Saturate the ground thoroughly each time you water by letting the hose run at the base of the plant. Use sprinklers in hot, dry weather to prevent the spread of spider mites.

Mulch After the soil warms up to 60°F (15.6°C) apply a mulch around the base of the plant. The mulch should come close to but not touch the canes. The most commonly used mulch is shredded (not whole) leaves. Place at least a 3-inch (7.5-cm) layer over the soil. If you have enough leaves to make a thicker mulch, so much the better. Leaves keep the soil moist and cool which encourages rapid root growth. The mulch feeds microorganisms and worms which enrich and aerate the soil. The mulch also inhibits weed growth and makes pulling any weeds that do sprout much easier. Other good mulches include pine needles and grass clippings. The latter should be only 2 inches (5 cm) deep, or they will heat the soil and may smell. These mulches are all inexpensive and effective. If you use chipped wood or shredded bark, apply additional fertilizer to the soil, since these rob the soil of nitrogen as they decompose. Because mulch is eaten by soil microorganisms and worms, it needs to be replenished regularly throughout the growing season. Remove and compost all mulch in the fall.

Fertilizing If combining the use of inorganic and organic fertilizers, be careful during the first year. Use inorganic fertilizer only after the plant has leafed out and is growing vigorously.

In subsequent years, as early as possible in spring, sprinkle 10-10-10 (or 5-10-5) granular fertilizer around the base of each plant. Use common sense in judging the amount, more for bigger plants, less for smaller. Hand sprinkle, never placing fertilizer against the stem of the plant. Water immediately to dissolve the granules and carry nutrients to the roots.

In both the first and subsequent years do the following: One to two weeks after the first feeding with 10-10-10 granules soak the base of each plant with a liquid containing 20-20-20 water soluble fertilizer following directions on the label. Use common sense, giving larger plants more, smaller plants less. This feeding will stimulate sensational bloom as the plant matures.

One week later soak the base of each plant with a liquid containing fish emulsion. Follow the directions on the label. Feed according to the size of the plant.

One week later give all plants the 20-20-20 treatment again. One week later, give all plants the fish emulsion soaking. Keep this rotation going until the middle of August. At this time, stop all use of fertilizers containing nitrogen, since nitrogen stimulates new growth which dies off in most winters. Use 0-10-10 fertilizer after mid-August if desired.

Additional fertilizing with Epsom salts (magnesium sulfate) is highly recommended. Give ⅓ cup (about 75 g) for larger plants, less for smaller, two times a year: in late May and in early July. Magnesium sulfate neutralizes the soil and makes it easier for the plant to take in nutrients. This stimulates the growth of new canes from the base of the plant.

Good fertilizers for organic growers are alfalfa meal (rabbit pellets), blood meal, bonemeal, compost, cow manure, fish emulsion, rotted horse manure, and Milorganite. Add bonemeal to the soil before plant-

ing. The others are effective added to the soil at planting time and as additional feedings on the surface of the soil throughout the season.

Weeding Before planting, kill off all perennial weeds with Roundup®. Keep the bed free of all weeds. They compete with the plants for water and nutrients. Mulch stops most annual weeds from sprouting. The few that do sprout are easy to pull from the soft mulch. Hand weed to prevent damage to the shallow root system.

Staking A few of the larger varieties of Hybrid Teas may be damaged in high winds. The wind jostles the plant, causing the roots to loosen. Prevent this by supporting taller plants with stakes. The best supports are made from aluminum conduit, available at most hardware stores in 10-foot (3-meter) lengths. Have these cut into two supports, one 6 feet (180 cm) high, the other 4 feet (120 cm), for varying plant heights. Cut the conduit with a hacksaw. Place the conduit into the ground early in the season when you plant the rose for the first time or when you lift it after protecting it during the winter. Aluminum is highly recommended because it does not get hot, is easy to cut, is easy to paint, and does not rust. It is also light and easy to work with. Attach the plant to the support with loose ties made of soft fabric. Use a figure-eight knot.

Disbudding Some growers remove flower buds in the first year, directing all energy to the plant itself. Pick off the buds just as they begin to form, for best results.

On mature plants, most growers do not remove any buds at all in an attempt to get as much bloom from the plant as possible.

A few growers remove all flower buds on a stem except for the one at the very top. The idea is to create one large, beautiful bloom. Snip off other buds on the same stem when they are just beginning to appear. The earlier you do this, the better. That way, they have no time to sap the terminal bud of its strength. They also come off easily at this stage. If you wait too long, the stem will often be scarred with a dark blotch. Either pinch the side buds off next to the main stem with your fingernails or snip them off with pruning shears.

Disbudding of this kind takes place all season long. It is not necessary for the health of the plant; it's simply an aesthetic decision.

Deadheading As blossoms fade, snip them off. This will encourage further bloom. Snip down to the next leaf with five leaflets. Make a diagonal cut about ¼ inch (6 mm) above the leaf. Hybrid Teas will repeat bloom, but they are not continuous bloomers.

Pruning During the first year do not prune, other than to remove dead, broken, or diseased cane. Snip off all flower buds, leaving as much foliage as possible. Removal of buds is optional, but it does help the plant develop a stronger root system. You can let the plant blossom late in the season as a way of inducing dormancy for the winter.

In subsequent years, prune after the last frost in spring. Prune just as buds begin to swell or earlier, but before leaves appear. Avoid all pruning in fall, since wounds may be infected when the plant is covered with soil. However, do remove any damaged or dead cane. If you're an organic gardener, paint the wound with Elmer's glue or a compound made for this purpose. Doing this is important because the plant can not cover its wound with a callus at low temperatures. Covering wounds is optional if you're using sprays during the season and a dormant spray before winter protection.

During the second year, leave basal stems alone, but remove all other spindly growth. Ideally, you want three to five healthy stems, each ½ inch (1.25 cm) or more in diameter. Cut out crossing canes. In this way you shape the plant so that it has an open center. This is good for air circulation to prevent disease and allows more light to strike the plant for good bloom. Cut out all damaged or dead cane, since these are easily infected with insects and disease. Finally, cut the good canes back according to their size. Cut larger canes back so that they have six to eight buds per

cane; cut smaller canes back so that they have two to four buds per cane.

This type of pruning is not an exact science. Some varieties withstand harder pruning than others. However, in colder climates the general rule is not to go overboard in pruning. High pruning makes the most sense. Cut off a smaller portion of cane than you would in more southerly areas. The growing season is short, so preserve as much healthy cane as possible.

During the third year, remove any cane that did not produce well in the previous season. Mark these each year for removal with a tag. If an older cane is producing strong laterals (side branches), leave it alone. The laterals often produce magnificent bloom. If a strong lateral appears only at the base of an old cane, keep it, but remove all cane above it.

Yellow-colored varieties resent pruning. Remove as little growth as possible to prevent damage to these more delicate plants.

Winter Protection Hybrid Tea roses need to be protected in cold climates. The best method is the Minnesota Tip Method described in detail in Part II. Without winter protection Hybrid Teas are hardy to about 26°F/−3.3°C (the mid-20s).

Problems

Insects Hybrid Teas are prone to the same insects as other roses. Use a regular spraying program every 7 to 10 days to protect your plants.

Disease Yellow and orange varieties are more prone to black spot than other Hybrid Teas. The following varieties are especially susceptible to anthracnose and powdery mildew: 'Christian Dior,' 'Chrysler Imperial,' 'Command Performance,' Double Delight®, Garden Party®, and 'Tropicana.' Use a regular spraying program every 7 to 10 days to prevent disease. Vary the fungicides used for best protection.

If you're an organic gardener, select varieties by their disease resistance, which correlates greatly with climate and growing conditions. Join a rose society in your area. Since resistance varies so much by area, this is the best way to get information on the best roses to grow in your exact location.

Propagation

You can start your own Hybrid Teas by budding or grafting plants, but few people do this because of the effort and time involved. Furthermore, these techniques are quite difficult. A few specialized growers are starting some Hybrid Teas on their own roots through cuttings. This works only with certain varieties and requires extreme skill developed over years of experimentation.

Special Uses

Cut Flowers Cut stems no longer than 6 inches (15 cm) from young plants. As plants mature, take longer stems. Make cuts ¼ inch (6 mm) above an outward- facing bud (found where a leaf joins the stem). Take cuttings early in the morning or in the evening. Always leave as much foliage on the plant as possible. A few outstanding Hybrid Teas for cutting are 'Bella'roma,' 'Bewitched,' 'Century Two,' 'Chrysler Imperial,' Folklore®, 'Fragrant Cloud,' Garden Party®, Honor™, 'Keepsake' (Esmeralda®), 'Marijke Koopman,' Mister Lincoln®, 'New Zealand, 'Olympiad®, Over the Moon™, Pascali®, 'Peace,' Pristine®, Secret®, Sheer Elegance®, 'Swarthmore,' and Touch of Class®.

If you want fewer flowers with long stems, hard prune these varieties in early spring. Cutting the cane back to a few buds increases the chance for a few long-stemmed flowers. However, it reduces the amount of bloom per plant. Frankly, unless you're into exhibiting Hybrid Teas, we would not advise this in light of the detrimental effects of hard pruning in colder areas.

Dried Flowers The individual blossoms of Hybrid Teas are among the most beautiful in the rose world and prized for dried flowers. However, colors often change dramatically during the drying process. Most people who dry flowers use silica. Experiment with different blossoms to see whether you like the end results. Pinks are among the favorites. Generally, dry only flowers, not stems. (But you can certainly try to dry both.) Stems can be replaced by wire available from florists.

Sources

Angel Gardens, P.O. Box 1106, Alachua, FL 32616, (352) 359-1133

Antique Rose Emporium, 9300 Lueckmeyer Rd., Brenham, TX 77833, (800) 441-0002

Burlington Rose Nursery, 24865 Rd. 164, Visalia, CA 93292, (559) 747-3624

Chamblee's Rose Nursery, 10926 US Hwy 69 N, Tyler, TX 75706, (800) 256-7673

Countryside Roses, 5016 Menge Ave., Pass Christian, MS 39571, (228) 452-2697

David Austin Roses Ltd., 15059 Hwy 64 W, Tyler, TX 75704, (800) 328-8893

Edmunds' Roses, 335 S High St., Randolph, WI 53956, (888) 481-7673

Garden Valley Ranch, 498 Pepper Rd., Petaluma, CA 94952, (707) 795-0919

Greenmantle Nursery, 3010 Ettersburg Rd., Garberville, CA 95542, (707) 986-7504

Heirloom Roses, 24062 Riverside Dr. NE, Saint Paul, OR 97137, (503) 538-1576

High Country Roses, P.O. Box 148, Jensen, UT 84035, (800) 552-2082

Hortico, Inc., 723 Robson Rd., RR# 1, Waterdown, ON L0R 2H1 Canada, (905) 689-6984

Inter-State Nurseries, 1800 E Hamilton Rd., Bloomington, IL 61704, (309) 663-6797

Jackson & Perkins, 2 Floral Ave., Hodges, SC 29653, (800) 872-7673

Jung Seed, 335 South High St., Randolph, WI 53957, (800) 297-3123

Kimbrew-Walter Roses, Inc., 2001 Van Zandt County Rd. 1219, Grand Saline, TX 75140, (903) 829-2968

Mary's Plant Farm & Landscaping, 2410 Lanes Mill Rd., Hamilton, OH 45013, (513) 894-0022

McKay Nursery Co., P.O. Box 185, Waterloo, WI 53594, (920) 478-2121

Northland Rosarium, 9405 S Williams Lane, Spokane, WA 99224, (509) 448-4968

Palatine Fruit & Roses, 2108 Four Mile Creek Rd., RR #3, Niagara-on-the-Lake, ON L0S 1J0 Canada, (905) 468-8627

Pickering Nurseries, 3043 County Rd. 2, RR #1, Port Hope, ON L1A 3V5 Canada, (905) 753-2155

Regan Nursery, 4268 Decoto Rd., Fremont, CA 94555, (800) 249-4680

Rogue Valley Roses, P.O. Box 116, Phoenix, OR 97535, (541) 535-1307

Rosemania, 4920 Trail Ridge Dr., Franklin, TN 37067, (888) 600-9665

Roses of Yesterday and Today, 803 Brown's Valley Rd., Watsonville, CA 95076, (831) 728-1901

Roses Unlimited, 363 N Deerwood Dr., Laurens, SC 29360, (864) 682-7673

S & W Greenhouse, Inc., P.O. Box 30, 533 Tyree Springs Rd., White House, TN 37188, (615) 672-0599

Two Sisters Roses, 1409 N Redbud Lane, Newcastle, OK 73065, (no phone by request)

Vintage Gardens (custom propagation), 4130 Gravenstein Hwy N, Sebastapol, CA 95472, (707) 829-2035

Wayside Gardens, 1 Garden Lane Hodges, SC 29695, (800) 213-0379

Wisconsin Roses, 7939 31st Ave., Kenosha, WI, 53142, (262) 358-1298

Witherspoon Rose Culture, 3312 Watkins Rd., Durham, NC 27707, (800) 643-0315

VARIETIES

Hybrid Tea roses with exhibition form display high centers. This is the classic modern rose you would expect to buy for a bouquet from a florist. These roses have pointed buds that unfurl slowly into elegant flowers with thick petals often exuding an exquisite fragrance. The roses Gemini™ and Moonstone™ are two great exhibition roses. The decorative form displays a different sort of beauty. It has a looser shape, often similar to a Peony or Dahlia. Yves Piaget® is a good example. It displays large, fragrant peony-shaped flowers. A number of Hybrid Teas are now being offered as plants growing on their own roots to improve winter hardiness. Examples are Aromatherapy™, Opening Night™, 'Pope John Paul II,' and 'Malibu.' The fragrance of Hybrid Teas varies from slight to strong. For fragrance Double Delight® and The McCartney Rose® stand out. Organic gardeners might want to try Elina® and Grande Amore®, designated All Deutschland Rose (ADR)—roses that have shown remarkable disease and insect resistance in Germany. When colors are separated by a forward slash (/) mark, the first color describes the upper surface of the petal and the color or colors after the (/) describe the undersides (reverse) of the petals.

VARIETIES	COLOR	FORM/PETAL COUNT	FRAGRANCE	HEIGHT
'Alabama'****	Deep rose pink	Exhibition 25—30	Moderate	36"
Alec's Red®****	Vivid medium red	Exhibition 45	Variable	30"
'Anastasia'****	Crystal white	Exhibition 30	Slight	36"
'Antigua'***	Apricot salmon	Decorative 30	Slight	48"
Aromatherapy™****	Medium pink	Exhibition 30—35	Strong	48"
Artistry™***	Orange pink	Exhibition 28—30	Slight	42"
Barbra Streisand™***	Lavender pink	Exhibition 32—35	Strong	48"
'Bella'roma'****	Yellow edged pink	Exhibition 30—35	Strong	42"
Betty White™****	Pink cream	Exhibition 60—65	Strong	42"
'Bewitched'****	Medium pink	Exhibition 28	Moderate	48"
'Big Ben'***	Dark red	Exhibition 35	Strong	48"

'Dainty Bess'

Double Delight®

'Fragrant Cloud'

Liebeszauber®

Princesse de Monaco®

Saint Patrick™

'Stainless Steel'

Touch of Class®

VARIETIES	COLOR	FORM/PETAL COUNT	FRAGRANCE	HEIGHT
Blue Moon®***	Light lilac mauve	Exhibition 40	Strong	48″
Brandy®****	Rich apricot	Exhibition 28–30	Moderate	42″
'Bride's Dream'****	Light satiny pink	Exhibition 28–30	Slight	36″
Brigadoon®***	Pink yellow edged dark pink	Exhibition 35–40	Spicy	36″
Cabana™***	Pink blotched white	Exhibition 25–30	Moderate	48″
'Cary Grant'***	Orange red	Exhibition 35–40	Spicy	40″
'Century Two'***	Medium pink	Exhibition 32	Moderate	48″
'Charlotte Armstrong'***	Deep pink	Decorative 35	Moderate	48″
Chicago Peace®***	Pink yellow	Exhibition 45–50+	Slight	36″
Chris Evert™***	Orange yellow	Exhibition 30	Moderate	36″
'Christian Dior'***	Bright red	Exhibition 45	Slight	48″
'Christopher Columbus'****	Orange red/yellowish	Decorative 25	Slight	48″
'Chrysler Imperial'****	Very dark red	Exhibition 45–50	Very strong	30″
'Color Magic'***	Salmon pink edged dark pink	Exhibition 28	Moderate	30″
'Command Performance'***	Orange red	Exhibition 25	Strong	48″
'Commonwealth Glory'****	Ivory peach	Decorative 32	Strong	48″
'Confidence'****	Pink tinged yellow	Exhibition 35	Moderate	48″
'Crimson Glory'****	Dark red	Exhibition 32	Very strong	30″
'Dainty Bess'*****	Light pink with red stamens	Decorative 5	Light to moderate	36″
'Dolly Parton'***	Vermilion red	Exhibition 35	Strong	36″
Double Delight®****	Creamy white edged red	Exhibition 32	Very strong	36″
'Elegant Beauty'****	Light yellow edged pink	Exhibition 20	Slight	48″
Elina® ('Peaudouce')*****	Light yellow to cream	Exhibition 32	Slight to moderate	36″
Falling in Love™****	Rich medium pink	Exhibition 28	Strong	48″
Firefighter™***	Dark red	Exhibition 42	Strong	42″
'First Love'***	Light pink	Exhibition 25	Slight	36″
'First Prize'****	Rose to light pink	Exhibition 28	Moderate	36″
Folklore®*****	Salmon/golden orange	Exhibition 44	Strong	60″
'Fragrant Cloud'****	Orange red	Exhibition 32	Very strong	36″
Friendship®***	Deep pink	Exhibition 28	Strong	48″
'Full Sail'***	White with pink tones	Exhibition 32	Strong	42″
Garden Party®***	White cream	Exhibition 28	Slight	48″
Gemini™*****	Pink cream edged dark pink	Exhibition 28	Slight	42″
Gentle Giant™**	Bright pink	Exhibition 30	Slight	42″
'Granada'****	Pinkish red yellow	Exhibition 20+	Strong	48″
Grande Amore®****	Medium red	Exhibition 32	Slight	42″
'Heirloom'***	Purplish mauve	Decorative 25	Strong	48″
Helmut Schmidt®****	Medium yellow	Exhibition 35	Light to moderate	36″
Honor™***	Cream white blushed pink	Exhibition 20	Slight	48″
In the Mood™****	Rich medium red	Exhibition 25–30	Slight	36″
Ingrid Bergman®***	Dark red	Decorative 35	Slight	36″
'John F. Kennedy'**	Snow white	Decorative 48	Moderate	36″
Just Joey®***	Coppery apricot	Exhibition 30	Moderate	36″

VARIETIES	COLOR	FORM/PETAL COUNT	FRAGRANCE	HEIGHT
'Keepsake' (Esmeralda®)****	Deep pink/yellowish/red edge	Exhibition 40	Moderate	48"
Konrad Henkel®****	Medium red	Exhibition 35	Moderate	40"
'Kordes' Perfecta'***	Deep pink with white base	Exhibition 68	Strong	48"
Lasting Love™***	Dark red	Decorative 25	Strong	48"
Legends™***	Medium ruby red	Decorative 30	Slight	42"
Let Freedom Ring™***	Dark red	Exhibition 17–25	Slight	42"
Liebeszauber®*****	Medium red	Decorative 32	Strong	40"
Love and Peace™****	Yellow pink cream	Exhibition 40	Slight	48"
'Malibu'***	Salmon orange	Exhibition 35	Spicy	30"
'Marijke Koopman'****	Medium pink	Exhibition 25	Slight	48"
Marilyn Monroe™**	Creamy apricot	Exhibition 32	Slight	36"
Medallion®****	Creamy pinkish apricot	Decorative 17–25	Moderate	40"
'Mellow Yellow'***	Soft yellow	Exhibition 40	Moderate	40"
Memorial Day™***	Pink lavender	Exhibition 35	Strong	42"
Michelangelo®***	Medium yellow	Decorative 42	Slight	48"
Midas Touch™***	Deep yellow	Exhibition 20	Moderate	36"
'Milestone'****	Medium red	Exhibition 40	Slight	48"
'Mirandy'***	Dark red	Exhibition 45	Strong	48"
'Miss All-American Beauty'****	Deep pink	Exhibition 55	Slight	36"
Mister Lincoln®****	Dark red	Exhibition 35	Strong	60"
Mon Cheri®***	Red with yellow tones	Exhibition 38	Spicy	48"
Moonstone™*****	White edged pink	Exhibition 32	Slight	42"
'Nantucket'****	Apricot pink	Exhibition 22–25	Slight	36"
Neptune™****	Pinkish mauve	Decorative 35	Strong	60"
'New Zealand'****	Light pink	Exhibition 32	Slight	48"
'Oklahoma'***	Dark red	Exhibition 48	Strong	36"
Olympiad®****	Medium red	Exhibition 32	Slight	48"
Opening Night™****	Bright dark red	Decorative 30	Slight	48"
'Oregold'***	Deep yellow	Exhibition 30–35	Slight	48"
'Osiria'***	Red/white	Exhibition 50	Strong	48"
Over The Moon™****	Apricot caramel	Exhibition 30+	Moderate	42"
Papa Meilland®***	Dark red	Exhibition 35	Strong	36"
Pascali®***	Creamy white	Exhibition 30	Slight	48"
'Peace'*****	Yellow pink white	Exhibition 43	Slight	40"
'Perfume Delight'***	Medium pink	Exhibition 28	Strong	40"
Peter Frankenfeld®***	Deep pink	Exhibition 40	Slight	48"
Peter Mayle™****	Dark pink	Exhibition 55	Strong	48"
'Pink Peace'***	Medium pink	Decorative 58	Strong	48"
Pink Promise™***	Light to dark pink	Exhibition 28	Moderate	42"
Polarstern® ('Polar Star')***	Creamy white	Exhibition 35	Slight	36"
'Pope John Paul II'***	Bright white	Decorative 50	Strong	42"
'Portrait'***	Light to dark pink	Exhibition 30+	Strong	48"

VARIETIES	COLOR	FORM/PETAL COUNT	FRAGRANCE	HEIGHT
Princesse de Monaco®****	White edged pink	Exhibition 35	Strong	40″
Pristine®***	White with light pink tones	Exhibition 30	Slight	40″
Proud Land®***	Dark red	Decorative 60	Moderate	60″
Queen Mary II™****	Pure white	Exhibition 36	Strong	36″
Red Masterpiece®***	Dark red	Exhibition 45	Strong	48″
Rio Samba™***	Yellow orange red	Exhibition 25	Slight	36″
Ronald Reagan™***	Rich red/whitish	Exhibition 35	Slight	42″
Rose Gaujard®***	Cherry red/silvery white pink	Exhibition 80	Slight	48″
'Rosie O'Donnell'***	Reddish orange/yellow white	Exhibition 32	Slight	36″
'Royal Highness'***	Light pink	Exhibition 40	Moderate	40″
'Rubaiyat'***	Deep pink	Exhibition 25	Strong	40″
Saint Patrick™****	Deep yellow tinted green	Exhibition 32	Slight	36″
'Seashell'***	Orange pink	Exhibition 48	Slight	48″
Secret®*****	Pink edged deep pink	Exhibition 32	Strong	36″
'Sedona'***	Coral orange red	Decorative 25–30	Slight	30″
'Sheer Bliss'***	White with pinkish tones	Exhibition 35	Spicy	60″
Sheer Elegance®***	Orange pink	Exhibition 32	Slight	40″
Signature®***	Deep pink	Exhibition 32	Slight	36″
'Simon Bolivar'***	Orange red	Exhibition 34	Slight	40″
'Soeur Thérèse'***	Gold yellow edged pink	Exhibition 25	Slight	40″
Solitaire®***	Vibrant yellow edged orange	Exhibition 25	Strong	40″
'South Seas'***	Orange pink	Decorative 48	Moderate	36″
Stainless Steel™****	Light lavender	Exhibition 26	Strong	40″
'Stephen's Big Purple'***	Pinkish purple crimson	Exhibition 60	Strong	36″
'Sue Hipkin'****	Apricot blushed pink	Exhibition 42	Strong	42″
'Suffolk'***	White edged pink	Exhibition 41+	Slight	36″
Summer Love™***	Medium yellow	Decorative 32	Slight	36″
Sunset Celebration™*****	Apricot cream orange	Exhibition 36	Moderate	36″
'Swarthmore'****	Red to deep pink	Exhibition 48	Slight	48″
Tahition Sunset™****	Light apricot pink	Exhibition 28	Moderate	42″
'Tallyho'***	Deep pink	Exhibition 35	Strong	48″
The McCartney Rose®*****	Medium pink	Decorative 22	Very strong	60″
'Tiffany'***	Pink with white tones	Exhibition 28	Strong	36″
Tineke®***	White cream	Exhibition 50	Slight	36″
Touch of Class®***	Orange pink	Exhibition 28	Slight	40″
Traviata™***	Dark red	Decorative 100	Slight	48″
'Tropicana'***	Orange red	Exhibition 32	Moderate	40″
Uncle Joe® ('El Toro')****	Dark red	Exhibition 80	Slight	48″
Veterans' Honor®***	Dark red	Exhibition 28	Slight	42″
'Voluptuous!'****	Dark pink	Exhibition 35	Slight	48″
Whisper™***	Cream white with peach tones	Exhibition 32	Moderate	42″
Yves Piaget®****	Dusky pink	Decorative 46–80	Strong	36″

Dee Bennett™ (Miniature)

MINIATURE AND MINIFLORA ROSES

Miniature roses grow from 8 to 18 inches (20 to 45 cm) tall with tiny blooms similar to those of Hybrid Teas and small leaves in proportion to their size. Their daintiness is their most desired quality. In 2000 a new class of rose was born to accommodate terrific roses that did not fall into already established categories. This new class was the Miniflora. It was too large to be called a Miniature, but it had wonderful qualities including larger blooms that could be seen from a distance and a smaller size that made it ideal for growers with limited space. In Europe these roses are called Patio Roses, a description that fits the plants perfectly. Some Miniature roses have now been reclassified as Minifloras. Miniatures are excellent for edging plants or ground covers, in rock gardens or beds, in window boxes or containers, on patios or decks, and as potted indoor plants. A few of them are ideal as Climbers or speci-

mens for hanging baskets. Some gardeners use them for bonsai or Tree roses (standards). Minifloras may be grown in pots or directly in small gardens and are a delightful addition to the rose garden with their lovely, colorful blooms.

How Miniature and Miniflora Roses Grow

Most Miniature roses sold today are grown on their own roots with the exception of Miniature Tree roses that are typically budded or grafted. Minifloras are often budded onto a different rootstock. True Miniatures have thin, almost wiry stems. Plants look like small, many branched bushes. Flowers, produced singly or in clusters, are typically less than 1 inch (2.5 cm) across, varying greatly in petal count. Leaves are usually ½ inch (12 mm) wide to 1 inch (2.5 cm) long. The space between leaves is quite short. Each leaf generally has from 3 to 5 leaflets. Minifloras could be called Miniature Floribundas. Plant, leaf, and flower size is larger than that of a Miniature, but smaller than most Floribundas. As with Miniatures, the flowers come in a wide range of form and colors. With some exceptions Miniatures lack a strong fragrance which is not the case for a number of Minifloras.

Where to Plant

Site and Light Plant in a sunny location. Best blooms occur if plants have at least 6 hours of direct sunlight a day. Sunlight also prevents many foliar diseases. Avoid planting under shade trees. Trees also compete for water and nutrients. Avoid planting under eaves as well unless you have lots of time to water the plants as the soil dries out. Also avoid planting under drip lines, since water running off the roof can damage the plants.

Soil and Moisture Miniatures and Minifloras need rich soil that retains moisture but drains freely. Replace clay or rock with loam purchased from a garden center or in bags as potting soil. Add lots of organic matter, such as compost, leaf mold, peat moss, or rotted manures. These keep the soil moist and cool during dry periods. They also help the soil drain freely, encourage the growth of beneficial soil microorganisms, and attract worms which keep the soil aerated and fertilized. Soil should be light, loose, and airy. This allows quick root growth and good drainage so that plenty of oxygen can get to the tender feeder roots. A handful of bonemeal or superphosphate mixed into the soil provides additional nutrients for the young plant. Proper soil preparation takes a little time, but it is critical to good growth.

If you grow these roses in pots, repot or pot up each year. Remove and replace some of the old soil.

Spacing Space according to the approximate height expected for the plant. Give enough space so that there is good air circulation around the branches to avoid disease. Space Miniatures about 12 to 18 inches (30 to 45 cm) apart. Give Minifloras more space. Miniature tree roses (standards) will need support.

Planting

You can start Miniatures and Minifloras either from bare root (dormant) plants or from roses already potted. Often when you order bare root Miniature roses, you get roses already potted in tiny pots. Removing these and planting them in larger and more appropriate pots is called *potting up*.

Bare Root Miniature and Miniflora roses grown on their own roots have a dense, fibrous root system. Budded bare root plants typically have fewer, larger, and longer roots. Both can be planted directly in the garden or in pots. Planting them in pots has several advantages: in cold climates the season is short, and the soil in pots warms up quickly in spring giving your plants a head start; it is easy to improve the soil each time you pot the plant up; you have tight control over the amount of fertilizer each plant gets; and, potted plants are easy to move around and winter protect. Still, especially with Minifloras, some gardeners prefer planting them in a set location in the landscape.

Miniatures often have thin canes, as many as 12 to 18 on the very best plants. Canes are usually 8 to 10 inches (20 to 25 cm) tall. Minifloras have fewer but larger canes.

Plant both types so that the lower portion of each cane is buried an inch or two (2.5 to 5 cm) below the soil. This encourages the rose to develop roots from the buried portion of the stem making the plants more vigorous and winter hardy. This is especially important if you buy budded or grafted plants (some Miniatures and many Minifloras are).

Potting Up Already Potted Plants Mail order Miniatures may arrive in small 2- to 3-inch (5- to 7.5-cm) pots. Some plants are completely dormant, with no sprouting buds or leaves. Other plants may have yellow, drooping leaves; remove these leaves immediately, and consider these stressed and denuded plants as fully dormant. If you have a plant that is covered with green, healthy leaves, don't strip off the leaves.

Regardless of what condition the plant is in, be sure to water the soil in the pot immediately. Then allow it to drain overnight in a cool place, such as your garage, before potting it up.

The next day pour 3 to 4 inches (7.5 to 10 cm) of potting soil into the bottom of a 7- to 10-inch (17.5- to 25-cm) plastic pot.

Tap the side of the little pot against something hard while you hold your fingers over the plant as it's turned upside down. The plant will pop right out of the pot. Keep the rootball intact.

Gently set the rootball on the soil in the larger pot. You want the lower portion of canes to end up buried an inch or two (2.5 to 5 cm) below the surface of the soil. The latter should end up 1 inch (2.5 cm) below the rim of the pot. You may have to add more soil underneath the plant to get it in the correct position. When doing this, gently scoop the plant from underneath to avoid breaking the rootball.

Pour 3 to 4 inches (7.5 to 10 cm) of soilless mix around the base of the rootball and firm it in place with your fingers.

Sprinkle about ¼ cup (55 g) of Milorganite on top of the mix. Using an organic fertilizer is recommended, since inorganic products could burn the plant's tender feeder roots. Some growers use coated, slow-release fertilizers instead. Organic gardeners will only use Milorganite.

Fill in the rest of the pot with potting soil. Then press down on the soil with your fingers to remove air pockets. Pour more soil in if necessary. The top of the soil should be about 1 inch (2.5 cm) from the rim of the pot.

Fill the pot to the brim with water. Do this several times until water begins to drain out the bottom holes. Do this again the following day so that soil stays consistently moist.

If the plant doesn't have leaves, keep it in your garage, where it will bud out over a period of days. Keep the soil moist but not soggy. Check moisture by feel, not by sight. Just push your finger into the soil to tell whether it's drying out.

Mist all canes twice daily to keep them moist. Continue doing this until the plant begins to form buds. You do not have to mist plants that have already leafed out.

Place any budding or fully leafed-out rose outdoors after all danger of frost. Place it first in partial shade. Gradually increase the amount of sun over a period of 10 days until the plant is in full sun. This is called *hardening off,* a gardening term for letting a plant get used to brighter light, varying temperatures, and lower humidity.

Buying Potted Plants Miniature (both bush and Tree) and Miniflora roses are becoming increasingly popular. The best-known varieties are often stocked in local garden centers or nurseries. Pick out plants with healthy canes and luxuriant foliage. Check plants carefully for any signs of disease or insect infestations to avoid buying infected plants. Avoid leaving the plant in the car while you do other errands, since the heat buildup can damage it. Also, if the plant will be exposed to wind, protect it well by wrapping it in thick plastic or paper. As soon as you get home, water the soil if it's at all dry. Plant already potted Miniature roses as outlined in Part II.

How to Care for Miniature and Miniflora Roses

Water Keep the soil evenly moist at all times throughout the season for best bloom. If the plant is kept in a pot, water it more frequently. This may mean watering twice or more a day. Most growers use plastic pots to cut down on watering, since clay ones dry out quickly. Water potted plants until water begins to pour out the drain holes. Then water again. This saturates the soil completely. If the soil moves away from the rim of the pot, it indicates that you need to water the plant more often. If this happens, push the soil back against the rim so that there is no gap between the soil and the pot. Saturate the soil immediately. Avoid letting this happen again because lack of water stresses plants badly.

Mulch After the soil warms up to 60°F (15.6°C) apply a mulch around the base of the plant. The mulch should come close to but not touch the canes. The most commonly used mulch is shredded (not whole) leaves. Place at least a 3-inch (7.5-cm) layer over the soil. If plants are growing in pots, simply cover the soil surface with pulverized leaves. Leaves keep the soil moist and cool which encourages rapid root growth. The mulch feeds microorganisms and worms which enrich and aerate the soil. The mulch also inhibits weed growth and makes pulling any weeds that do sprout much easier. Other good mulches include pine needles and grass clippings. The latter should be only 2 inches (5 cm) deep, or they will heat the soil and may smell. These mulches are all inexpensive and effective. If you use chipped wood or shredded bark, apply additional fertilizer to the soil, since these rob the soil of nitrogen as they decompose. Because mulch is eaten by soil microorganisms and worms, it needs to be replenished regularly throughout the growing season. Remove and compost all mulch in the fall.

Cocoa bean hulls make an excellent mulch around planted Miniatures. Many growers place groups of potted Miniatures on a bed of cocoa bean hulls. Used in both ways, the small hulls are an attractive light brown and exude a fragrance of chocolate. A layer 3 inches (7.5 cm) deep is fine. If cocoa bean hulls are too thick, they will sometimes develop a white, crusty look. Rake them lightly to get rid of this discoloration.

Fertilizing If combining the use of inorganic and organic fertilizers, be careful during the first year. Use inorganic fertilizer only after the plant has leafed out and is growing vigorously.

In subsequent years, as early as possible in spring, sprinkle 10-10-10 (or 5-10-5) granular fertilizer around the base of each plant. Use common sense in judging the amount, more for bigger plants, less for smaller. Hand sprinkle, never placing fertilizer against the stem of the plant. Water immediately to dissolve the granules and carry nutrients to the roots.

In both the first and subsequent years do the following: One to two weeks after the first feeding with 10-10-10 granules soak the base of each plant with a liquid containing 20-20-20 water soluble fertilizer following directions on the label. Use common sense, giving larger plants more, smaller plants less. This feeding will stimulate sensational bloom as the plant matures.

One week later soak the base of each plant with a liquid containing fish emulsion. Follow the direc-

tions on the label. Feed according to the size of the plant.

One week later give all plants the 20-20-20 treatment again. One week later, give all plants the fish emulsion soaking. Keep this rotation going until the middle of August. At this time, stop all use of fertilizers containing nitrogen, since nitrogen stimulates new growth which dies off in most winters. Use 0-10-10 fertilizer after mid-August if desired.

Additional fertilizing with Epsom salts (magnesium sulfate) is highly recommended. Give 1 tablespoon (15 g) for larger plants, less for smaller, two times a year: in late May and in early July. Magnesium sulfate neutralizes the soil and makes it easier for the plant to take in nutrients. This stimulates the growth of new canes (basal breaks) from the base of the plant.

Many people like to grow Miniatures and Minifloras in flower pots throughout the plant's entire lifetime. In such cases, adding iron to the soil is an excellent idea to keep the foliage healthy. Sprint is one of the best products for this. Follow the directions on the label exactly, matching the amount you use to the size of the pot.

Good fertilizers for organic growers are alfalfa meal (rabbit pellets), blood meal, bonemeal, compost, cow manure, fish emulsion, rotted horse manure, and Milorganite. Add bonemeal to the soil before planting. The others are effective added to the soil at planting time and as additional feedings on the surface of the soil throughout the season.

Weeding Weeds compete with roses for nutrients and water. They also act as hosts for harmful insects. Destroy all perennial weeds before planting. Use an herbicide such as Roundup®. Inhibit annual weeds with the use of mulch. Pull by hand any weeds that appear through the mulch. Avoid hoeing, since this can harm the shallow root system.

If you're growing Miniatures in pots, pull out all weeds. Clean up all dead leaves and fallen petals from the soil, since they act as breeding areas for disease spores and insect eggs.

Staking Miniature Tree roses (standards) require support. Place a metal rod to the side of the stem at the time of planting. A typical support would be a 36-inch (90-cm) piece of ¼-inch (6-mm) stainless steel rod or aluminum conduit. Wind electric tape around both the stem and support. Do this 3 inches (7.5 cm) from the base of the plant and just under the upper branches. You do not want the plant to jiggle in the soil during heavier winds; this disturbs the root system and harms the plant. If the plant is exposed to heavy winds, tie individual upper canes to the support with polyester twine. This may seem fussy, but it will often prevent canes from snapping in wind gusts. In effect, it can save your plant.

Disbudding Some growers remove flower buds in the first year, directing all energy to the plant itself. Remove buds just as they appear, for best results.

Removing buds on mature plants is rarely done, since the goal is to have bushes covered in a shower of blossoms.

Deadheading When flowers fade, remove them immediately by snipping off the blossoms to a point just above the nearest leaf below. This encourages the plant to form more blossoms. It also keeps the area around the plant tidy and prevents disease and insect infestations. You can also snip off growth to keep the plant at exactly the desired shape and size. This could fall under the category of pruning or pinching back. Don't overdo it, or you will harm the plant.

Pruning During the first year, do no pruning at all, other than to remove diseased or damaged cane with sharp shears. Cut back to a point ¼ inch (6 mm) above an outward-facing bud. Also, remove all flower buds until the middle of summer. Let the plant bloom lightly to help it go into winter dormancy.

During the second year, remove any dead or damaged cane as you lift the plant from the ground. Remove especially spindly growth. Some gardeners cut Miniatures back to one-half their original size. Cutting back is most helpful on vigorous growers; however, it is not necessary on all plants and not nec-essary at all on Minifloras (although done in warmer climates).

During the third and subsequent years, remove dead canes. Also remove canes that cross so that there is good air circulation in the center of the plant. Prune to shape the plant for aesthetic reasons. Leave at least six healthy canes on each plant (or more on Miniatures). Pruning encourages bushiness and fuller bloom.

Plants growing on their own roots often produce new cane off to the side of the parent plant. Do not remove these, since they will produce flowers identical to the original plant (unlike many roses grown on a different rootstock).

Miniatures can be trimmed into bonsai specimens. Since this is a complicated process, you should try to join a local group that specializes in bonsai for specific tips.

Winter Protection Most Miniatures and Minifloras are hardy only to 26°F (−3.3°C). The Minnesota Tip Method provides excellent winter protection. If you prefer to leave roses standing, protect them with the Hass Teepee method. If planted in containers, bury them alongside tipped roses or protect them using the Hass Mound Method (see pp. 193–199).

Some growers with a limited number of potted plants, place them in the garage during the winter. When it gets close to freezing outdoors, clean debris off the soil around the base of the plants and snip off all buds and flowers. Moisten the soil thoroughly until water drains out the bottom drain holes. Spray the plants with a dormant spray soaking leaves, canes, and the soil. Tie the canes together. Transfer the pots into the garage. Place a section of moistened newspaper in the bottom of a sturdy, black plastic garbage bag. Set the bag on a board or similar object to give support but keep the bottom off the cement. Set two or three pots in the bag, depending on their size. Tie the bag with a twist'em. Naturally, place the pots in the most protected spot in your garage, away from opening and closing doors, and keep your garage door closed during the winter. *Unless your garage is heated somewhat, this kind of winter protection is a gamble.*

Few guides recommend Miniature roses as indoor plants. However, they can do well if given lots of attention (more than the average person is willing to do). They will not bloom continually, but they will bloom intermittently throughout the winter. Note that indoor conditions are not ideal for growing these plants—the air is dry and the sun often shrouded—so, you'll have to make up for these poor growing conditions.

At the end of the season, after the plant has gone through a chilling period outdoors, repot it. Remove some of the old soil and add new. Some growers cut all cane back to half its length at this time (strictly optional). The container must drain freely. Dip the newly potted rose into a bucket of water just over the rim of the pot. When the bubbling stops, lift it out of the water and let it drain. Keep it in a cool, shaded area (such as your garage) for 2 to 4 weeks. This allows the plant to go through a pseudo-dormant period before going indoors; this mimics a mini-winter.

After this period, bring it indoors. Set it on a plastic dish (sold in most garden centers) so that excess water will drain out the bottom of the pot and collect in the dish. Place it in a sunny window (southern exposure is recommended). Provide good artificial light (grow lights or fluorescent light 14 to 16 hours per day, 8 inches [20 cm] above the top of the plant).

Moisture is critical. Keep the soil evenly moist at all times. High humidity helps but can be difficult to provide indoors during the winter. Placing plants on a tray filled with pebbles and water helps (the bottom of the pot should not be in water). Misting is also recommended.

Feed once a month until the plant blooms. Then feed the rose twice a month. Remove spent blossoms as you would outdoors by cutting back to a leaf below the flower. This encourages additional bloom.

If the plant gets infested with insects or spider mites, water it well. Then take it out of direct light, enclose it in a plastic bag, and spray it with an insecticide or miticide through the opening at the top of the bag. Spray quickly, and immediately close the bag tightly by first twisting the plastic and then tying the bag shut with a twist'em. Keep the bag sealed for at least an hour while the spray drifts inside and settles on the plant and soil. Open the bag in a ventilated area after an hour. If you work quickly, you can do this outdoors in just a few seconds. Then place the plant back where it was growing in full light.

Some growers place pots in the crisper of the refrigerator during January and February to force dormancy—fine if you have only one or two plants; otherwise, impractical. After the forced dormancy, the procedure is to prune the plant to half its original size (if this was not already done in the fall), then continue to water and feed the plant as recommended earlier. Forced dormancy is not necessary.

Problems

Insects Follow a regular routine of spraying every 7 to 10 days to ward off insect infestations. During hot, dry weather watch out for spider mites. Begin using a miticide immediately if any of these appear.

Disease Follow a 7- to 10-day spraying routine to prevent common rose diseases, including powdery mildew and black spot. These are much easier to prevent than to control. There are many effective products available in local garden centers. If you're an organic gardener, join a local rose society or a garden club. Members will tell you which plants have done well in your area. This is especially helpful since disease resistance varies greatly throughout the cold-climate region.

Propagation

Miniatures are among the easiest groups of roses to propagate. *It is illegal to propagate any plant still under patent.* Patented plants have symbols behind their name, such as actual plant patent numbers or PPAF (Plant Patent Applied For).

Division Plants grown on their own roots may produce shoots off to the side. Cut these off in early spring as close to the crown as possible, and plant them immediately as if bare root. Keep the shoots consistently moist to encourage them to take root.

Cuttings When leaves have already formed, cut off the tip of a vigorous new cane. The tip should be 3 to 8 inches (7.5 to 20 cm) long and contain four nodes (places where leaves join the stem). Cut the stem straight across ¼ inch (6 mm) below a leaf node with a razor-sharp knife (pruning or utility knife). Remove all leaves except two at the end of the cutting. Dip the entire cutting in a disinfectant. Dip the cut end in rooting hormone. Tap off any excess powder. Place the cutting in a moist, sterile rooting medium (a combination of peat moss, perlite, and vermiculite is good) so that most of the stem is buried. Keep the medium consistently moist. Bottom heat from heating pads or cables is highly recommended, especially for varieties that are the most difficult to grow from cuttings.

Most Miniatures, such as 'Rise 'n' Shine,' are extremely easy to grow from cuttings, while others are quite difficult. It's possible to grow many by putting them in a dark (opaque) container filled with water. Place the container in indirect light. Roots form underwater in the darkness of the container. Replace water regularly to keep it oxygenated. Many indoor gardeners have done this with cuttings from houseplants, such as geraniums and ivies. It's exactly the same process. When the young cutting has roots 3 to 4 inches (7.5 to 10 cm) long, plant it as you would a bare root plant in the garden or preferably in a pot.

Special Uses

Cut Flowers Flowers of these roses are delightful floating in small bowls or placed on cakes for decoration. Rose flowers are edible, but do not eat them or place them on anything that will be eaten if they have been sprayed with chemicals.

Dried Flowers Miniature roses are very popular in arrangements, plaques, collages, and locket jewelry. Dry the flowers in silica according to recommendations on the package.

Sources

Angel Gardens, P.O. Box 1106, Alachua, FL 32616, (352) 359-1133

Antique Rose Emporium, 9300 Lueckmeyer Rd., Brenham, TX 77833, (800) 441-0002

Burlington Rose Nursery, 24865 Rd. 164, Visalia, CA 93292, (559) 747-3624

Chamblee's Rose Nursery, 10926 US Hwy 69 N, Tyler, TX 75706, (800) 256-7673

Classic Miniature Roses, P.O. Box 2206, Sardis, BC V2R 4L4 Canada (Canada only)

Countryside Roses, 5016 Menge Ave., Pass Christian, MS 39571, (228) 452-2697

Edmunds' Roses, 335 S High St., Randolph, WI 53956, (888) 481-7673

Garden Valley Ranch, 498 Pepper Rd., Petaluma, CA 94952, (707) 795-0919

Heirloom Roses, 24062 Riverside Dr. NE, Saint Paul, OR 97137, (503) 538-1576

High Country Roses, P.O. Box 148, Jensen, UT 84035, (800) 552-2082

Hortico, Inc., 723 Robson Rd., RR# 1, Waterdown, ON L0R 2H1 Canada, (905) 689-6984

Kimbrew-Walter Roses, Inc., 2001 Van Zandt County Rd. 1219, Grand Saline, TX 75140, (903) 829-2968

Northland Rosarium, 9405 S Williams Lane, Spokane, WA 99224, (509) 448-4968

Palatine Fruit & Roses, 2108 Four Mile Creek Rd., RR #3, Niagara-on-the-Lake, ON L0S 1J0 Canada, (905) 468-8627

Pickering Nurseries, 3043 County Rd. 2, RR #1, Port Hope, ON L1A 3V5 Canada, (905) 753-2155

Pine Tree Miniature Roses, 5283 N Shore Dr., Duluth, MN 55804, (218) 525-4794

Regan Nursery, 4268 Decoto Rd., Fremont, CA 94555, (800) 249-4680

Rogue Valley Roses, P.O. Box 116, Phoenix, OR 97535, (541) 535-1307

Rosemania, 4920 Trail Ridge Dr., Franklin, TN 37067, (888) 600-9665

Roses Unlimited, 363 N Deerwood Dr., Laurens, SC 29360, (864) 682-7673

S & W Greenhouse, Inc., P.O. Box 30, 533 Tyree Springs Rd., White House, TN 37188, (615) 672-0599

Two Sisters Roses, 1409 N Redbud Lane, Newcastle, OK 73065, (no phone by request)

Vintage Gardens (custom propagation), 4130 Gravenstein Hwy N, Sebastapol, CA 95472, (707) 829-2035

Wells Midsouth Roses, 471 Lucy Kelly, Brighton, TN 38011, (901) 476-6064

Wisconsin Roses, 7939 31st Ave., Kenosha, WI, 53142, (262) 358-1298

Witherspoon Rose Culture, 3312 Watkins Rd., Durham, NC 27707, (800) 643-0315

VARIETIES

Choose varieties by color, disease resistance, flower form, fragrance, and size. Also, consider how you'll be using them. 'Candy Cane,' 'Jeanne Lajoie,' 'Red Cascade,' and 'Valentine's Day' are excellent climbing varieties if given support. Particularly good in baskets are Denver's Dream™, 'Party Girl,' and Sweet Chariot™. Fragrance varies from none to strong. 'Jennifer' and Sweet Chariot™ are two exceptional choices for fragrance. Arcanum®, 'Butter Cream,' and Lady E'owyn™ produce lovely roses for exhibition, but are not free flowering in cold climates.

Miniatures are susceptible to powdery mildew. Among the more resistant plants are Baby Masquerade®, 'Bee's Knees,' Cupcake™, Gizmo™, 'Gourmet Popcorn,' 'Jeanne Lajoie,' Magic Carrousel®, 'Popcorn,' Rainbow's End™, Starina®, Ty™, and 'Valentine's Day.'

The roses listed are Miniatures unless designated (MinFl) after the height, which stands for Miniflora. A number of roses once classified as Miniatures are now listed as Minifloras. When colors are separated by a forward slash (/), the first color describes the upper surface of the petal and the color or colors after the (/) describe the undersides (reverse) of the petals. The slash may also be followed by information on the center or eye of the flower, which may contrast nicely to the color of the petals surrounding it.

VARIETIES	COLOR	FORM/PETAL COUNT	FRAGRANCE	HEIGHT
'Abby's Angel'****	Yellow edged red	Exhibition 26—40	Slight	36" (MinFl)
'All American Girl'***	Light to medium pink	Exhibition 17—25	Slight	28" (MinFl)
Ambiance™****	Pastel peach	Exhibition 40	Slight	30" (MinFl)
'Amy Grant'***	Light pink	Exhibition 17—25	Variable	24" (MinFl)
'Andie MacDowell'****	Orange red	Exhibition 17—25	None	24" (MinFl)
Antique Rose™***	Medium to dark pink	Decorative 38	Slight	18"
Apricot Twist™****	Apricot pink	Exhibition 15—22	Slight	18"
Arcanum®****	Apricot edged red	Exhibition 17—25	None	30"
'Ashton'****	Deep to light pink	Exhibition 17—25	None	30" (MinFl)
'Autumn Bliss'**	Mustard edged orange red	Exhibition 26—40	Slight	30" (MinFl)
'Autumn Splendor'****	Gold yellow orange	Exhibition 26—40	Slight	30" (MinFl)
'Avandel'***	Yellow apricot pink	Decorative 20—25	Moderate	18"
'Baby Betsy McCall'****	Light pink	Decorative 20	Moderate	10"
Baby Boomer™***	Medium pink	Exhibition 17—25	None to slight	24"
'Baby Katie'****	Pink white tones	Exhibition 28	Slight	10"
Baby Masquerade®***	Yellow pink red	Decorative 17—25	Slight	10"
'Baldo Villegas'****	White edged red	Exhibition 25—35+	Slight	36" (MinFl)
'Beauty Secret'****	Medium red	Decorative 17—25	Strong	12"
'Bee's Knees' (Bees Knees™)*****	Yellow edged pink	Exhibition 25—35+	Slight	30"
'Best of 04'***	Yellow orange	Exhibition 17—25	None	18"
'Billie Teas'***	Dark red	Decorative 26—40	None	14"
Black Jade™****	Dark red	Exhibition 30—35	None	18"
'Butter Cream'*****	Medium yellow	Exhibition 26—40	None	30" (MinFl)
'Café Olé'***	Russet	Decorative 40—50	Moderate	20"
'Caledonia'****	Light yellow	Exhibition 26—40	None	30" (MinFl)
'Camden'****	Deep pink red	Exhibition 17—25	Slight	36" (MinFl)
'Candy Cane' (Cl)**	Pink red splashed white	Decorative 13	Slight	Variable
'Carolyn's Passion'***	Medium yellow	Exhibition 17—25	Moderate	24" (MinFl)
'Charismatic'***	Red white edged red	Exhibition 17—25	None	24" (MinFl)
Child's Play™***	White edged pink	Exhibition 20	Moderate	30"

Black Jade™ (Miniature)

'Conundrum' (Miniflora)

'Leading Lady' (Miniflora)

Little Paradise™ (Miniature)

VARIETIES	COLOR	FORM/PETAL COUNT	FRAGRANCE	HEIGHT
'Cinderella'****	Light pink to white	Decorative 35—50+	Moderate	12″
Class of '73™***	Cream pink edged deep pink	Exhibition 17—25	None	30″ (MinFl)
'Conundrum'*****	Yellow edged red	Exhibition 17—25	Slight	24″ (MinFl)
'Corsage'**	White pink tones	Decorative 17—25	Slight	24″ (MinFl)
Cupcake™****	Medium pink	Exhibition 45—50+	None	18″
'Daddy Frank'*****	Medium red	Exhibition 17—25	None	30″ (MinFl)
Daddy's Little Girl™***	Rose pink	Decorative 17—25	Slight	18″
Debidue™***	Deep pink	Decorative 26—40	Slight	12″
Dee Bennett™****	Orange gold red	Exhibition 25	Fruity	18″
'Déjá Blu'***	Mauve magenta	Exhibition 26—40+	Slight	30″ (MinFl)
Denver's Dream™***	Orange red	Exhibition 26—40	None	12″
'Dr. John Dickman'*****	Mauve edged red	Exhibition 26—40	Moderate	36″ (MinFl)
'Dr. Troy Garrett'**	Bright red	Exhibition 26—40	Moderate	30″ (MinFl)
Double Gold™***	Light golden yellow	Exhibition 28—36	Strong	24″ (MinFl)
'Double Take'****	Red white	Exhibition 17—25	None	30″ (MinFl)
'Easter Morning'***	White	Decorative 60—70	Slight	12″
'Emily Louise'***	Yellow edged pink	5	Slight	24″ (MinFl)
'Equinox'***	Reddish orange/light white	Exhibition 26—40	None	30″ (MinFl)
'Fairhope'*****	Light yellow	Exhibition 16—28	Slight	24″
'First and Foremost'***	Deep pink/white/pink	Exhibition 17—25	None	30″ (MinFl)
'First Choice'***	Cream edged pink	Exhibition 17—25	Slight	30″ (MinFl)
'Fitzhugh's Diamond'****	Yellow edged deep pink	Exhibition 26—40	None	30″ (MinFl)
'Flawless'*****	Medium pink	Exhibition 17—25	Strong	30″ (MinFl)
'Focal Point'***	Apricot pink to deep pink	Exhibition 26—40	Slight	30″ (MinFl)
'Foolish Pleasure'****	Pink cream	Exhibition 26—40	None	30″ (MinFl)
Gingerbread Man™***	Apricot orange	Decorative 35+	Slight	18″
Gizmo™****	Orange red/white eye	5	Slight	20″
'Glowing Amber'***	Red yellow/yellow	Exhibition 26—40	Slight	24″
'Gourmet Popcorn'***	White	Decorative 9—16	Slight	18″
'Grace Seward'*****	White/gold center	5	Strong	24″
'Green Ice'***	White green	Decorative 17—25+	Slight	24″
'Harm Seville'***	Dark red	Decorative 17—25	Slight	24″ (MinFl)
'Heartbreaker'***	Pink cream	Exhibition 17—25+	Slight	24″
'Herbie'***	Mauve purple	Exhibition 25—30	Slight	12″
'High Ambition'****	Light apricot cream	Exhibition 26—40	Strong	30″ (MinFl)
Hot Tamale™****	Orange yellow edged red	Exhibition 26—40	Slight	14″
'Hula Girl'***	Salmon orange	Exhibition 35—45	Moderate	12″
Incognito™****	Red mauve	Exhibition 15—25	Slight	30″
'Ingrid'***	Red/yellow	Exhibition 26—40	Moderate	36″ (MinFl)
Innocence™****	White	Exhibition —25+	Slight	20″
'Irresistible'*****	White tinged pink	Exhibition 43—45	Moderate	18″
Jean Kenneally™*****	Apricot pink	Exhibition 22	Slight	18″
'Jeanne Lajoie'(Cl)****	Medium pink	Decorative 35—40	Slight	Variable
'Jennifer'****	Pink cream	Exhibition 25—35	Strong	18″
'Jerry Lynn'****	Apricot	Exhibition 26—40	None	24″ (MinFl)
Jerry-O™***	Orange pink/gold center	Decorative 25	Spicy	18″
Josh Alonso™****	Orange peachy pink	Exhibition 25—35	None to slight	24″

VARIETIES	COLOR	FORM/PETAL COUNT	FRAGRANCE	HEIGHT
'Joy'***	White edged pink	Exhibition 17—25	None	24"
'Joyful'**	Pink/cream	Exhibition 17—25	Slight	24" (MinFl)
'Judy Fischer'***	Medium pink	Decorative 17—25	Slight	14"
Julie Ann™***	Orange red	Decorative 20	Moderate	12"
'Just For You'***	Deep pink	Exhibition 35	Slight	12"
'Kismet'***	Yellow red	Exhibition 17—25	None	30" (MinFl)
Kristin™ ('Pirouette')****	White edged red	Exhibition 27—30	None	24"
Lady E'owyn™****	Pink white edged pink	Exhibition 26—40	None	24" (MinFl)
'Leading Lady'***	White pink/white	Exhibition 17—25	Moderate	36" (MinFl)
'Lemon Drop'***	Medium yellow	Decorative 35+	Slight	30"
Little Jackie™***	Orange red	Exhibition 20	Strong	10"
'Little Linda'***	Light yellow edged red	Decorative 17—25	Slight	8"
Little Paradise™***	Pinkish lavender edged red	Exhibition 17—25	Slight	18"
'Little White Lies'****	White	5	None	10"
'Lo & Behold'****	Deep clear yellow	Exhibition 17—25	Slight	30" (MinFl)
'Loyal Vassal'***	Orange/yellow orange	Exhibition 26—40	Slight	30" (MinFl)
'Luis Desamero'****	Light yellow	Exhibition 28	Fruity	24"
'Luscious Lucy'****	White yellow edged red	Exhibition 17—25	None	30" (MinFl)
Magic Carrousel®****	White edged red	Exhibition 17—25+	Slight	18"
'Magic Show'***	Red white/white	Exhibition 17—25	None	24"
'Mariam Ismailjee'***	Red yellow edged red	Exhibition 25—35	Slight	30" (MinFl)
'Mary Marshall'***	Pink orange tints	Exhibition 17—25+	Moderate	12"
Maurine Neuberger™****	Rich medium red	Exhibition 30	Moderate	12"
Memphis King™****	Dark red	Exhibition 26—40	Slight	30" (MinFl)
Memphis Queen™****	White blushed pink	Exhibition 50—60+	None to slight	24
Merlot™***	Red/white	Exhibition 17—25	Slight	24"
Minnie Pearl®*****	Pink salmon	Exhibition 17—25+	Slight	18"
Miss Flippens™****	Medium red	Exhibition 17—25	None	24"
'Moonlight Scentsation'***	White with lavender	Decorative 50—55	Strong	36" (MinFl)
'Mother's Love'***	Dark pink to light pink	Exhibition 20—25	Slight	24"
'My Inspiration'***	Red/white	Exhibition 26—40	Slight	30" (MinFl)
'My Sunshine'****	Bright clear yellow	5	Moderate	20"
Neon Cowboy™****	Bright red/golden center	5	Slight	16"
'Orange Honey'***	Pale orange	Decorative 20—23	Slight	12"
'Over the Rainbow'***	Red/yellow orange	Exhibition 17—25	Slight	10"
Overnight Scentsation™*****	Bright medium pink	Exhibition 41+	Strong	36" (MinFl)
Pacesetter®*****	Creamy white	Exhibition 35—45+	Moderate	14"
Pacific Serenade™***	Medium yellow	Decorative 17—25	Moderate	30"
Palmetto Sunrise™***	Orange red yellow	Decorative 20—25	None	18"
'Party Girl'****	Apricot pink yellow	Exhibition 23	Moderate	12"
Peach Delight™***	Light apricot peach	Decorative 35—60+	Strong	28" (MinFl)
'Peaches n' Cream'***	Pink creamy white	Exhibition 50—52	Slight	10"
Pierrine™****	Orange pink	Exhibition 30—40+	Slight	12"
'Popcorn'***	White/yellow center	Decorative 13	Strong	12"
'Power Point'***	Medium cherry red	Exhibition 26—40	Slight	24" (MinFl)
Powerhouse™***	Orange red	Exhibition 17—25	None	30" (MinFl)

VARIETIES	COLOR	FORM/PETAL COUNT	FRAGRANCE	HEIGHT
Pride 'n' Joy™****	Orange/yellow	Exhibition 30—35	Slight	14″
Rainbow's End™*****	Yellow edged pink red	Exhibition 30—35	None	12″
'Ready'****	Yellow edged red	Exhibition 26—40	None	24″ (MinFl)
'Red Beauty'****	Dark red	Exhibition 35	Slight	12″
'Red Cascade' (Cl)****	Dark red	Decorative 26—40	Slight	Variable
'Red Imp'****	Dark red	Decorative 54	Slight	10″
'Regina Lee'****	Red cream/white	Exhibition 26—40	Slight	30″ (MinFl)
'Renegade'****	Red white	Exhibition 17—25	None	30″
'Rise 'n' Shine'****	Medium yellow	Exhibition 35	Moderate	14″
'Rocky Top'***	Orange red	Exhibition 17—25	Slight	30″ (MinFl)
Ruby Ruby™***	Medium red	Exhibition 26—40	Slight	12″
'Sassy Cindy'****	Red/white	Exhibition 26—40	Slight	30″ (MinFl)
Scentsational™****	Mauve edged pink	Exhibition 24—30	Strong	30″
'Shameless'***	White edged red	Exhibition 26—40	Slight	40″ (MinFl)
'Show Stopper'****	Apricot pink	Exhibition 45—50	Slight	30″ (MinFl)
'Shrimp Hit' (Shrimp™)***	Orange red salmon	Decorative 15—25	None	24″ (MinFl)
'Simplex'***	White/yellow center	5	Slight	10″
'Simply Beautiful'***	Lavender pink	Exhibition 17—25	Strong	24″ (MinFl)
Sleeping Beauty™***	Salmon orange	Exhibition 26—40	Strong	22″ (MinFl)
Solar Flair™****	Yellow edged red	Exhibition 17—25+	Slight	36″ (MinFl)
'Soroptimist International'*****	Pink red yellow	Exhibition 26—40	Slight	24″
'Spirit Dance'***	Butterscotch orange/yellow	Exhibition 26—40	Slight	30″ (MinFl)
'Spring's A Comin'***	Pink white edged pink	Exhibition 17—25+	Slight	30″ (MinFl)
Starina®****	Orange scarlet red	Exhibition 23—28+	Slight	12″
Strawberry Swirl®***	Red white stripes	Decorative 48	Slight	24″
Sun Sprinkles™***	Dark yellow	Exhibition 25—30	Slight	20″
'Sunglow'***	Yellow pink	5	Strong	30″ (MinFl)
'Sweet Arlene'***	Lavender	Exhibition 26—40	Strong	36″ (MinFl)
Sweet Chariot™****	Mauve purplish red	Decorative 40+	Strong	12″
Sweet Diana™****	Golden yellow	Exhibition 25+	Slight	18″
'Tennessee Sunrise'***	Deep yellow pink	Exhibition 26—40	Moderate	30″ (MinFl)
'Tennessee Sunset'***	Yellow orange	Exhibition 26—40	Strong	30″ (MinFl)
'The Guthrie Rose'***	Red yellow/light red	Exhibition 17—25	Slight	36″ (MinFl)
'Tiffany Lynn'***	Pink edged darker pink	Exhibition 15+	Slight	24″ (MinFl)
'Top Contender'****	Deep yellow	Exhibition 25+	Slight	30″ (MinFl)
'Top Secret'****	Medium red	Exhibition 17—25+	Slight	12″
Ty™*****	Dark yellow	Exhibition 17—25	None	30″
'Unbridled'**	Cream yellow edged peach	Exhibition 17—25	None	36″ (MinFl)
'Valentine's Day' (Cl)****	Dark red	Decorative 26—40	Slight	36″+ (MinFl)
Vista™****	Light purplish mauve	Exhibition 25+	Slight	16″
'Warm & Fuzzy'****	Red/gold center	Decorative 17—25	Slight	14″
'Water Lily'***	White pink tones	Exhibition 17—25	Slight	24″ (MinFl)
'Watercolor'***	Medium pink	Decorative —25+	Slight	12″
'White Quill'***	White	Decorative 17—25+	Slight	30″ (MinFl)
Winsome®*****	Pinkish purple	Exhibition 25—30	Slight	24″
'Wonderful'**	White edged pink	Decorative 26—40	Strong	30″ (MinFl)
'Yellow Bird'**	Bright yellow	Decorative 25—30	None to slight	30″ (MinFl)

'Königin von Dänemark' (*Alba*)

OLD GARDEN ROSES

In 1867 the first Hybrid Tea was hybridized, ushering in the so-called modern era of roses. Roses in existence before that date are classified rather arbitrarily into a group known as Old Garden Roses, although many of them are quite different from each other. Many Old Garden Roses (but not all) are quite hardy and long-lived. They often are equally fragrant (great for potpourris) and produce colorful hips, popular with cooks, floral arrangers, and birds. Old Garden Roses make good landscape or specimen plants. They are, generally, resistant to insects and disease. However, their cultural requirements are quite varied and require a certain amount of specialized skill, as outlined in this section. Old Garden Roses are very fashionable at the moment. Sometimes their drawbacks are minimized or ignored. This wonderful group deserves attention but also appropriate words of caution, as given throughout this section.

How Old Garden Roses Grow

Old Garden Roses are varied in their growth patterns, since the classification covers a wide range of plants. The essential secret to growing Old Garden Roses in colder climates is to buy ones grown on their own roots. These grow more vigorously, bloom more prolifically, and are generally much hardier than budded or grafted roses. They also do not produce unwanted suckers from the rootstock which are a nuisance to remove and may sap the plant of its strength.

Most Old Garden Roses bloom only once during the season. Others have recurring or intermittent bloom. And others have true repeat (remontant) bloom. Flowers vary in shape from single or semi-double to double or even very double. Many Old Garden Roses produce large clusters or trusses of roses for a truly spectacular floral display. Although many of them do much better in the South than in colder climates, some cold-climate rose growers find them challenging and worth the extra effort. Possibly, that's because so many of the older roses have such a wonderful aroma.

Where to Plant

Site and Light Most of the Old Garden Roses thrive in full sun—the brighter the light, the better. Bright light increases plant vigor, produces better bloom, and helps prevent disease. The Alba roses, however, not only tolerate some shade but also grow well in it. They are an exception.

Soil and Moisture Most Old Garden Roses need rich soil that retains moisture but still drains freely. Replace clay or rock with loam purchased from a garden center or in bags as potting soil. Add lots of organic matter, such as compost, leaf mold, peat moss, or rotted manures. All of these keep the soil moist and cool during dry periods. They also help the soil drain freely, encourage the growth of beneficial soil microorganisms, and attract worms which keep the soil aerated and fertilized. Soil should be light, loose, and airy. This allows quick root growth and good drainage so that plenty of oxygen can get to the tender feeder roots. A handful of bonemeal or superphosphate mixed into the soil will provide additional nutrients for the young plant. Proper soil preparation takes a little time, but it is critical to good growth.

One group thrives in poor soil: the Foetidas. They must not be given too much fertilizer, or they will produce poor bloom.

Spacing Give each plant as much space as possible, according to the projected height and width of the plant. Some get quite large and need lots of space, but others will die back each year and require less space than what might be indicated by catalogs. Spacing is important for sufficient sunlight, adequate soil moisture and nutrients, and good air circulation to prevent disease.

Planting

Old Garden Roses are available primarily through mail-order sources. Order them well in advance. You'll get bare root (dormant) plants. Buy plants growing on their own roots if possible. These will be hardier than budded plants, ones with upper cane different from the rootstock. Unfortunately, it is getting more difficult to find plants that have not been budded. But, at least, ask when ordering plants whether they have been budded or not. Plants grown on their own roots are sometimes a little smaller than budded stock, but they quickly catch up in size.

Bare Root Plant bare root Old Garden Roses directly in the garden as soon as the ground can be worked in spring. Follow the exact steps outlined in Part II.

Each plant should have healthy canes. The number, length, and size of canes vary with the type of rose ordered. Alba, Bourbon, Centifolia, Centifolia muscosa, Damascena, Gallica, Hybrid Perpetual, and Portland roses usually have three solid canes and a root system similar to Hybrid Teas. The Foetida roses often have more, but spindlier canes. The Noisette roses often are twiggy. The Spinosissima roses may look like a thorny shrub.

If you should get a budded rose, place the bud union 3 to 4 inches (7.5 to 10 cm) below the soil. This may stimulate root growth from the upper budded portion of cane. This helps to increase the winter hardiness of the plant.

Potted Plants Old Garden Roses are hard to find in most garden centers and nurseries. Most of the time you'll have to start with bare root plants. Sometimes, you'll find them potted at special sales, such as at arboretums or garden clubs.

If potted roses are available, get ones with at least three strong canes. The number varies with the class of Old Garden Rose. Check the leaves carefully for any sign of disease or insect infestation. Take the potted rose home immediately. Keep it out of extreme heat, as in the trunk of a car. If the plant will be exposed to wind, cover the stem and upper canes with thick plastic or paper. Remove this as soon as you get home, and make sure the soil is moist. Plant following the steps outlined in Part II.

How to Care for Old Garden Roses

Water Old Garden Roses thrive on frequent waterings. Keep the soil moist at all times. Spring waterings are particularly important. You want the plant to create lots of new cane, thick foliage, and many buds. Continue deep watering throughout the season, especially if there is a dry period. Drought in late summer and early fall is common in cold climates.

Mulch After the soil warms up to 60°F (15.6°C) apply a mulch around the base of the plant. The mulch should come close to but not touch the canes. The most commonly used mulch is shredded (not whole) leaves. Place at least a 3-inch (7.5-cm) layer over the soil. If you have enough leaves to make a thicker mulch, so much the better. Leaves keep the soil moist and cool which encourages rapid root growth. The mulch feeds microorganisms and worms which enrich and aerate the soil. The mulch also inhibits weed growth and makes pulling any weeds that do sprout much easier. Other good mulches include pine needles and grass clippings. The latter should be only 2 inches (5 cm) deep, or they will heat the soil and may smell. These mulches are all inexpensive and effective. If you use chipped wood or shredded bark (lovely around the larger plants), apply additional fertilizer to the soil, since these rob the soil of nitrogen as they decompose. Because mulch is eaten by soil microorganisms and worms, it needs to be replenished regularly throughout the growing season. Remove and compost all mulch in the fall.

Fertilizing Old Garden Roses are a diverse group. Some need lots of fertilizer to do well, while others perform better when given far less. Most prefer organic fertilizer.

Bourbon, China, Hybrid Perpetual, and Noisette roses respond best to a combination of inorganic and organic fertilizers. During the first year, use chemical fertilizer only after the plant has leafed out and is growing well.

In subsequent years, as early as possible in spring, sprinkle 10-10-10 (or 5-10-5) granular fertilizer around the base of each plant. Use common sense in judging the amount, more for bigger plants, less for smaller. Hand sprinkle, never placing fertilizer against the stem of the plant. Water immediately to dissolve the granules and carry nutrients to the roots.

One week later, soak the base of each plant with a liquid containing fish emulsion. Follow the directions on the label. Feed according to the size of the plant.

Alba, Centifolia, Damask, Foetida, Gallica, and Portland roses grow best and flower more profusely if fed organic fertilizers only. This is especially true of the Foetida group that does poorly if overfertilized. This has nothing to do with a nonorganic-versus-organic debate; these roses simply prefer organic fertilizers.

Feed these roses once a year in spring just as plants start to show signs of growth. Match the amount of fertilizer to the size of the plant. If a plant blooms vigorously, give the same amount of organic fertilizer the following year. If it grows lush foliage but produces little bloom, cut down on fertilizer the following spring or give none at all. If a plant appears spindly with poor foliage, give more fertilizer immediately.

Good organic fertilizers are alfalfa meal (rabbit pellets), blood meal, bonemeal, compost, cow manure, fish emulsion, fish meal, rotted horse manure, and Milorganite. Add bonemeal to the soil before planting. The others are effective added to the soil at planting time and as additional feedings on the surface of the soil in subsequent seasons.

Weeding Old Garden Roses, like all other woody plants, don't want to compete with weeds or other plants for nutrients and soil moisture. Kill all perennial weeds before planting. Use Roundup® if necessary. The use of mulch helps a great deal, since mulch retards most annual weed growth. The few weeds that do appear are easy to pull from the moist, loose earth. Old Garden Roses make good specimen plants. However, if the plants are surrounded by lawn, the grass can act like a weed. Keep grass away from the base of your plants.

Staking Some of the bushes become quite large and will whip back and forth in heavy winds. This can cause root damage or pull the shank of the plant from the soil. Stake bushes that are not protected from heavy winds. The easiest way to do this is with electric conduit cut into appropriate lengths. Stake plants as early in the season as possible.

Pegging Among the Old Garden Roses, Hybrid Perpetuals often produce long, arching canes that are suitable for pegging. For better repeat bloom, you can bend the canes over and attach them to a stake or piece of wire in the ground. Avoid crimping the base of each cane. This will cause the canes to produce laterals (side branches) and sublaterals (branches off side branches), which in turn produce profuse bloom.

Some Bourbon roses also respond well to pegging. If canes reach 6 to 7 feet (180 to 210 cm), then try this same method to increase bloom. Bourbons are not truly hardy in colder climates, so canes are unlikely to reach this length early in the season unless plants have been protected using the Minnesota Tip Method.

Disbudding Some growers remove flower buds in the first year, directing all energy to the plant itself. Remove buds just as they begin to form, for best results. Do this throughout the season if you plan to bury the roses using the Minnesota Tip Method. If you intend to leave the plants standing without protection, let a few blossoms form late in the season to induce dormancy in the rose to protect it during the winter.

Don't disbud mature plants, since Old Garden Roses are grown for their abundant floral display.

Deadheading Remove spent blossoms on any of the repeat-flowering varieties. When removing spent blossoms, snip the stem back to a leaf below the faded bloom. If you don't, the amount of repeat bloom will be limited. The plants will form hips and stop blooming. If plants bloom only once, removing spent blossoms will simply stop the plant from forming hips, so few gardeners take the time to remove blossoms on one-time-blooming plants.

Pruning Old Garden Roses need little pruning. Remove dead cane. Shape the plant for aesthetic reasons by removing canes that cross or seem too dense. Some growers cut healthy cane back by one-third to induce the formation of laterals, which produce more bloom.

Hybrid Perpetuals bloom heavily in spring. Immediately after bloom, cut branches or laterals back to the second or third bud. This will encourage another round of bloom later in the season. (See "Pegging" earlier in the section for tips on getting even more bloom.)

Winter Protection Bourbon, China, Hybrid Perpetual, Noisette, and Portland roses require excellent winter protection. Using the Minnesota Tip Method is highly advised for maximum winter protection of these less-than-hardy roses (see pp. 192–195). However, tipping these plants is an arduous chore and is one reason why many rose growers do not bother with them in colder climates. We suggest trying the simpler and less tiring Hass Teepee Method of winter protection (see p. 198). If canes are properly protected, these roses survive and produce incredible roses. Always begin protecting these roses after the first hard frost in late fall.

Problems

Insects Old Garden Roses are a varied group. In general, they are quite resistant to attacks by insects. However, if you see a problem developing, don't hesitate to begin spraying with an appropriate insecticide. If you see any spider mites, kill them immediately with a miticide before they have a chance to colonize. Using a preventive spraying program as you would for modern roses certainly can't hurt, but it is usually unnecessary.

Disease Centifolia and Gallica roses are prone to powdery mildew. Spray them early in the season with a fungicide. It is easier to prevent than to control this disease. Centifolia and Foetida roses are particularly vulnerable to black spot. Use a scheduled spray program for prevention. Alba roses are extremely disease resistant. Spinosissima roses are quite resistant. Just keep an eye on your roses, and spray accordingly.

Propagation

Division Some Old Garden Roses send out underground stems from the parent plant. These stems (stolons) produce little plants called suckers. Once the suckers are growing well, sever them from the parent plant by digging down through the underground stolon with a spade. The best time to do this is in early spring before the plants leaf out. Keep as much soil around the sucker as possible. Plant immediately as you would a bare root plant. Water well. Keep the soil moist at all times to promote vigorous growth.

Cuttings You may want to experiment with growing Old Garden Roses from softwood cuttings. If so, you'll have the most success with Alba, Bourbon, and Gallica roses.

Take cuttings from the center of a firm cane in late June or July. The cane should be mature, not green. Refer to Chapter 7 in Part II for the exact steps in taking softwood cuttings.

Special Uses

Cut Flowers Old Garden Roses are a diverse group. Experiment with cut flowers by taking some from mature plants. Keep in mind that some Old Garden Roses have such a beautiful scent that they can be overpowering. It's best to put them somewhere other than on a dining room table at dinnertime.

Always leave as much foliage on a plant as possible when taking cut flowers. Cut just above an outward-facing growth bud (place where a leaf joins the stem).

Dried Flowers Experiment with drying different varieties to see whether you can retain the form and color desired. Dry flowers in silica according to directions on the label. Many retain a delicate scent.

Potpourris Many of the Old Garden Roses produce heavily scented blossoms whose petals are ideal for potpourris. Place petals loosely in a box only one layer deep. Do this in a dark, dry place. If the petals are too deep, they'll often get moldy. Let them dry out slowly. When they're completely dry, put them in a bowl with your favorite spices. Or combine different roses for an exquisite scent. See Chapter 8 for detailed instructions on making and storing potpourris.

Sources

Angel Gardens, P.O. Box 1106, Alachua, FL 32616, (352) 359-1133

Antique Rose Emporium, 9300 Lueckmeyer Rd., Brenham, TX 77833, (800) 441-0002

Burlington Rose Nursery, 24865 Rd. 164, Visalia, CA 93292, (559) 747-3624

Chamblee's Rose Nursery, 10926 US Hwy 69 N, Tyler, TX 75706, (800) 256-7673

Corn Hill Nursery Ltd., 2700 Rte. 890, Corn Hill, New Brunswick E4Z 1M2, Canada, (506) 756-3635

Countryside Roses, 5016 Menge Ave., Pass Christian, MS 39571, (228) 452-2697

David Austin Roses Ltd., 15059 Hwy 64 W, Tyler, TX 75704, (800) 328-8893

Edmunds' Roses, 335 S High St., Randolph, WI 53956, (888) 481-7673

ForestFarm, 990 Tetherow Rd., Williams, OR 97544, (541) 846-7269

Fraser's Thimble Farms, 175 Arbutus Rd., Salt Spring Island, BC V8K 1A3 Canada, (250) 537-5788

Fritz Creek Gardens, P.O. Box 15226, Homer, AK 99603, (907) 235-4969

Garden Valley Ranch, 498 Pepper Rd., Petaluma, CA 94952, (707) 795-0919

Goodness Grows, Inc., P.O. Box 311, 332 Elberton Rd., Lexington, GA 30648, (706) 743-5055

Greenmantle Nursery, 3010 Ettersburg Rd., Garberville, CA 95542, (707) 986-7504

Heirloom Roses, 24062 Riverside Dr. NE, Saint Paul, OR 97137, (503) 538-1576

High Country Roses, P.O. Box 148, Jensen, UT 84035, (800) 552-2082

Hortico, Inc., 723 Robson Rd., RR# 1, Waterdown, ON
L0R 2H1 Canada, (905) 689-6984

Kimbrew-Walter Roses, Inc., 2001 Van Zandt County
Rd. 1219, Grand Saline, TX 75140, (903) 829-2968

Linda's Antique Roses, 405 Oak Ridge Dr., San Marcos,
TX 78666, (512) 353-3220

Mary's Plant Farm & Landscaping, 2410 Lanes Mill Rd.,
Hamilton, OH 45013, (513) 894-0022

North Creek Farm, 24 Sebasco Rd., Phippsburg, ME
04562, (207) 389-1341

Northland Rosarium, 9405 S Williams Lane, Spokane,
WA 99224, (509) 448-4968

Palatine Fruit & Roses, 2108 Four Mile Creek Rd.,
RR #3, Niagara-on-the-Lake, ON L0S 1J0 Canada,
(905) 468-8627

Pickering Nurseries, 3043 County Rd. 2, RR #1, Port
Hope, ON L1A 3V5 Canada, (905) 753-2155

Regan Nursery, 4268 Decoto Rd., Fremont, CA 94555,
(800) 249-4680

Rogue Valley Roses, P.O. Box 116, Phoenix, OR 97535,
(541) 535-1307

Rose Fire, Ltd., 09394 State Rte. 34, Edon, OH 43518,
(419) 272-2787

Rosemania, 4920 Trail Ridge Dr., Franklin, TN 37067,
(888) 600-9665

Roses of Yesterday and Today, 803 Brown's Valley Rd.,
Watsonville, CA 95076, (831) 728-1901

Roses Unlimited, 363 N Deerwood Dr., Laurens, SC
29360, (864) 682-7673

S & W Greenhouse, Inc., P.O. Box 30, 533 Tyree Springs
Rd., White House, TN 37188, (615) 672-0599

Two Sisters Roses, 1409 N Redbud Lane, Newcastle, OK
73065, (no phone by request)

Vintage Gardens (custom propagation), 4130 Gravenstein
Hwy N, Sebastopol, CA 95472, (707) 829-2035

Wayside Gardens, 1 Garden Lane, Hodges, SC 29695,
(800) 213-0379

Witherspoon Rose Culture, 3312 Watkins Rd., Durham,
NC 27707, (800) 643-0315

VARIETIES

Following are the better varieties of Old Garden Roses for cold-climate gardeners. Some are easy to grow, others more difficult. The reward for growing some of the more tender roses can be incredible. The size, beauty, and fragrance of many of these roses can only be appreciated in person. The descriptive word "quartered" means that any given flower looks like it is divided into four sections each with its own set of petals. The petal count in the following table is approximate. The number of petals varies greatly by culture and overall climatic conditions! When a specific group of Old Garden Roses is not fully hardy, we have selected the hardiest within that group. Tipping these roses can be a chore, so we advise placing three black plastic bags filled with whole leaves around the more tender plants to winter protect them (see p. 198). You're not just protecting them from cold but desiccating winter winds as well. So you can never go wrong with winter protection no matter what the hardiness rating given for these plants here, in catalogs, or on the Internet. You do not want canes to die back significantly because some of the roses will only form blossoms on the previous year's growth (old wood). *Comments on dieback (death of cane) are simply a warning to winter protect the canes as best you can to get good bloom.*

Rosa alba (Alba rose)

(ROW-suh AL-buh)

These are tall, dense plants, growing quickly. The Albas produce clusters of extremely fragrant blossoms once a year (June), mostly on side branches (laterals). Foliage is downy on both bottom and top. It's a soft, gray green color sometimes with a bluish cast. Canes on most varieties are thorny, but a few varieties are not prickly at all. Albas often produce large, scarlet hips (especially on plants producing flowers with fewer petals). Albas are quite resistant to powdery mildew, mildly susceptible to black spot. They are one of the few roses to tolerate partial shade. They sometimes respond well to pegging (bending canes over and attaching them to stakes in the ground) to increase bloom. Hardy to about −25°F (−32°C).

VARIETIES	COLOR	PETAL COUNT	FRAGRANCE	HEIGHT
'Alba Semi-plena'****	Clear white	8—12	Strong	60″

Rather lax, lanky growth. Will sucker. Clusters of single to semi-double flowers up to 3 inches (7.5 cm) wide with bright yellow stamens and lovely fragrance (used in making attar of roses), both sweet and lemony. Grayish green leaves on long, arching canes. Very thorny. Abundant red hips. Possibly the true 'White Rose of York.'

'Belle Amour'***	Light pink	30—60	Moderate	60″

Upright, dense. Will sucker. Clusters of cupped camellia-like flowers up to 3 inches (7.5 cm) across, often with a salmon tinge and prominent yellow stamens. Myrrh- to musky scent. Medium green foliage. Thorny cane. Round, red hips possible. Severe dieback expected.

'Céleste' (see 'Celestial')

'Celestial'****	Light clear pink	17—25	Strong	48″

Upright. Exquisite 3 inch (7.5) flowers with lovely gold stamens, beautifully formed on neat bushes with bluish gray green foliage. Somewhat resistant to black spot. Shade tolerant. Severe dieback to death possible.

'Chloris'****	Light pink	41—60+	Slight to moderate	60″

Upright arching. Will sucker. Flowers just under 3 inches (7.5 cm) across and exquisitely delicate with a greenish button eye. Almost thornless cane with a deep red tinge. Deep bluish green, leathery foliage. No hips. Resistant to black spot. Dieback common.

'Cuisse de Nymphe Émue' (see 'Maiden's Blush')

'Félicité Parmentier'****	Soft shell pink/white	41—80+	Very strong	60″

Dense, stiff, upright. Forms clusters of lightly quartered, flatish flowers up to 3 inches (7.5 cm) across and fading to white. Long bloom period. Exquisite scent. Grayish, green foliage on thorny cane. Susceptible to black spot. Severe dieback to death possible.

'Jeanne d'Arc'***	White	41—50	Strong	60″

Dense, upright. Flowers with pink hue and nearly 3 inches (7.5 cm) across. Deep gray green foliage on thorny cane. Susceptible to black spot. Dieback common.

'Königin von Dänemark'***	Light to medium pink	41—80	Very strong	48″

Dense to open, mounded. Forms clusters of flowers, like many petalled cups (sometimes quartered), almost 3 inches (7.5 cm) across with carmine hue and button eye. Lovely, sweet scent. Dark bluish gray green foliage. Very thorny. Large red hips are possible. Resistant to black spot. Quite hardy when grown on its own roots. Dieback common.

'Mme. Legras de St. Germain'****	White	41—80	Moderate to strong	48″

Upright, open to arching. Buds striped red open into clusters of flowers roughly 3 inches (7.5 cm) across, occasionally very double with a light yellow tinge in the center. Long bloom period. Light gray, green foliage on arching, nearly thornless canes. Severe dieback to death possible.

'Mme. Plantier'****	Light pink to white	41—60	Moderate	60″

Dense, arching. Plant spreads rapidly. Forms clusters of delicately colored, pompon flowers from 2 to 3 inches (5 to 7.5 cm) wide with yellow eye. Long bloom. Delicate light gray green foliage on nearly thornless cane. Susceptible to black spot. Severe dieback expected.

'Petite de Hollande' (*Centifolia*)

'Henri Martin' (*Centifolia muscosa*)

'Saint Nicholas' (*Damascena*)

'Austrian Copper' (*Foetida*)

'Tuscany Superb'/'Superb Tuscan' (*Gallica*)

'Mrs. John Laing' (Hybrid Perpetual)

'Empress Josephine' (Miscellaneous)

'Nastarana' (Noisette)

VARIETIES	COLOR	PETAL COUNT	FRAGRANCE	HEIGHT
'Maiden's Blush'****	White pink	41–60	Variable	60″

Upright, slightly arching. Technically, **Rosa alba incarnata**. Lovely 2-inch (5-cm) and larger flowers with nearly perfect form and pinkish tint. Fragrance varies from moderate to very strong. Some plants sold under this name are light to medium pink. Bluish tinge to foliage. Used as a Climber in warmer areas. Susceptible to black spot. Severe dieback to death possible. A gem!

'Queen of Denmark' (see 'Königin von Dänemark')

'Semi-plena' (see 'Alba Semi-plena')

VARIETIES	COLOR	PETAL COUNT	FRAGRANCE	HEIGHT
'White Rose of York'***	White	30–60	Strong	60″

Lovely, sweet-smelling flowers with yellow center. Lovely contrast to smooth, medium gray green foliage and light green slightly thorny cane. Scarlet hips. Frequently confused with **Rosa** 'Alba Semi-plena.' The debate about which is the true 'White Rose of York' will never be settled. So ask whether the plant you are ordering has a high petal count.

Apothecary rose (see **Rosa gallica**)

Bourbon rose

These roses are prone to extreme dieback and even death in severe winters. In more mild winters they will typically die back to the snow line. Some bloom more than once with lovely, fragrant, many-petalled flowers that look like large globes. The petals are stunning in both translucence and hue. The spring bloom is most spectacular; the fall bloom not as reliable or heavy. Cane varies in thorniness by variety, with some very prickly. They are quite susceptible to black spot. Few are grown on their own roots since they often grow better when budded. Hardy to 0°F (−18°C) without protection. A few will make it to −20°F (−29°C) without winter protection. *We suggest winter protecting all of them.*

VARIETIES	COLOR	PETAL COUNT	FRAGRANCE	HEIGHT
'Commandant Beaurepaire'***	Pink lilac cream	45–50	Strong	60″

Scarlet flowers up to 3 inches (7.5 cm) across, often striped or splashed with pink and white. Lemony scent. Blooms once in June. Light green foliage on thorny cane. One of the hardier roses in this group. Some dieback.

VARIETIES	COLOR	PETAL COUNT	FRAGRANCE	HEIGHT
'Honorine de Brabant'****	Pale pink spotted white	30–40+	Moderate to strong	60″

Small clusters of lilac-tinged flowers just under 3 inches (7.5 cm) across, bloom once in June. Sometimes spotted or striped purple. Raspberry scent. Lush, light green foliage on cane with few thorns. Potential for repeat bloom. Will die back.

'La Reine Victoria' (see 'Reine Victoria')

VARIETIES	COLOR	PETAL COUNT	FRAGRANCE	HEIGHT
'Louise Odier'****	Deep pink	60+	Very strong	60″

Vigorous. Lovely clusters of rose-colored, cupped flowers just under 3 inches (7.5 cm) across. Repeat bloom possible. Light to mid green foliage on mildly prickly canes. May set hips. Severe dieback.

VARIETIES	COLOR	PETAL COUNT	FRAGRANCE	HEIGHT
'Mme. Ernst Calvet'**	Soft to medium pink	41–60	Very strong	60″

Scraggly, arching. Small clusters of saucer-shaped flowers just under 3 inches (7.5 cm) across with intense fragrance. Repeat bloom possible. Thorny cane. An unattractive plant with remarkable flowers. Also sold as 'Ernest Calvet.' Severe dieback.

VARIETIES	COLOR	PETAL COUNT	FRAGRANCE	HEIGHT
'Mme. Pierre Oger'****	Pearly pinkish white	41–60	Moderate to strong	Variable

Slender, erect. Cup-shaped flowers just over 2 inches (5 cm) across with lilac rose hue. Repeat bloom possible. A delicate plant that needs excellent winter protection for any chance of survival. Will die out completely in severe weather.

VARIETIES	COLOR	PETAL COUNT	FRAGRANCE	HEIGHT
'Reine Victoria'****	Lilac pink/silver	60+	Very strong	60"

Slim, erect. Clusters of lovely, cupped blossoms up to 4 inches (10 cm) across with an extraordinary fragrance. Repeat bloom possible. Flowers easily damaged by rain. Pale green foliage on cane with few thorns. Excellent cut flower. Will die out completely in severe weather.

VARIETIES	COLOR	PETAL COUNT	FRAGRANCE	HEIGHT
'Souvenir de la Malmaison'****	Light pink to cream	35—60+	Variable	36"

Upright. Moderately vigorous. Flowers up to 4 inches (10 cm) across or larger. Blooms once in June with quilled or quartered blossoms easily damaged by rain. Fragrance varies from light to strong. Large, glossy leaves. Prune this plant as little as possible! An extraordinary rose but needs extreme winter protection. Severe dieback and possible death in severe weather.

VARIETIES	COLOR	PETAL COUNT	FRAGRANCE	HEIGHT
'Variegata di Bologna'****	Pale pink to white	60+	Moderate	60"

A vigorous plant. Many-petalled 3-inch (7.5 cm) flowers are striped red or even purple. Petals are ideal for potpourris. Collect before a rain since they are easily damaged. Medium green, glossy foliage. Needs excellent winter protection. Will die out completely in severe winters.

VARIETIES	COLOR	PETAL COUNT	FRAGRANCE	HEIGHT
'Zéphirine Drouhin'***	Cherry pink	25—30	Moderate	60"

Semi-climber that is fairly vigorous. Lovely blooms up to 3 inches (7.5 cm) across and occasionally streaked white. Light green foliage on truly thornless cane, tinged purple when immature. Reblooms under good conditions. Requires extreme winter protection. Will die out completely in severe winters.

Burnet rose (see *Rosa spinossissima*)

Cabbage rose (see *Rosa centifolia*)

Rosa centifolia (Cabbage or Provence rose)
(ROW-suh sen-teh-FOH-lee-uh)

The flowers of these roses look like large globes with many overlapping petals. The fragrant blossoms flower on the end of slender, arching branches. The scent of this rose is "the beauty of life itself." Classifying the scent as "strong" may at times be misleading. It's sometimes the beauty of the scent rather than its intensity that merits the label. The roses bloom once in June. Blossoms are susceptible to rain damage. Rarely do these roses set hips. Some canes may need support. Most are prone to both powdery mildew and black spot. They are vigorous growers, but do suffer dieback. Even if all cane dies back, the plants are generally crown hardy and will spring back to life. Hardy to −25°F (−32°C) or colder, but winter protect anyway.

VARIETIES	COLOR	PETAL COUNT	FRAGRANCE	HEIGHT
'Bullata'****	Medium pink	41—60	Variable	48"

Upright, open. Flowers are about 3 inches (7.5 cm) across and often dark in the center. Fragrance varies from moderate to very strong. The plant has crinkly looking leaves which resemble lettuce, turning from bronzy to green as they mature.

VARIETIES	COLOR	PETAL COUNT	FRAGRANCE	HEIGHT
'Fantin-Latour'****	Light pink white	60—80	Moderate	60"

Lanky, arching. Vigorous. Small clusters of peony-like flowers 3 inches (7.5 cm) across and flattish with a button eye. Train the plant horizontally to increase bloom. Dark green, semi-glossy foliage on canes with few but large thorns (when present). Wonderful fragrance in a group noted for this characteristic.

VARIETIES	COLOR	PETAL COUNT	FRAGRANCE	HEIGHT
'Juno'***	Light silvery pink	60—80	Strong	48"

Lax, open, arching. Flowers up to 4 inches (10 cm) across and sometimes deep pink to light red with a quartered appearance. Bright green foliage with a hint of gray. Often classified in different categories depending upon the source.

VARIETIES	COLOR	PETAL COUNT	FRAGRANCE	HEIGHT
'Lettuce Leafed Rose' (see 'Bullata')				
'Petite de Hollande'****	Medium pink	60–80	Moderate to strong	Variable

Dense, bushy. Will sucker. Clusters of smallish, globular flowers about 2 inches (5 cm) across on thorny cane. Flowers often display a rosy coloration. Small light green leaves are glossy and serrated along the edges. Most plants are compact and small, but a few will grow taller.

'Rose à Feuilles de Laitue' (see 'Bullata')				
'Rose de Meaux' ('De Meaux')****	Light pink	60–80	Variable	36″

Upright, dense. Smallish flowers, about 1 inch (2.5) across. They are tight, frilly, almost like buttons. Blossoms easily damaged by rain. Scent varies from light to strong. Foliage dainty as well, but wonderfully dark. Stems often short, almost twiggy, and thorny. Plant needs exceptionally good soil to thrive.

'Rose des Peintres'***	Medium pink	80–100	Strong	60″

Upright, open, arching. Vigorous. Cupped, clear pink globes about 3 inches (7.5 cm) across contrast beautifully to the medium green foliage on prickly cane. Very popular once with painters, from which it derived its name.

'Tour de Malakoff'****	Purplish pink	41–60	Moderate	Variable

Lax, sprawling. Will sucker. Flowers up to 4 inches (10 cm) across with papery petals ranging from pink to deep purple. Very unusual bloom color and form. Smallish smooth medium green leaves.

Rosa centifolia muscosa (Moss rose)
(ROW-suh sen-teh-FOH-lee-uh muss-KOH-suh)

These roses get their name from the mossy texture of buds, leaves, and stems, often described as "sticky hairs." The moss varies in color with each variety and may exude a distinctive aroma. The flowers are extremely fragrant globes. Some varieties bloom only once in June; others repeat bloom (very rarely). All are susceptible to disease, especially black spot. The plants are hardy, but prone to severe dieback. Rarely does a plant die out completely. Hardy to about −25°F (−32°C) at best. Winter protection is highly recommended.

VARIETIES	COLOR	PETAL COUNT	FRAGRANCE	HEIGHT
'Alfred de Dalmas'****	Light pink/darker pink	41–60	Moderate	36″

Upright, spreading. Suckers, but is not particularly vigorous. Exquisite cupped flowers just under 3 inches (7.5 cm) across with creamy tones. Often described as having the scent of honeysuckle or sweet peas. Possible repeat bloom (deadhead and cut back cane slightly). Small grayish green leaves. Thorny cane. Severe dieback possible.

'Communis' ('Common Moss')****	Light to medium pink	50–60	Moderate	48″

Upright, bushy. Semi-vigorous growth. Will sucker. Flowers up to 3 inches (7.5 cm) across. Light green foliage on prickly cane. Severe dieback possible.

'Crested Moss' ('Cristata')****	Medium pink	50–70	Strong	48″

Upright, arching. Vigorous growth, but flops over. Flowers 3 inches (7.5 cm) across, bloom only in June. Very unusual look when in bud. The moss extends out from underneath the flower like wings, giving rise to its name. Medium green foliage. Severe dieback possible.

'Deuil de Paul Fontaine'***	Crimson purplish red	50–70	Variable	48″

Upright, open. Cupped flowers up to 4 inches (10 cm) across with scent varying from light to strong. Light green foliage often with a dark reddish edge. Thorny cane. Severe dieback possible.

'Gloire des Mousseuses' (see 'Gloire des Mousseux')

'Gloire des Mousseux'****	Bright medium pink	41–60	Moderate to strong	48"

Upright, dense. A vigorous plant. Flowers up to 4 inches (10 cm) across with clear color and deeper-toned center. Foliage is a light green and aromatic, but susceptible to disease. Although vulnerable to dieback, this is an excellent choice in this group. Severe dieback possible.

'Henri Martin'****	Deep pink to light red	35–45	Moderate to strong	60"

Lax, arching. Very vigorous. Abundant, slightly bristly buds burst into clusters of crimson flowers roughly 3 inches (7.5 cm across). Long bloom period. Lovely, medium green foliage. Cane is not thorny. Small round red hips. Severe dieback common. A gem worth the extra effort to protect.

'Jeanne de Montfort'***	Medium pink	41–60+	Strong	60"

Dense, lanky, sprawling. Buds and flower stalks covered with brown green moss. Wonderful red buds burst into clusters of blossoms up to 4 inches (10 cm) wide, often silvery in tone and quite loose, with prominent yellow stamens. Medium green foliage. Severe dieback possible.

'Mme. de la Rôche-Lambert'***	Pinkish mauve	41–60	Strong	60"

Upright, dense. Cupped to globular lowers up to 3 inches (7.5 cm) across, often reddish purple. Dark green foliage. Susceptible to powdery mildew. Severe dieback possible.

'Mousseline' (see 'Alfred de Dalmas')

'Nuits de Young'	Reddish purple	41–60+	Moderate to strong	48"

Upright, dense. Will sucker. Mossy stems and buds. Loose, velvety purplish flowers about 2 inches (5 cm) wide with golden stamens once fully open. Deep green foliage. Dieback common.

'Salet'****	Medium pink/light	41–60+	Moderate to strong	48"

Upright, arching, open. Flowers up to 3 inches (7.5 cm) across with a lovely reddish hue may be damaged by rain. Divine fragrance. Medium green foliage. Canes are quite thorny. Repeat bloom possible, even probable in most seasons. Severe dieback possible.

'Striped Moss' ('OEillet Panachée')***	White pink/marked red	41–50+	Moderate to strong	60"

Upright, dense. Unique flowers up to 3 inches (7.5 cm) across striped or flecked red. Flower is shaped like cup at first but opens flat, with quilled petals and golden eye. Very unusual form. Prone to black spot. Severe dieback possible.

'William Lobb'***	Purplish/lilac pink	41–60+	Moderate to strong	60"

Upright, arching. Nicknamed 'Old Velvet Moss.' Flowers (sometimes quartered) up to 3 inches (7.5 cm) across and almost purple. Lovely for potpourris. Medium grayish green foliage on long, thorny stems that may need support. Prone to powdery mildew. Severe dieback common.

Rosa chinensis (China rose)
(ROW-suh cheye-NEN-siss)

Chinas (occasionally called Bengal or Indica roses) are delicate plants more suitable for mild climates. The plants grow to about 24 inches (60 cm) in cold climates and bloom intermittently throughout the season. Blooms tend to be small and loose in form with only a mild fragrance. Plants are potentially susceptible to powdery mildew and/or black spot. These need excellent winter protection to survive severe winters. Use the Minnesota Tip Method or grow them in pots to be buried in late fall. Hardy to approximately 26°F/−3.3°C (mid twenties) without protection.

VARIETIES	COLOR	PETAL COUNT	FRAGRANCE	HEIGHT
'Archduke Charles'***	Pink red to crimson	26—40	Slight	24″

Twiggy, bushy. Noted for changing colors throughout the season. Blossoms typically about 3 inches (7.5 cm) wide with deep pink and red petal colorations. Often becomes wholly crimson as blossoms age. Dark, glossy green foliage. Very tidy growth and almost continuous bloom under ideal conditions.

'Hermosa'**	Light pink	26—40	Moderate	24″

Twiggy, tidy. Flowers tinged lilac pink with a noticeable tea fragrance. Flowers often cupped or globular and just over 2 inches (5 cm) wide. Distinctive bluish gray light green foliage on thin stems. Does well grown in a pot.

'Old Blush'****	Medium pink	20—30	None	24″

Buds are a deep pink red and burst into lighter coloration. Loose clusters of flowers just under 3 inches (7.5 cm) across. Decent repeat bloom throughout the season. Some people insist this rose has a slight, fruity scent. Medium green foliage tinged red when emerging. Slender canes slightly prickly. Small orange red hips.

Rosa chinensis viridiflora (see 'The Green Rose')

'The Green Rose'****	Green	25—30	None to slight	24″

Upright. 'The Green Rose' looks like a 1½ inch (4 cm) green pin cushion tinged brown. Flowers are really colored sepals, not flower petals. A true oddity favored more for cutting and dried flowers than garden performance. Some growers detect a spicy scent at times.

Rosa damascena (Damask rose)
(ROW-suh dam-uh-SEEN-uh)

Damask roses are strong, vigorous plants. With winter protection they could grow up to 60 inches (150 cm) tall (taller in the South). They are generally upright plants, some of which will sucker if grown on their own roots. Flowers tend to have an intense fragrance. The oils of these flowers are used in making attar of roses, an essential ingredient for perfumes. Flowers vary in number of petals but often form lovely clusters in ideal conditions. Canes are typically thin and have a tendency to arch over. They can be very prickly. Foliage has a downy feel on both upper and lower leaf surfaces. Damasks fall into two categories: Summer and Autumn (the latter will bloom twice in more southerly locations; rarely more than once in the North). Autumn damask roses have funnel-shaped, rather than round, hips. While as a group they tend to be disease resistant, plants are certainly susceptible to infection, especially black spot, on occasion. They also need lots of attention, including frequent watering and proper feeding to make them capable of withstanding cold winters. They will die out completely if neglected. Hardy to 0°F (−18°C) and occasionally to −10°F (−23C) without protection. Crown hardy to −25°F (−32°C) with protection.

VARIETIES	COLOR	PETAL COUNT	FRAGRANCE	HEIGHT
'Autumn Damask' (see *Rosa damascena semperflorens*)				

'Celsiana'****	Light clear pink	16—25	Moderate to strong	36″

Pale red buds open into clusters of clear pink flowers 2 to 4 inches (5 to 10 cm) across with delicate, yellow stamens. Loose, open look. Long bloom period. Grayish green foliage. Cane with small thorns. Severe dieback at times.

'Kazanlik' (see *Rosa damascena trigintipetala*)				

'Léda'****	White	30—50	Variable	36″

Will sucker. Sometimes referred to as 'Painted Damask.' Dark crimson buds open into clusters of white to light pink flowers edged red and up to 3 inches (7.5 cm) across. Scent varies from light to strong. Lovely dark gray-green foliage. Deep red hips in fall. Some dieback.

VARIETIES	COLOR	PETAL COUNT	FRAGRANCE	HEIGHT
'Mme. Hardy'*****	White	60+	Very strong	60"

Will sucker. Described as white but often cream-colored or pink-tinged blossoms ranging from 2 to 3 inches (5 to 7.5 cm) wide with a green button eye. Fragrance of lemons. Medium green foliage on thorny light green cane. Some dieback common, but it is the hardiest of the group.

*Rosa damascena semperflorens****	Medium pink	40—60+	Moderate to strong	48"

Dense. Will sucker. Clusters of blossoms are often clear pink, but occasionally white, and up to 3 inches (7.5 cm) across. Exquisite scent. Excellent for potpourris or sachets. Light gray green foliage. Expect dieback.

*Rosa damascena trigintipetala*****	Medium to deep pink	30	Strong	48"

Clusters of flowers up to 3 inches (7.5 cm) across with exquisite scent. One of the best roses for potpourris and sachets. Used in making attar of roses. Foliage is light to dark green with gray tones. Canes tend to be slender and thorny. May form hips. It is suggested by many that this should be considered a group of similar plants rather than an individual species (would explain variance in foliage coloration). Expect some dieback.

*Rosa damascena versicolor****	Pink/white	17—25	Moderate to strong	48"

May have both white and pink bloom or only one color per plant. Most blossoms under 3 inches (7.5 cm) across. Light bluish gray green foliage. Its aroma and foliage are its strong points. Tends to spread out. Expect dieback.

'Rose de Rescht'****	Deep pink	50+	Very strong	36"

Compact. Will sucker. Flowers really reddish to lilac and very rich. Most are just under 3 inches (7.5 cm) wide. Long-lasting bloom. Thick, wrinkly, medium green foliage. Red hips. Some consider this compact plant a Portland rose. Expect dieback.

'Rose des Quatre Saisons' (see *Rosa damascena semperflorens*)

'Rose of Castile' (see *Rosa damascena semperflorens*)

'Saint Nicholas'***	Deep pink lilac	15—20	Fruity	36"

Will sucker. Vigorous grower. Large clusters of flowers about 3 inches (7.5 cm) across with rich coloration contrasting nicely to yellow stamens. Open, loose look. Foliage is slightly glossy and light green. Round, red hips. Expect dieback.

'Trigintipetala' (see *Rosa damascena trigintipetala*)

'York and Lancaster Rose' (see *Rosa damascena versicolor*)

Rosa foetida
(ROW-suh FEH-tee-duh)

The Foetidas are vigorous growers. Flowers will bloom along the entire cane for several weeks (usually in June), and then the flush ends. The plants prefer poor soil (the exception to the rule). Never prune except to remove dead wood, since flowers appear on old wood. Black spot is a common problem with all of these roses. See the section on Shrub roses for hybrid Foetidas, including 'Harison's Yellow.' Hardy to −30°F (−34°C) without winter protection.

'Austrian Copper' (see *Rosa foetida bicolor*)

'Persian Yellow' (see *Rosa foetida persiana*)

VARIETIES	COLOR	PETAL COUNT	FRAGRANCE	HEIGHT
Rosa foetida *****	Medium yellow	5	Unpleasant	72″

'Austrian Briar' or 'Austrian Yellow' is a stunning plant with many blooms just larger than 2 inches (5 cm) across. Nice green foliage. Attractive round hips almost maroon in color. Highly prone to black spot and usually requires spraying to survive. Very hardy.

VARIETIES	COLOR	PETAL COUNT	FRAGRANCE	HEIGHT
Rosa foetida bicolor *****	Red orange yellow	5	None	60″

Profuse bloom along entire cane. Blossoms coppery red on the outside, rich yellow underneath. Prone to black spot. Now often sold on budded cane. Depending upon the rootstock, this once hardy plant may now need winter protection. *Try to get plants grown on their own roots.*

VARIETIES	COLOR	PETAL COUNT	FRAGRANCE	HEIGHT
Rosa foetida persiana *****	Rich golden yellow	20—25	Slight	72″

Flowers like globes with wonderful coloration. Plant has sprawling habit and often thin, very thorny branches. Very susceptible to black spot.

French rose (see *Rosa gallica*)

Rosa gallica (Apothecary, French, or Provins rose)
(ROW-suh GAL-li-kuh)

Gallicas bloom once in spring (June) with a nice fragrance. Many have quartered blooms. Plants sometimes look spindly (wiry) but have upright growth and may sucker prolifically (they form thickets in the wild). They often grow wider than they are tall. Some varieties have nearly thornless cane, while others have slightly thorny cane. Many respond well to pegging (arching canes over and attaching them to stakes in the ground). Some produce lovely hips. Gallicas are prone to powdery mildew. Keep plants open and airy to cut down on disease. Removal of older cane that is flowering poorly is highly recommended. Most varieties are hardy to very hardy, but dieback is common. Hardy to at least −25°F (−32°C).

VARIETIES	COLOR	PETAL COUNT	FRAGRANCE	HEIGHT
'Alain Blanchard' ***	Reddish mauve/gold	9—16+	Moderate	48″

Upright, wiry. Fairly vigorous. Will sucker. Flowers up to 4 inches (10 cm) across with purplish to black hue and beautiful gold stamens. Medium green foliage. Prickly stems. Lovely, round, red hips. Quite hardy.

VARIETIES	COLOR	PETAL COUNT	FRAGRANCE	HEIGHT
'Alika' ***	Deep pink to red	12+	Moderate to strong	60″

Upright. Will sucker. Striking landscape or specimen plant. Flowers are up to 3 inches (7.5 cm) across, have a brilliant coloration, and lovely gold stamens. Sweetly scented. Sparsely prickled cane. Excellent hips. Very hardy.

'Apothecary's Rose' (see *Rosa gallica officinalis*)

VARIETIES	COLOR	PETAL COUNT	FRAGRANCE	HEIGHT
'Belle de Crécy' ***	Red to purple/green	41—80	Strong	48″

Arching. Striking variation in color as the small clusters of flowers just under 3 inches (7.5 cm) open and finally fade. Often quartered in appearance with a memorable fragrance. Dark green foliage on cane with reddish thorns. Hardy.

VARIETIES	COLOR	PETAL COUNT	FRAGRANCE	HEIGHT
'Belle Isis' ***	Light to medium pink	41—60	Strong	40″

Upright, compact. Will sucker. Cupped flowers (often quartered) up to 3 inches (7.5 cm) across with a greenish eye. Unusual scent most often described as myrrh-like. Grayish, light green foliage (small leaves) on prickly cane. Some dieback common.

VARIETIES	COLOR	PETAL COUNT	FRAGRANCE	HEIGHT
'Camaieux' ('Camaieu') ***	Pink streaked white	41—60	Moderate	30″

Upright, compact, vigorous. Often twiggy in colder climates. Will sucker. Deep pinkish red flowers just over 3 inches (2.5 cm) across, spotted with crimson and pink. Unusual frilly appearance. Unique scent. Medium, glossy green foliage. Severe dieback possible.

VARIETIES	COLOR	PETAL COUNT	FRAGRANCE	HEIGHT
'Cardinal de Richelieu'****	Wine burgundy/white	41–60	Moderate	48"

Will sucker. Crimson purple buds open into small clusters of exquisite flowers just under 3 inches (7.5 cm) across, almost velvety purple in color under ideal growing conditions. May have so many petals the flower appears quartered. Unusual, subtle scent. Small, dark green leaves on nearly thornless canes. Excellent cut flower. Severe dieback possible.

VARIETIES	COLOR	PETAL COUNT	FRAGRANCE	HEIGHT
'Charles de Mills'*****	Purplish red	41–80	Variable	48"

Erect, arching. Vigorous. Will sucker. Large flowers up to 4 inches (10 cm) wide or larger, almost maroon in color. So many petals that the rose may appear quartered. Fragrance can be faint to intense. Bush has good form, dark green foliage, and few thorns. Listed as disease resistant, but is susceptible to black spot. Severe dieback possible.

VARIETIES	COLOR	PETAL COUNT	FRAGRANCE	HEIGHT
'Complicata'***	Pink/white base	5	Variable	60"

Upright, arching to climbing. Clusters of wild rose-like flowers up to 5 inches (12.5 cm) wide with white to rose pink eyes and lovely, yellow stamens. Soft, sweet scent. Grayish, bright green foliage on nearly thornless stems. Stunning round orange to scarlet hips. Could be used as a Pillar rose in more southerly areas. Dieback possible, but basically hardy.

VARIETIES	COLOR	PETAL COUNT	FRAGRANCE	HEIGHT
'Désirée Parmentier'***	Bright medium pink	60–80	Strong	60"

Upright. Moderately vigorous. Will sucker. Vivid pink flowers, often quartered, fade as they age. Medium to dark green foliage may turn orange red by fall. Expect severe dieback.

VARIETIES	COLOR	PETAL COUNT	FRAGRANCE	HEIGHT
*Rosa gallica officinalis*****	Deep pink	9–24	Moderate	48"

Lovely 2- to 3-inch (5- to 7.5-cm) flowers (often reddish with yellow stamens) contrasting to rich grayish, green foliage on nearly thornless cane. Favorite for potpourris and sachets. Large, round, red hips. Spreads freely. Quite hardy.

VARIETIES	COLOR	PETAL COUNT	FRAGRANCE	HEIGHT
*Rosa gallica versicolor*****	Pink striped red or white	9–24	Moderate to strong	40"

Arching. Vigorous. Spreads freely. Reddish flowers just over 3 inches (7.5 cm) across with white stripes (or the reverse) and prominent yellow stamens. Medium green foliage on nearly thornless cane. Also known as 'Rosa Mundi.' May be a sport of *Rosa gallica officinalis* and may revert to that plant. Severe dieback possible.

'Rosa Mundi' (see *Rosa gallica versicolor*)

'Rose de la Maître-École' (see 'Rose du Maître d'École')

VARIETIES	COLOR	PETAL COUNT	FRAGRANCE	HEIGHT
'Rose du Maître d'École'***	Lavender pink mauve	41–70	Very strong	60"

Upright, bushy. Will sucker. Flowers up to 5 inches (12.5 cm) across (sometimes quartered), varying from pink to lilac with a green center. May actually bend canes over from their weight. Bright medium green foliage. Cane with few thorns. Severe dieback possible.

'Superb Tuscan' (see 'Tuscany Superb')

VARIETIES	COLOR	PETAL COUNT	FRAGRANCE	HEIGHT
'Tuscany Superb'****	Maroon crimson purple	41–70	Variable	48"

A gem nicknamed 'Old Velvet Rose.' Will sucker. Flowers up to 3 inches (7.5 cm) across varying in color from intense crimson to velvety maroon or mauve with golden center when fully open. Faint to strong sweet scent. Lush, medium green foliage. Not thorny. Possible orange red hips. Severe dieback possible.

Hybrid Perpetual rose

The Hybrid Perpetuals are difficult to grow in cold climates. Flowers are lovely (cupped) and fragrant. They bloom prolifically early in the season with marginal repeat bloom on new wood. They are very vigorous growers. However, most plants are susceptible to severe dieback and even death in cold winters. Nevertheless, some cold-climate growers love these plants, so here are a few secrets: Proper pruning and care is very important. During the first year, prune as little as possible, removing only flower buds until late summer. Winter protect. The following year, cut out any weak, dead, or diseased cane. Create an open, vase-shaped plant. Leave as much cane on the plant as possible within those guidelines. Winter protect. When the plant is mature, cut canes back to 4 feet (120 cm). Remove 3-year-old cane if the plant is growing well. If it's weak, remove no canes at all. Arch the longest canes over and attach them to stakes. This forces many laterals to form along the cane, so that what might appear to be a spindly plant suddenly becomes bushy. Cut each of these laterals back to 3 eyes (buds). These will then bloom in a magnificent display. Deadhead regularly. This is a lot of work, suited only to the avid cold-climate rose grower. Hardy to −10°F (−23°C) without protection (more conservative estimates are 0°F/−18°C). Varieties vary slightly in hardiness and susceptibility to disease. Winter protect using either the Minnesota Tip or Hass Teepee methods.

VARIETIES	COLOR	PETAL COUNT	FRAGRANCE	HEIGHT
'American Beauty'*****	Deep pink	50	Very strong	48″

Flowers up to 4 inches (10 cm) across and extremely fragrant. Color somewhere between deep pink and light red. Dark green foliage and prickly stems. Very susceptible to disease. Severe dieback possible. Despite its flaws, still one of the loveliest roses on the market.

'Baron Girod de l'Ain'**	Dark red edged white	30—40+	Slight	36″

Stunning flowers up to 3 inches (7.5 cm) across. Often crimson-colored with white, ragged edges. Light green leaves on thorny cane. Severe dieback expected.

'Baronne Prévost'***	Medium pink	60—80	Moderate to strong	36″

Flowers up to 4 inches (10 cm) across, almost purple or deep rose at times. Can appear quartered under ideal conditions. Medium bright green foliage. Prickly cane. Severe dieback possible.

'Frau Karl Druschki'****	White	25—35	None to slight	48″

Pointed pink buds open into lovely white blooms more than 5 inches (12.5) across. Strong, vigorous plant which under ideal conditions may be used as a Climber. Medium to dark green foliage on thorny cane. May occasionally form red hips. Severe dieback possible.

'Général Jacqueminot'****	Red to deep pink	25—30	Strong	60″

Bright crimson red buds. Flowers up to 4 inches (10 cm) across with very clear coloration. Medium bright green foliage on prickly stems. Severe dieback possible.

'Georg Arends'***	Medium pink	25—30	Very strong	36″

Especially vigorous. Long buds open into high-centered flowers up to 4 inches (10 cm) across with a silvery pink hue. Light green foliage on canes with few thorns. Severe dieback expected.

'Mme. Ferdinand Jamin' (see 'American Beauty')				
'Magna Charta'****	Bright pink/silver	26—40+	Moderate	36″

Clear pink flowers up to 3 inches (7.5 cm) across with lights shades of red and silvery undersides. Dull deep green, leathery foliage. Spreads rapidly. Severe dieback expected.

'Marchesa Boccella'****	Light to medium pink	60+	Moderate to strong	36″

Often confused with 'Jacques Cartier,' a Portland rose. Both are exquisite. Flowers roughly 3 inches (7.5 cm) across float on the tips of canes. Light to medium green foliage. Severe dieback possible.

VARIETIES	COLOR	PETAL COUNT	FRAGRANCE	HEIGHT
'Mrs. John Laing'****	Medium pink	45	Moderate	36″

Silvery tinted flowers up to 4 inches (10 cm). Loosely formed but capable of withstanding rain. Large dark green leaves with a matte finish on nearly thornless cane. Vigorous, but expect severe dieback.

'Paul Neyron'****	Warm pink/silver	50—55+	Moderate	60″

Lovely cupped flowers up to 5 inches (12.5 cm) across. Flowers tinged lilac almost look like Peonies. Stunning cut flower. Rich medium green foliage on very vigorous plant. Severe dieback to death possible.

'Reine des Violettes'****	Mauve	60—75	Very strong	60″

Cupped (sometimes quartered) flowers up to 4 inches (10 cm) across. Listed as mauve, but more a violet red. Light grayish, green foliage on nearly thornless cane tinged red. Severe dieback possible.

'Roger Lambelin'***	Red edged white	30—40	Strong	36″

Compact, vigorous. Crimson flowers edged and striped white. Very unusual petal form, almost feathery or spiked. Light green foliage. Severe dieback expected.

Miscellaneous Old Garden Rose

No one is *certain* about the parentage of this rose although it is typically listed as a Gallica or sold as ***Rosa × francofurtana***. It dates back to well before 1824 and pays homage to a woman who did more for rose culture than any person ever has. It has very little scent; a good clue that it is not a true Gallica. It will form orange to red hips. Differing roses may be sold under this name which explains the wide variability in petal count and descriptions of scent. Cover the crown with potting soil and a thick layer of leaves in late fall. Or, use the Minnesota Tip Method or Hass Teepee Method for optimal winter protection. Crown hardy to −25°F (−32°C) or lower with winter protection.

VARIETY	COLOR	PETAL COUNT	FRAGRANCE	HEIGHT
'Empress Josephine'***	Deep pink	30—60	Variable	30″

Dense, lax. Flowers up to 4 inches (10 cm) across, often shaded purple and papery in texture. Long bloom period, but blossoms damaged easily by rain. Plants vary in fragrance (the true rose has slight scent). Light green leaves on almost thornless canes. Orange hips like mini turnips. Expect dieback.

Moss rose (see *Rosa centifolia muscosa*)

The term *Moss rose* may be used for other roses as well, but in this guide we're covering the one that has the best chance of doing well in colder climates.

Noisette rose

Definitely not hardy in colder climates. In southern climates this class of rose produces clusters of loose, fragrant blooms on climbing canes. The following Noisette is worth the gamble if you use the Minnesota Tip Method or Hass Teepee Method of winter protection. You can also try mounding the crown with soil and a thick layer of whole leaves in late fall. This should protect it to −10°F (−23°C) and lower with an early snowfall. One final option is to grow it in a container to be buried in late fall (often survives this way). The plant is worth the work. Hardy to approximately 20°F (6.7°C) without protection.

VARIETY	COLOR	PETAL COUNT	FRAGRANCE	HEIGHT
'Nastarana'****	Pale pink to white	9—16+	Strong	36″

The 'Persian Musk Rose.' Flowers just over 2 inches (5 cm) across, tinged pink with exquisite creamy gold-colored stamens. Medium green foliage, sometimes tinted blue or gray. May form slender hips. Will bloom profusely on new wood. Warning: many different plants are sold under this name.

Portland rose

Portlands tend to be sturdy, erect, and bushy in appearance. Flowers are fragrant, many petalled, and will repeat bloom if deadheaded. Leaves have light green, narrow leaflets. In our opinion, 'Jacques Cartier' stands out as a gamble for cold climates. It's light pink, 50+ petals, very fragrant, and will grow to 36 inches (90 cm) or so. It is often mixed up in the trade with 'Marchesa Boccella,' a lovely Hybrid Perpetual. Give it extreme protection with the Minnesota Tip or Hass Teepee method to preserve as much cane as possible, since the plant does poorly if it has to regenerate a great deal of new wood. Consider growing it in a container to be buried in late fall alongside tipped roses or protected using the Hass Mound Method. Hardy to 0°F (−18C) without protection.

Provence rose (see *Rosa centifolia*)

Provins rose (see *Rosa gallica*)

Scotch rose (see *Rosa spinosissima*)

Rosa spinosissima (Burnet rose or Scotch rose)
(ROW-suh speye-noh-SISS-i-muh)

These are hardy roses growing up to 72 inches (180 cm) tall in colder climates. Profuse bloom in early spring. Most produce flowers with relatively few petals and varying degrees of fragrance. Arching, densely branched canes with ferny foliage and prickly thorns. Fall foliage color can be good, varying from yellowish orange to purplish red, depending upon the year. Salt tolerant, making them good for informal, roadside hedges. Their hybrids are listed under Shrub roses. The species here are also listed under Species Roses. Hardy to at least −20°F (−29°C) and lower.

VARIETIES	COLOR	PETAL COUNT	FRAGRANCE	HEIGHT
Rosa spinosissima***	White cream	5	Variable	36″

The 'Scotch Brier Rose.' Suckers freely. Parent to a number of fine hybrids. Large, cream-colored blossoms. Very thorny. Reddish cane bears leaves with tiny leaflets. Forms small, black or purple hips. Very hardy.

VARIETIES	COLOR	PETAL COUNT	FRAGRANCE	HEIGHT
Rosa spinosissima var. *altaica*****	White	5	Variable	60″

Blooms in clusters of blossoms up to 2 inches (5 cm) across. Open faintly yellow, turning to cream. Ferny foliage. Extremely spiny canes. Lovely and delicious maroon purple to black hips. Vigorous. Hardy.

VARIETIES	COLOR	PETAL COUNT	FRAGRANCE	HEIGHT
Rosa spinosissima plena****	White cream	26—41+	Moderate	72″

The plant spreads freely forming a large clump. Commonly known as the 'White Rose of Finland.' The true variety is often hard to find because it is difficult to propagate. It has long, arching canes covered with small 2-inch (5-cm) blossoms in early summer, each with bountiful, bright yellow stamens and sweet scent. Fairly hardy (winter protect). A good replacement is Polstjärnan under Shrub roses (it too lays claim to the title 'White Rose of Finland').

Tea rose

Tea roses get their name from the faint tea-like scent of their foliage. They are repeat bloomers. Plants will grow to about 36 inches (90 cm) in cold climates but require extreme winter protection to survive. A few diehards grow Teas in colder areas by raising them in pots and protected them like Miniatures grown the same way. Really a waste of time and money when compared to the results you'll get from other types of roses recommended throughout this guide. However, they are highly recommended in warmer climates.

'China Doll'

POLYANTHA ROSES

Polyanthas are delightful, versatile roses. They are quite compact, usually under 3 feet (90 cm) and produce clusters of cupped, 1-inch (2.5-cm) or slightly larger blossoms. Plants will bloom more than once if spent blossoms are removed. They are among the most free-flowering of roses providing season long interest. They provide nice color when planted in a large group as bedding plants. Polyanthas blend beautifully with perennials and make fine container plants. Admittedly, a number of them are susceptible to powdery mildew. And, most need winter protection.

How Polyantha Roses Grow

Some Polyanthas are grown on their own roots; others are budded or grafted on rootstock of another rose. These roses often grow as wide as they are tall. The hardier varieties may get quite tall. Polyanthas tend to spread out more in warmer climates, less so in colder areas. New canes often push up from the base of the plant as older canes mature and die back. You simply remove older canes to keep the plant healthy.

Where to Plant

Site and Light Polyanthas require lots of sun to do well. Plant them in full light. This will encourage larger flushes of bloom. Bright light also helps them grow vigorously and discourages disease.

Soil and Moisture Polyanthas need rich soil that retains moisture but drains freely. Replace clay or rock with loam purchased from a garden center or in bags as potting soil. Add lots of organic matter, such as compost, leaf mold, peat moss, or rotted manures. These keep the soil moist and cool during dry periods. They also help the soil drain freely, encourage the growth of beneficial soil microorganisms, and attract worms which keep the soil aerated and fertilized. Soil should be light, loose, and airy. This allows quick root growth and good drainage so that plenty of oxygen can get to the tender feeder roots. A handful of bonemeal (or superphosphate) mixed into the soil provides additional nutrients for the young plant. Proper soil preparation takes a little time, but it is critical to good growth.

Spacing Give more space than you might think necessary from the size of the bare root plants. Polyanthas spread out in warmer climates, sometimes several feet (about a meter). In colder climates they don't grow as wide and need less space. 'The Fairy' is an exception and does spread out fairly wide even in colder climates. Proper spacing helps circulate air around plants to prevent mildew.

Planting

Most Polyanthas are sold as bare root plants. You'll generally get them through a mail-order source.

Bare Root Plant Polyanthas as soon as the ground can be worked in spring.

Each plant should have 8 to 10 canes about 12 to 15 inches (30 to 37.5 cm) high. The number of canes varies greatly by supplier. The length and width of canes vary by variety.

Traditional guidelines are to place the bud union one-half below and one-half above the soil. Deeper planting causes no harm and may be advisable for plants such as 'The Fairy' if you don't plan to winter protect it.

Potted Plants Polyanthas were once hard to find in local garden centers. More recent introductions are now commonly available as potted plants. Pick out plants with healthy canes and luxuriant foliage. Check plants carefully for any signs of disease or insect infestations to avoid buying infected plants. Avoid leaving the plant in the car while you do other errands, since the heat buildup can damage it. Also, if the plant will be exposed to wind, protect it well by wrapping it in thick plastic. As soon as you get home, remove the plastic, and water the soil if it's at all dry. Plant exactly as outlined in Part II.

How to Care for Polyantha Roses

Water Water Polyanthas frequently. If the soil is loose and drains freely, it is almost impossible to over-water these plants. Water encourages bushy growth, lots of foliage, and excellent repeat bloom. These plants will bloom almost continuously if grown under ideal conditions. Never let the soil dry out. Keep it evenly moist all season long.

Mulch After the soil warms up to 60°F (15.6°C) apply a mulch around the base of the plant. Since Polyanthas fan out, it's fine to distribute the mulch under the cane. The most commonly used mulch is shredded (not whole) leaves. Place at least a 3-inch (7.5-cm) layer over the soil. If you have enough leaves to make a thicker mulch, so much the better. Leaves keep the soil moist and cool which encourages rapid root growth. The mulch feeds microorganisms and worms which enrich and aerate the soil. The mulch also inhibits weed growth and makes pulling any weeds that do sprout much easier. Other good mulches include pine needles and grass clippings. The latter should be only 2 inches (5 cm) deep, or they will heat the soil and may smell. If you use chipped wood or shredded bark, apply additional fertilizer to the soil, since these rob the soil of nitrogen as they decompose. Because mulch is eaten by soil micro-organisms and worms, it needs to be replenished regularly throughout the growing season. Remove and compost all mulch in the fall.

Fertilizing If combining use of inorganic and organic fertilizers, be careful during the first year. Use inorganic fertilizer only after the plant has leafed out and is growing vigorously.

In subsequent years, as early as possible in spring, sprinkle 10-10-10 (or 5-10-5) granular fertilizer around the base of each plant. Use common sense in judging the amount, more for bigger plants, less for smaller. Hand sprinkle, never placing fertilizer against the stem of the plant. Water immediately to dissolve the granules and carry nutrients to the roots.

In both the first and subsequent years do the

following: One to two weeks after the first feeding with 10-10-10 granules soak the base of each plant with a liquid containing 20-20-20 water soluble fertilizer following directions on the label. Use common sense, giving larger plants more, smaller plants less. This feeding will stimulate sensational bloom as the plant matures.

One week later soak the base of each plant with a liquid containing fish emulsion. Follow the directions on the label. Feed according to the size of the plant.

One week later give all plants the 20-20-20 treatment again. One week later, give all plants the fish emulsion soaking. Keep this rotation going until the middle of August. At this time, stop all use of fertilizers containing nitrogen, since nitrogen stimulates new growth which dies off in most winters. Use 0-10-10 fertilizer after mid-August if desired.

Additional fertilizing with Epsom salts (magnesium sulfate) is highly recommended. Give ⅓ cup (about 75 g) for larger plants, less for smaller, two times a year: in late May and in early July. Magnesium sulfate neutralizes the soil and makes it easier for the plant to take in nutrients. This stimulates the growth of new canes (basal breaks) from the base of the plant.

Good fertilizers for organic growers are alfalfa meal (rabbit pellets), blood meal, bonemeal, compost, cow manure, fish emulsion, rotted horse manure, and Milorganite. Add bonemeal to the soil before planting. The others are effective added to the soil at planting time and as additional feedings on the surface of the soil throughout the season.

Weeding Kill all perennial weeds before planting. Weeds compete with roses for nutrients and water and also act as hosts for harmful insects that carry disease. Inhibit annual weed growth with the use of a mulch.

Polyanthas often have spreading branches close to the ground, so wear thick leather gloves while placing mulch underneath them. Many annual weed seeds need light to germinate; without it they won't sprout. Mulch stops most of these seeds from growing. A few may sprout anyway; pull them up by hand as soon as they are noticeable. A layer of mulch makes pulling weeds easy, as the soil stays moist and loose.

Avoid hoeing, which can harm the shallow root system or wound the low-lying branches.

Staking Because Polyanthas are relatively low-growing plants, they require no staking whatsoever.

Disbudding Some growers remove flower buds in the first year, directing all energy to the plant itself. Remove buds just as they begin to form, for best results. Continue removing buds all season long if you intend to bury the plant for winter protection. If you will leave the plant standing, let it bloom lightly in the end of the season to induce dormancy in order to help protect it during the winter.

On mature plants no one removes buds. Flowers are extremely small. It is the amount and frequency of bloom on these plants that is attractive.

Deadheading Some growers remove all flower buds during the first year of growth, letting plants bloom only late in the season. Others aren't so fussy or don't have the time. Bud removal does encourage better root growth, but it is not essential. In following years, removing spent blossoms will encourage repeat bloom, but Polyanthas are noted for their strong intermittent to repeat bloom anyway. Frankly, deadheading Polyanthas is often bypassed by the average rose grower for this reason.

Pruning Let the plant grow freely in the first year. Remove only broken or diseased portions of cane and flower buds (if you have the energy and time). The following spring, cut 1-year-old canes back by one-third. Cut back to a dormant, outward-facing bud. The following year, cut 2-year-old canes back by two-thirds. Remove all 3-year-old canes in subsequent years. The removal of old cane stimulates the plant to grow new canes from its base. These will eventually produce spectacular bloom. Polyanthas tend to send up many spindly canes from the base of the plant. Prune out all but the healthiest and thickest to replace old canes. Keep the plant bushy but as open and widespread as possible. The more light to the buds, the better.

However, if plants are not growing vigorously, prune as little as possible. The pruning method outlined here assumes excellent growth patterns and little winter dieback.

Winter Protection Protect the more tender varieties by covering their crown with potting soil and a thick layer of whole leaves. If grown in containers, bury them with tipped roses. Hardier varieties typically die back to the snow line.

Problems

Insects Polyanthas are prone to the same insect infestations as other roses. Use a preventive spray program for best results. If any mites appear, kill them immediately with a miticide to prevent colonization.

Disease Polyanthas also may get infected with powdery mildew. Proper care helps avoid this, but chemical treatment may be necessary in some years if the disease gets out of hand.

Propagation

Most Polyanthas sold to the general public are on budded stock. If you can buy plants growing on their own roots, they will sometimes send out underground roots (stolons) which produce little plantlets off to the side of the mother plant. Dig these up in early spring and plant as bare root plants.

If you buy budded plants, you can try bending over one of the branches and following the steps outlined under "Ground Layering" in Chapter 7 (p. 212). This is frequently successful. Taking cuttings is also possible, but the soil layering (ground layering) method is easier and more often successful.

Budding is best left in the hands of professionals. Considering the amount of work compared with the foregoing methods, it doesn't make much sense except for commercial growers.

Special Uses

Cut Flowers Flowers are quite small. Most people relish the display in the garden and leave the plants alone.

Dried Flowers Although the flowers of Polyanthas are generally small, some contain many petals and are quite attractive dried. Pick them just before they peak and dry them quickly.

Sources

Angel Gardens, P.O. Box 1106, Alachua, FL 32616, (352) 359-1133

Antique Rose Emporium, 9300 Lueckmeyer Rd., Brenham, TX 77833, (800) 441-0002

Burlington Rose Nursery, 24865 Rd. 164, Visalia, CA 93292, (559) 747-3624

Chamblee's Rose Nursery, 10926 US Hwy 69 N, Tyler, TX 75706, (800) 256-7673

Countryside Roses, 5016 Menge Ave., Pass Christian, MS 39571, (228) 452-2697

David Austin Roses Ltd., 15059 Hwy 64 W, Tyler, TX 75704, (800) 328-8893

Dutch Gardens, P.O. Box 2999, 4 Currency Dr., Bloomington, IL 61702, (800) 944-2250

Edmunds' Roses, 335 S High St., Randolph, WI 53956, (888) 481-7673

Garden Valley Ranch, 498 Pepper Rd., Petaluma, CA 94952, (707) 795-0919

Heirloom Roses, 24062 Riverside Dr. NE, Saint Paul, OR 97137, (503) 538-1576

High Country Roses, P.O. Box 148, Jensen, UT 84035, (800) 552-2082

Hole's Greenhouses & Gardens Ltd., 101 Bellerose Dr., St. Albert, AB T8N 8N8 Canada, (780) 419-6800 (Canada only)

Hortico, Inc., 723 Robson Rd., RR# 1, Waterdown, ON L0R 2H1 Canada, (905) 689-6984

Jackson & Perkins, 2 Floral Ave., Hodges, SC 29653, (800) 872-7673

Linda's Antique Roses, 405 Oak Ridge Dr., San Marcos, TX 78666, (512) 353-3220

Mary's Plant Farm & Landscaping, 2410 Lanes Mill Rd., Hamilton, OH 45013, (513) 894-0022

Pickering Nurseries, 3043 County Rd. 2, RR #1, Port Hope, ON L1A 3V5 Canada, (905) 753-2155

Regan Nursery, 4268 Decoto Rd., Fremont, CA 94555, (800) 249-4680

Rogue Valley Roses, P.O. Box 116, Phoenix, OR 97535, (541) 535-1307

Rose Fire, Ltd., 09394 State Rte. 34, Edon, OH 43518, (419) 272-2787

Roses of Yesterday and Today, 803 Brown's Valley Rd., Watsonville, CA 95076, (831) 728-1901

Roses Unlimited, 363 N Deerwood Dr., Laurens, SC 29360, (864) 682-7673

Orange Triumph®

Oso Happy™ Candy Oh!

'Sigrid'

'The Fairy'

S & W Greenhouse, Inc., P.O. Box 30, 533 Tyree Springs Rd., White House, TN 37188, (615) 672-0599

Spring Valley Roses, P.O. Box 7, Spring Valley, WI 54767, (715) 778-4481

Two Sisters Roses, 1409 N Redbud Lane, Newcastle, OK 73065, (no phone by request)

Vintage Gardens (custom propagation), 4130 Gravenstein Hwy N, Sebastapol, CA 95472, (707) 829-2035

Wayside Gardens, 1 Garden Lane, Hodges, SC 29695, (800) 213-0379

White Flower Farm, P.O. Box 50, Litchfield, CT 06759, (800) 503-9624

Witherspoon Rose Culture, 3312 Watkins Rd., Durham, NC 27707, (800) 643-0315

VARIETIES

Many Polyanthas date back to the early part of the 20th century. A few have been developed more recently. If you buy budded Polyanthas, place the bud union 3 inches (7.5 cm) below the ground to improve winter hardiness. Many of the more tender varieties will die back to the ground during the winter. As long as they are healthy going into winter dormancy, they usually regrow well in spring. There is no harm in covering their crowns with a thick layer of whole leaves or marsh hay in late fall for optimal winter protection. Many of the zone 6 plants have survived in zone 4 with proper care and protection. Early and deep snow provides the best protection, with some of the plants only dying back to the snow line. 'Snowbelt' and 'Zenaitta,' both bred by Paul Jerabek, are tender, but well worth winter protecting because they flower so freely. Oso Happy™ Candy Oh! ('Candy Oh! Vivid Red') bred by David Zlesak, and 'Lena,' 'Ole,' 'Sigrid,' and 'Sven,' bred by Kathy Zuzek are recent introductions which are hardy. 'Lena' and 'Ole' are good choices for organic gardeners because they are quite disease resistant as is the more tender 'Marie Daly.' Polyanthas suited for containers are 'China Doll' and 'White Pet.' For optimal winter protection bury these potted roses along with tipped roses or protect them using the Hass Mound Method. Note that the most hardy plants listed below may get considerably taller than the heights listed.

VARIETIES	COLOR	PETAL COUNT	FRAGRANCE	HEIGHT	HARDINESS
'Britannia'***	Red/white/gold	5	Slight	20"	Zone 6
'China Doll'***	Bright pink/yellow base	25	Slight	18"	Zone 6
Fairy Dance®****	Fuchsia red/gold	17+	Slight	24"	Zone 4
'Katharina Zeimet'****	White to cream	26—40	Moderate	24"	Zone 4—5
'La Marne'****	Light pink/white/gold	10 (variable)	Slight	24"	Zone 6
'Lena'****	Pink white	5	Slight	30"	Zone 3
'Margo Koster'***	Orange salmon	25—35	None to slight	24"	Zone 6
'Marie Daly'***	Medium pink	17—25	Moderate to strong	24"	Zone 5—6
'Marie-Jeanne'***	Pink cream fading to white	26—40	Slight to moderate	24"	Zone 6
'Marie Pavié'***	White pink/pink/gold	17—25	Strong	24"	Zone 5—6
'Ole'****	Pink to ivory	16—	Strong	36"	Zone 3
Orange Triumph®**	Reddish orange	17—25	Slight	24"	Zone 6
Oso Happy™ Candy Oh!****	Medium red to crimson	5	Slight spicy	36"	Zone 4
'Sigrid'****	Deep red	9—11	Slight	30"	Zone 4
'Snowbelt'****	White/gold center	15—25	Slight	24"	Zone 6
'Sven'****	Purple to pink/yellow	16—40	Strong	24"	Zone 3
'The Fairy'****	Light to deep pink	35—40	None to slight	20"	Zone 4
'Verdun'***	Rich carmine red	41+	Slight	24"	Zone 5
'White Pet'**	Creamy white pink blush	41+	Moderate	18"	Zone 6
Yesterday®**	Lilac pink/gold center	13	Slight	30"	Zone 4
'Zenaitta'****	Orange red/white base	20	None	24"	Zone 5

'Lillian Gibson' (*Blanda*)

SHRUB ROSES
(HYBRID SHRUB ROSES)

All roses are shrubs, but the catchall group known today as *Shrub roses* is booming in popularity as more and more varieties become available to home growers. In general, Shrub roses are hardy, easy to care for, vigorous, and generally resistant to disease and insects. They are extremely versatile plants to work with. Use them as specimen or accent plants, for hedges, in borders or beds, or as Climbers or ground covers. While many Shrub roses are hardy, some are not suited to colder climates and will die out completely if not fully winter protected. Others are so tender that they die out even if fully protected. So, the term *Shrub rose* is not synonymous with the term *winter-hardy rose* as many have been told. Many Shrub roses promoted as fully hardy in colder climates are not. Check with local growers and arboretums to see which Shrub roses are thriving in your area.

How Shrub Roses Grow

Shrub roses are hybrids, crosses between various Species (wild) roses. Many Shrub roses grow on their own roots, which increases their chance of survival in colder climates. Others are budded or grafted (buds or scions are joined to rootstock) but, even if they die back, will regrow and produce good bloom. Budded plants are often less hardy than roses grown on their own roots. Budded plants will also produce suckers from the rootstock which must be removed, so they are not as hassle-free as Shrub roses grown on their own roots. However, by planting the bud union 3 to 4 inches (7.5 to 10 cm) below the soil, you may get roots growing from the buried portion of canes. Shrub roses vary in height and growth habit. A few of them spread out, producing plantlets off to the side of the parent plant. Shrubs such as these may become many times as wide as they are tall.

Where to Plant

Site and Light You must plant all Shrub roses in an open, sunny area away from trees or other woody plants. They have an extensive root system and need lots of space to grow freely. Sun dries off morning dew which helps the plants resist disease, although for the most part, they are not prone to many diseases. However, disease resistance varies by individual plant and variety.

Avoid low-lying areas, which are frost pockets in spring and fall. Avoid planting Shrubs under eaves unless you plan to water them frequently. Also avoid planting them under drip lines where they can be damaged by falling snow in the winter or heavy rains in the summer.

Soil and Moisture Shrub roses thrive in moist, well-drained soil. If possible, prepare beds in the fall. If you are planting a single plant, dig a large hole, perhaps as much as 36 inches (90 cm) wide and 36 inches deep. Remove all poor soil, including clay and rocks. Add loam mixed with peat, compost, rotted leaves, or any other organic matter. The organic matter should make up about one-third of the soil. Soil should be almost loose enough to dig into with your bare hand. It should drain freely—avoid low-lying areas with boggy soil or any spot that collects water after a rain. Many Shrub roses do not need the amount of fertilizer of other roses. However, some growers mix alfalfa meal (rabbit pellets), blood meal, bonemeal, Milorganite, or a combination of these into the soil and claim to double the number of flowers. Superphosphate may replace bonemeal for non-organic growers.

Spacing Space Shrub roses according to the potential height of each plant. Most plants will grow almost as wide as, or wider than, they are tall. Place the rose in its permanent spot as you would any tree or large shrub. You don't want to move it at a later date. Shrub roses form an extensive and difficult-to-dig root system. Note, too, that some Shrub roses sucker profusely and will spread rapidly over a period of years.

Planting

You can start Shrub roses either from bare root plants or from roses already potted. To get specific varieties, you may have to buy bare root plants. Generally, these must be ordered from catalogs. Ask for plants grown on their own roots whenever possible. As noted, these tend to withstand cold winters better than those grown on budded stock.

Bare Root Plant bare root Shrub roses directly in the garden as soon as the ground can be worked in spring. Follow the exact steps outlined in Part II.

Each plant should have healthy canes. The number and size of the canes will vary considerably by supplier and variety ordered. The length and width of canes also vary by variety.

If the plant is budded, place the bud union 3 to 4 inches (7.5 to 10 cm) below the soil. This may stimulate root growth from the buried portion of canes, which will make the plant hardier in cold climates.

Potted Plants You'll often find an assortment of Shrub roses in local garden centers or nurseries. Pick out plants with healthy canes and luxuriant foliage. Check plants carefully for any signs of disease or insect infestations to avoid buying infected plants. Avoid leaving the plant in the car while you do other errands, since the heat buildup can damage it. Also, if the plant will be exposed to wind, protect it well by wrapping it in thick plastic or paper. As soon as you get home, remove the protective covering, and water the soil if it's at all dry. Plant as outlined in Part II.

How to Care for Shrub Roses

Water Mature Shrub roses tolerate drought better than other roses. However, in the early stages of growth keep the soil evenly moist at all times. Just because a plant tolerates drought doesn't mean that it likes it. If you want lush, vigorous plants, water frequently. This stimulates lots of new growth from the base of the plant. Watering frequently means watering every day if necessary. Soak the soil around the plant. Just let the hose run at its base for 10 minutes or so. Saturate the soil to a depth of 18 inches (45 cm) or more.

Continue watering through late summer into fall until the first freeze. In cold climates it is common to have a dry spell and even a drought during this period. Water stress at this time can damage roses. Late watering was originally believed to stimulate new growth susceptible to die back during the winter. Dieback is preferred over extreme stress.

Mulch After the soil warms up to 60°F (15.6°C), apply a mulch around the base of the plant. The mulch should come close to but not touch the cane. The most commonly used mulch is shredded (not whole) leaves. Place at least a 3-inch (7.5-cm) layer over the soil. If you have enough leaves to make a thicker mulch, so much the better. Leaves keep the soil moist and cool which encourages rapid root growth. The mulch feeds microorganisms and worms which enrich and aerate the soil. The mulch also inhibits weed growth and makes pulling any weeds that do sprout much easier. Other good mulches include pine needles and grass clippings. The latter should be only 2 inches (5 cm) deep, or they will heat the soil and may smell. These mulches are all inexpensive and effective. If you use chipped wood or shredded bark (lovely around larger plants), apply additional fertilizer to the soil, since these rob the soil of nitrogen as they decompose. Because mulch is eaten by soil microorganisms and worms, it needs to be replenished regularly throughout the growing season. Remove and compost all mulch in the fall.

Fertilizing Place bonemeal or superphosphate in the base of the planting hole as outlined earlier. These nutrients must be close to the roots in order to be effective.

In subsequent years, sprinkle 10-10-10 granular fertilizer around the base of each plant just as buds appear in spring. Use common sense in judging the amount, more for bigger plants, less for smaller. Hand sprinkle, never placing fertilizer against the stem of the plant. Water immediately to dissolve the granules and carry nutrients to the root zone.

Good organic fertilizers are alfalfa meal (rabbit pellets), blood meal, bonemeal, compost, cow manure, fish emulsion, rotted horse manure, and Milorganite.

Many Shrub roses prefer organic fertilizers, which contain lower doses of essential nutrients than chemical fertilizers.

Foetidas do well in poor soil. Avoid overfertilizing them, especially with products containing high amounts of nitrogen.

Shrub roses do not need frequent feedings. Stop all fertilizing by mid-June. Feed only once each season at the most. If mature plants form lots of foliage but little bloom, do not feed the following spring. If plants are not forming lush, full foliage, then feed them immediately.

Weeding Kill off all perennial weeds when preparing a bed or planting hole in late summer or early fall. Roundup® is extremely effective. It will disintegrate by the following spring. You can also prepare a bed in spring, but if you use an herbicide to kill perennial weeds, follow the directions on the label carefully, and plant only after the recommended waiting period.

Shrub roses are tough, durable plants. However, they do best when not competing with grass or weeds. Mulch is the best method of suppressing annual weeds, since it keeps soil moist and cool at the same time. When weeding around the base of the plant, hand pull all weeds.

Disbudding Some growers remove flower buds in the first year, directing all energy to the plant itself. Remove buds just as they begin to form, for best results. Let a few buds bloom late in the season to encourage the plant to go into dormancy for the winter season.

Most growers remove no buds on mature plants.

Some rose growers suggest removing a portion of buds early in spring on repeat bloomers to prolong the bloom season. At best, this would be a tiresome chore, and whether or not it will promote more bloom is questionable. We do not advise removal of any buds whatsoever. Prolong bloom by removing spent blossoms as follows.

Deadheading If a plant blooms only once, do not remove spent blossoms. If a plant is a repeat bloomer, remove all spent blossoms. If you do not remove these, the plant's energy will go into forming hips, and you will have less repeat bloom. If you want hips, do not remove blossoms after mid-summer. In all instances, stop removing spent blossoms by September. This encourages the plant to form some hips, which initiates a process that better protects the plant from winter damage.

When removing blossoms, use pruners to snip the plant back to a point just above a set of leaves. New growth will sprout from the growth bud there to produce a new branch with flower buds.

Staking (on plants that bloom in spring only) Sometimes, one-time-flowering plants get leggy—long and spindly. If this happens, bend the longest cane over and tie it to a stake. This induces new growth from the base of the plant. The new canes are called *basal breaks*. The new canes make the plant very bushy. Once the new canes are growing vigorously, cut the original cane off at ground level. This usually takes at least a year. Always be careful not to crimp the cane when bending it over. Staking works especially well with 'Harison's Yellow.'

Pegging (on plants that bloom more than once) If repeat-blooming Shrubs produce little bloom, arch the canes over and peg them to the ground with a piece of wire. Bent canes produce more laterals (branches off the main cane) and sublaterals (branches off other branches). These in turn produce more bloom. When doing this, be gentle but firm. Avoid crimping the canes, especially at their base.

Pruning Little or no pruning is generally required in the first year. If wood is damaged or dead, snip it off. Otherwise, do not prune.

In subsequent years, cut out dead wood each spring. Shape the plant according to the form and size desired. Many growers cut out crossing or inward-growing canes to open the plant up and give it a nicer look. Crossing canes sometimes rub against each other, creating wounds that become infected with disease. Cutting out crossing canes also allows light and air into the center of the plant. This keeps interior cane growth healthy. To encourage the growth of laterals (branches) on older canes, cut them back by one-third. Never remove the canes (basal breaks) growing from the base of the plant, since these will form new cane with heavy bloom. Always prune before plants flower (never prune after flowering other than to remove spent blossoms). Shrub roses require little pruning, which is one of their major advantages.

Never prune Shrub roses late in the season. This will encourage new growth which will die back during severe winters. The dead cane is often infected with disease or infested with insects.

Winter Protection Shrub roses vary in their hardiness. The term *Shrub rose* does not mean "hardy!"

A very hardy rose is one that stands up to almost any winter condition. The tips of some stems may die, but these can be snipped off. The death of part or all of a stem is known as *dieback*.

Some plants die back to the ground each year but will regrow and form a lovely plant in one season. These plants are said to be *crown hardy*. They require minimal to no winter protection because they act like a perennial.

A few of the roses listed in this section need winter protection to survive. Some may even require the Minnesota Tip or Hass Teepee method to survive. The Minnesota Tip Method is quite difficult for larger roses, so you may choose the less strenuous Hass Teepee Method (pp. 193–199).

Hardiness varies from year to year. Plants do best if there is a gradual cooling-off period before a severe cold. This allows them to go dormant before winter sets in. Heavy snow cover is an excellent winter protection. Plants fully exposed to winter winds and cold often suffer dieback, even when properly cared for. In very rare circumstances, plants that are normally very hardy suffer severe dieback. The exact cause is still unknown.

Problems

Insects Japanese beetles have become a serious problem. Ask local growers about plants that don't attract them in your area (in ours it is 'Sunrise Sunset'). Rose stem girdles can also be a major nuisance, as can aphids and spider mites. To control these see Appendix A (pp. 238–243).

Disease Powdery mildew and black spot are the most common problems. Many Shrub roses are quite resistant to them. Problems with these diseases are noted in the varietal information that follows. If a Shrub is commonly infected with mildew or black spot, you should use a preventive program. If not, don't spray at all. Never spray plants if you plan to use hips for food. Note, too, that certain types of Shrubs, especially *Rugosas*, are allergic to fungicides.

Propagation

Division Some Shrub roses grow suckers, or pups, off to the side of the mother plant. When removing these, follow the underground root (stolon) as far back to the mother plant as possible, and make your cut there. This helps the plantlet take root immediately and survive the trauma of being separated from the mother plant. Use a sharp spade to dig down and sever the root. Dig up the sucker with as much soil around it as possible. Plant it immediately as if it were a bare root plant. Keep it well watered until growing vigorously.

Cuttings Some Shrub roses grow well from softwood cuttings, generally taken in late June or early July from the center of a firm cane. Firm canes are mature, not too green. Refer to Chapter 7 in Part II for specific tips on propagation with softwood cuttings.

Special Uses

Cut Flowers Most people mistakingly think of Shrub roses as second-rate cut flowers. But in fact, many Shrub roses produce lovely sprays of scented flowers that are stunning in indoor arrangements. By growing a wide variety of repeat-blooming types, you can have cut flowers all summer long. It is true, however, that they are not as long-lasting as the more popular Hybrid Teas.

Dried Flowers Shrub roses are not generally used for dried flowers, although the petals can be dried and used in potpourris.

Hips Many Shrub roses produce abundant hips. These attract birds and are also extremely beautiful on plants during the winter season. Hips are extraordinarily high in vitamin C. They make excellent jellies. For an excellent recipe see Chapter 8 (p. 225).

Sources

Angel Gardens, P.O. Box 1106, Alachua, FL 32616, (352) 359-1133

Antique Rose Emporium, 9300 Lueckmeyer Rd., Brenham, TX 77833, (800) 441-0002

Bergeson Nursery, 4177 County Highway 1, Fertile, MN 56540, (218) 945-6988

Burlington Rose Nursery, 24865 Rd. 164, Visalia, CA 93292, (559) 747-3624

Chamblee's Rose Nursery, 10926 US Hwy 69 N, Tyler, TX 75706, (800) 256-7673

Corn Hill Nursery Ltd., 2700 Rte. 890, Corn Hill, New Brunswick E4Z 1M2, Canada, (506) 756-3635

Countryside Roses, 5016 Menge Ave., Pass Christian, MS 39571, (228) 452-2697

David Austin Roses Ltd., 15059 Hwy 64 W, Tyler, TX 75704, (800) 328-8893

Edmunds' Roses, 335 S High St., Randolph, WI 53956, (888) 481-7673

ForestFarm, 990 Tetherow Rd., Williams, OR 97544, (541) 846-7269

Fraser's Thimble Farms, 175 Arbutus Road, Salt Spring Island, BC V8K 1A3 Canada, (250) 537-5788

Fritz Creek Gardens, P.O. Box 15226, Homer, AK 99603, (907) 235-4969

Garden Valley Ranch, 498 Pepper Rd., Petaluma, CA 94952, (707) 795-0919

Goodness Grows, Inc., P.O. Box 311, 332 Elberton Rd., Lexington, GA 30648, (706) 743-5055

Greenmantle Nursery, 3010 Ettersburg Rd., Garberville, CA 95542, (707) 986-7504

Heirloom Roses, 24062 Riverside Dr. NE, Saint Paul, OR 97137, (503) 538-1576

High Country Roses, P.O. Box 148, Jensen, UT 84035, (800) 552-2082

Hortico, Inc., 723 Robson Rd., RR# 1, Waterdown, ON L0R 2H1 Canada, (905) 689-6984

Jackson & Perkins, 2 Floral Ave., Hodges, SC 29653, (800) 872-7673

Knud Pedersens Planteskole, Tastrup Sovj 1, Harlov DK-8462, 86-94-18-66

Mary's Plant Farm & Landscaping, 2410 Lanes Mill Rd., Hamilton, OH 45013, (513) 894-0022

McKay Nursery Co., P.O. Box 185, Waterloo, WI 53594, (920) 478-2121

North Creek Farm, 24 Sebasco Rd., Phippsburg, ME 04562, (207) 389-1341

Northland Rosarium, 9405 S Williams Lane, Spokane, WA 99224, (509) 448-4968

Palatine Fruit & Roses, 2108 Four Mile Creek Rd., RR #3, Niagara-on-the-Lake, ON L0S 1J0 Canada, (905) 468-8627

Pickering Nurseries, 3043 County Rd. 2, RR #1, Port Hope, ON L1A 3V5 Canada, (905) 753-2155

Raintree Nursery, 391 Butts Rd., Morton, WA 98356, (800) 391-8892

Regan Nursery, 4268 Decoto Rd., Fremont, CA 94555, (800) 249-4680

Rogue Valley Roses, P.O. Box 116, Phoenix, OR 97535, (541) 535-1307

Rose Fire, Ltd., 09394 State Rte. 34, Edon, OH 43518, (419) 272-2787

Rosenhof Schultheis, Bad Nauheimer Str. 3, 61231 Bad Nauheim-Steinfurth, Germany 06032 92528-0

Roses of Yesterday and Today, 803 Brown's Valley Rd., Watsonville, CA 95076, (831) 728-1901

Roses Unlimited, 363 N Deerwood Dr., Laurens, SC 29360, (864) 682-7673

St. Lawrence Nurseries, 325 State Hwy 345, Potsdam, NY 13676, (315) 265-6739

Select Plus International Lilac Nursery, 1510 Pine Rd., Mascouche, QC J7L 2M4 Canada, (450) 477-3797

S & W Greenhouse, Inc., P.O. Box 30, 533 Tyree Springs Rd., White House, TN 37188, (615) 672-0599

Spring Valley Roses, P.O. Box 7, Spring Valley, WI 54767, (715) 778-4481

Two Sisters Roses, 1409 N Redbud Lane, Newcastle, OK 73065, (no phone by request)

Vintage Gardens (custom propagation), 4130 Gravenstein Hwy N, Sebastopol, CA 95472, (707) 829-2035

Wayside Gardens, 1 Garden Lane, Hodges, SC 29695, (800) 213-0379

White Flower Farm, P.O. Box 50, Litchfield, CT 06759, (800) 503-9624

Witherspoon Rose Culture, 3312 Watkins Rd., Durham, NC 27707, (800) 643-0315

VARIETIES

Shrub roses vary greatly in winter hardiness. Some survive extremely cold winters with little damage, while others die back to the ground but produce lovely plants anyway. The latter are said to be *crown hardy*. Still others must be protected from the cold and drying winter winds to survive at all (see pp. 192–199 on winter protection). Since winter protection takes energy and time, hardiness may affect which plants you buy. Most plants that suffer dieback still produce excellent bloom from emerging cane, so don't be afraid to buy them. In the fall plants go through a process called acclimation. Water slowly moves out of the cells making them less vulnerable to rupture from cold (like a can of soda in a freezer). The technical term for this is *supercooling*. If there is a sudden cold snap before the canes have acclimated, cells rupture and cane is damaged. In the spring the process is reversed in deacclimation. Cells take in water and expand. Again, if cells have expanded too early and there is a sudden cold snap, they can rupture causing damage to the cane. Some Shrub Roses acclimate and deacclimate better than others. This explains why roses may come through an extremely severe weather without damage, but in another year suffer considerable damage even though mid-winter temperatures are relatively mild (fall or spring cold snaps cause the damage in these years).

General zone hardiness ratings are helpful to some degree. However, you have a number of microclimates or mini zones in your yard. For example, if you place a Shrub rose so that it is covered by drifting snow each year, it may survive much colder temperatures than you might expect. *Also, we strongly suggest that you buy roses growing on their own roots whenever possible. These plants tend to be hardier than budded roses.*

Shrubs vary greatly in their resistance to disease. You can improve disease resistance with good culture—proper placement and spacing, good soil preparation, adequate watering, mulching, and correct pruning. However, diseases can be spread by the wind or by bringing infected plants into the garden. Black spot is one of the most common diseases in cold climates. There are many types (races) of black spot. Specific Shrub roses may be resistant to most races, but not to a specific one. This is called *vertical resistance*. Other specific roses can be infected by the disease, but fight it off fairly well. This is known as *horizontal resistance*. As with other diseases, black spot mutates so that new races appear over a period of time. Breeders are working hard to develop roses with good disease resistance. A number of new introductions will be on the market in coming years. We have tried extremely hard to give you accurate information on a plant's disease resistance. But things change. 'Martin Frobisher,' once very disease resistant, has become extremely susceptible to certain races of black spot.

For organic gardeners, our notes on disease resistance will be helpful. However, it's beneficial to join a local chapter of the rose society to get even more information on this problem. Disease and insect problems are often related to a specific area and its climate. Local rose growers will help you choose named varieties that have done well in your specific location without spraying. We cannot match this kind of area-specific knowledge in a guide aimed at the entire cold-climate region. We rely on personal observation and input from colleagues across the country to make generalized statements on disease. By the way, some very fine roses susceptible to disease are grown without problems in a number of rose gardens, including those of organic gardeners.

Following is a list of possible choices for Shrub roses in cold climates. Ratings may be affected by the plant's hardiness or general susceptibility to disease (especially black spot and powdery mildew). Some plants that are gems in southern areas are rated quite low in this section because they may die off in colder temperatures. Choose roses by plant form and height, vigor, flower color, petal count, the expected time of bloom, repeat bloom, fragrance, hardiness, susceptibility to disease, summer and fall foliage color, and production of hips. Following are the general categories of Shrub roses which appear in the Varieties table:

Austin (English roses): These hybrid roses are crosses by David Austin between Old Garden Roses and modern Hybrid Teas and Floribundas. He has been able to retain the form and fragrance of the older roses and acquire the repeat bloom capacity of the newer ones. Most have great first bloom and limited repeat bloom. A number of these are gems and have been listed because *they are worth the extra effort to protect using the Minnesota Tip Method.* Cutting the plants back to 12 inches (30 cm) and then mounding soil over the crown also protects the plant. The loss of cane does affect their performance the following year. These plants are hardy to the mid-20s (−3°C) without winter protection. Note that David Austin has bred other roses not listed under the category English roses.

Bailey roses

Bailey Nurseries has introduced a number of roses including those in the The Easy Elegance® series. The latter was developed primarily through the breeding of Ping Lim. *The roses selected for this guide are grown on their own roots and are among the most disease resistant of the series.* They tend to be vigorous with good bloom. Plants will die back to the snow line or more without winter protection. However, they are crown hardy to −20°F (−29°C) or lower. Still, covering the crowns with potting soil and a thick layer of whole leaves is suggested for optimal winter protection.

Graham Thomas® (Austin)

'David Thompson' (Explorer)

'Persian Yellow' (*Foetida*)

Carefree Beauty™ (Buck)

'Black Boy' (*Kordesii*)

Carefree Wonder™ (Meilland)

'Marguerite Hilling' (*Moyesii*)

'Morden Fireglow' (Parkland)

'White Pavement'/'Snow Owl' (Pavement)

'Grootendorst Supreme' (*Rugosa*)

'Wildenfels Gelb' (*Spinosissima*)

'Assiniboine' (*Suffulta*)

Blanda hybrids: 'Lillian Gibson' is the only one in the following table. These hybrids are known for the reddish coloration of their stems and are usually hardy to about −30° to −40°F (−34° to −40°C) without winter protection. This one-time bloomer is becoming hard to find, but can be budded. Hopefully, more mail-order sources will offer it in the future as it is an exceptional rose.

Buck roses: The 88 roses bred by Griffith Buck grow well on the southern edge of the cold-climate region and should be considered marginally hardy at best with some notable exceptions such as 'Amiga Mia' and 'Applejack.' However, they are lovely and often fragrant. Many of them (not all) are crown hardy, which means that the cane dies back to the ground each winter and is removed in the spring. New shoots sprout from the ground to form a lovely plant. In effect, these roses act more like perennials than woody plants. The toughest varieties are crown hardy to −10°F (−23°C) without winter protection, but protection is strongly advised for all of these roses. Cover the crowns with potting soil and a thick layer of leaves in late fall for optimal winter protection. Death of these plants can occur in especially cold winters without adequate snow.

Canadian Artist Series

Canadian rose breeders are working together to develop tough new plants suited to cold-climate gardening. The first two introductions named after famous artists are 'Emily Carr' and 'Felix Leclerc.' More are expected to follow. We have not given them stars yet since they have not been fully field-tested by us.

Explorer series: This group of roses, named after famous Canadian explorers, tend to be quite hardy. Even if they die back, the plants sprout new growth that blooms profusely. Most of these plants are hardy to −30° to −40°F (−34° to −40°C) without winter protection. However, there are exceptions as noted in the varieties Table. These will do best with minimal to more extreme winter protection. In recent years this series has shown susceptibility to at least one type (race) of black spot, which is a shame, considering how valuable these plants are.

Floribunda roses: A few Floribundas are so hardy that they are included in catalogs as if they were shrubs. 'Chuckles' and 'Nearly Wild' fall into this category. They require mild winter protection to survive in severe climates.

Foetida hybrids: 'Harison's Yellow' is listed here under Old Garden Roses (p. 75) and under Species Roses (p. 129). These plants produce true and lasting yellow blossoms. The plants are graceful and quite vigorous if given plenty of space. They grow well in poor soil and are very hardy. However, they are prone to rust and may have brought black spot into the rose world. If they are not routinely sprayed with a fungicide, they usually defoliate and may die out completely. They are hardy to −30°F (−34°C) without winter protection.

Grandiflora roses: We have included a few Grandiflora roses in this section because they are quite hardy, resemble shrubs, and are often sold under the Shrub Rose category. They are also included in the Grandiflora section.

Kordes roses: These roses are being produced by a German company of the same name. They are most noted for their disease resistance. The Shrubs are presently working their way from Europe into the north American market. Included in these are the Vigorosa™ Series. Many of these are not reliably hardy in cold climates. If hardier varieties are introduced, we will include them in future editions of this guide.

Kordesii hybrids: These plants have glossy foliage. They have good repeat bloom in most cases. They are resistant to disease and fairly hardy, although prone to some dieback and even death if not winter protected. We suggest tipping some of them because there is extreme variation in their parentage. Some of these hybrids will die out when temperatures dip below the mid-20s (−3°C), while others like 'Robusta' are crown hardy to −30°F (−34°C). When crown hardy, they spring back to life and show vigorous new growth.

Meidiland roses These roses are not truly hardy in cold climates. They often suffer severe dieback and even death. Some growers successfully protect them using the Minnesota Tip Method. We strongly advise you to plant the bud union 6 or more inches (15 cm) below the ground so that the buried portion of budded canes form roots. This increases the chances of the plant's surviving. These roses are hardy to 10°F (−12.2°C) without winter protection. With winter protection they often do fine.

Meilland roses Meilland is the breeder of Meidiland® roses. He is also responsible for the Drift® Series. The Drift® Series, seven excellent ground cover roses, are grown on their own roots for added winter hardiness. Cover their crowns with soil and a thick layer of whole leaves in late fall for optimal winter protection. Included in the table is one of the finest Apricot Drift®.

Miscellaneous: Some roses don't fall into any formal classification because no one really can be certain of their parentage. In fact, the Shrub rose group began more or less as a catchall category. While somewhat complex, it has become better defined. The Varieties table makes navigating this maze much easier. A number of fine breeders have created some superb plants in this "miscellaneous" category. Among these breeders are Bedard, Carruth, and Moore.

Morden (see **Parkland series**)

Moyesii **hybrids:** Plants in this group form lovely shrubs up to 6 feet (180 cm) tall. They grow vigorously, form beautiful hips, and are fairly hardy with severe dieback possible, although plants generally spring back to life. Some bloom only once in spring; others such as 'Marguerite Hilling' and 'Nevada,' are marginal repeat bloomers. Most shrubs in this group are hardy to −10°F (−23°C) without winter protection. 'Marguerite Hilling' and 'Nevada' are hardier.

Nitida **hybrids:** Although we have included 'Corylus' and 'Metis' from this group, they are rarely grown in home gardens. They have nice form with rich green foliage and fit beautifully into perennial borders. They bloom once in June but form lovely hips for late-season interest. With adequate snowfall these roses are hardy to −20°F (−29°C) without winter protection. Covering the crowns with potting soil and a thick layer of whole leaves in late fall is highly recommended.

Oso Easy™ Rose series: This small group of roses make good ground cover or container plants. They grow up to 24 inches (60 cm), have varied foliage, flower form (petal count), color, and fragrance. These plants were developed with the organic gardener in mind and are noted for their resistance to black spot and powdery mildew. They are said to be hardy to −20°F (−29°C). However, reports indicate that they are surviving in colder regions, especially 'Fragrant Spreader,' 'Paprika,' and 'Peachy Cream.' For optimal winter protection cover the crowns with potting soil and a thick layer of leaves in late fall. We do not have enough personal experience to rate these yet. In the **Oso Happy™ Rose series** we have included one in the Polyantha section p. 86 ('Oso Happy™ Candy Oh!').

Parkland or **Morden series:** These roses are grown on their own roots. If plants die back during the winter, they normally send up new shoots in the spring. They are especially popular, since they make good bedding plants. Most of them are low growers. Parentage varies greatly so that the plants may be hardy from 10° to −30°F (−12.2° to −34°C) without winter protection. Predicting hardiness in this group is obviously difficult, but they are worth the gamble. However, they are susceptible to black spot.

Pavement series: Technically, these are *rugosa* hybrids. They tend to be small shrubs with dense foliage. They stand out for fragrance, good repeat bloom, and large red hips. These Shrub roses generally survive to −30° F (−34°C) without winter protection. However, in recent years they have shown susceptibility to black spot. Still, they are worth growing for their many good qualities.

Poulsen roses

Poulsen has introduced a number of plants in the Renaissance® and Towne and Country® Series. These are not yet widely distributed. We have included one ('Sophia Renaissance') without a star rating because it appears to be extremely promising.

Radler Roses

Bill Radler is the creator of the Knock Out® series of roses, which offers a nice variety of sizes and colors. They have proven to be quite disease resistant. They are said to be hardy to −20°F (−29°C), but, in our opinion, are marginally hardy and need good winter protection to survive. Cover the crown with potted soil and a thick layer of whole leaves. *Also, these are best purchased as potted roses rather than bare root.* The latter are often waxed and don't grow as well. These roses are well worth growing and would be rated much higher if they were hardier.

Rugosa **hybrids:** These tough, vigorous plants have wrinkled (rugose) leaves. The foliage is somewhat coarse but nicely colored, usually dark green. Plants are salt tolerant, leading to the common names "Beach Rose" or "Shore Pears." Rugosas tend to be large plants with good repeat bloom. Flowers are often large as well and sometimes fragrant. Most Rugosas are hardy in cold climates. They have become increasingly susceptible to disease. *Even if infected, they should not be sprayed with a fungicide. Rugosas are allergic to these. Spraying results in foliage burn.* Many (not all) produce abundant hips and lovely fall foliage. The more wrinkled (rugose) the leaf, the hardier the plant tends to be. Many of the shrubs in this category are hardy to −30°F (−34°C) without winter protection. A few are less hardy and may die back to the ground, but come back to life in spring with vigorous new growth. They need neutral to acidic soil to thrive.

Spinosissima **hybrids:** These plants are quite hardy. Flowers are lovely and, sometimes, fragrant. Foliage is usually deep green and clean, but stems do tend to be quite thorny. The plants are hardy to −20°F (−29°C) without winter protection. Some of these roses are getting hard to find ('Haidee,' for example). You may have to order them from Europe (see Knud Pedersens Planteskole in the source list). Although one-time bloomers, they have a faithful following among ardent rose growers.

Suffulta **hybrids:** These are parents to the Parkland series. They rarely grow very tall and are noted for their resistance to drought and heat. Most of these hybrids are extremely hardy surviving to −30°F (−34°C) and even to −40°F (−40°C) on occasion without winter protection. At the very least, they are crown hardy with cane dying back to the ground after a severe winter, but soon replaced by abundant new growth.

VARIETIES	COLOR	PETAL COUNT	HEIGHT	BLOOM
'Adelaide Hoodless'**	Deep pink to medium red	12–16	40″	Repeat

A Parkland rose. Spreading, arching. Vigorous growth but lax stems. Clusters of lovely 2½-inch (6.25-cm) flowers with slight fragrance and varying coloration. Fairly good repeat bloom. Dark glossy foliage. Very susceptible to black spot. Expect some dieback (zone 3).

VARIETIES	COLOR	PETAL COUNT	HEIGHT	BLOOM
'Agnes'**	Light yellow	30–40	60″	May to June

A *rugosa* hybrid. Dense, upright, vase. Often bare at bottom. Large 3-inch (7.5-cm) flowers with an amber glow. Quite fragrant and fruity. Petals tend to fall easily, especially after rain. Long and early bloom period. Green, crinkly foliage. Thorny cane. Potential for some orange red hips. Susceptible to black spot. Occasionally, dies back in severe weather, but generally regarded as fairly hardy. Cover the crown with potting soil and a thick layer of leaves in late fall during the first few years (zone 4 once established).

VARIETIES	COLOR	PETAL COUNT	HEIGHT	BLOOM
'Aïcha'***	Lemon yellow to cream	5—30+	72″	June

A *spinosissima* hybrid. Tall, arching. Will climb if given support. Suckers freely. Deep yellow buds open into flowers just under 4 inches (10 cm) across with prominent yellow stamens. Varying fragrance. Cane very thorny and covered with light green foliage. Quite hardy (zone 4 to 5).

VARIETIES	COLOR	PETAL COUNT	HEIGHT	BLOOM
'Alexander MacKenzie'***	Medium red	40—50	48″	Intermittent

An Explorer rose. A *kordesii* hybrid. Extremely elegant, upright growth that arches over. Clusters of raspberry-scented, cup-shaped flowers just under 3 inches (7.5 cm) across with a yellowish tinge at their base. Shiny, almost waxy, foliage matures from red tint to glossy light to mid green. Fairly disease resistant. Very good cane hardiness (zone 3).

VARIETIES	COLOR	PETAL COUNT	HEIGHT	BLOOM
'All the Rage'***	Apricot pink	10—12	36″	Repeat

A Bailey rose in the Easy Elegance® series. Upright, rounded. Coral buds open into large 3-inch (7.5-cm) flowers with bright yellow centers. Flowers freely but has no fragrance. Deadhead to encourage more bloom. Foliage is a glossy dark green and quite disease resistant. Thorny cane. Crown hardy, but benefits by having the crown covered with potting soil and a thick layer of leaves for optimal winter protection in the far north (zone 4 to 5).

VARIETIES	COLOR	PETAL COUNT	HEIGHT	BLOOM
'Amélie Gravereaux'**	Medium red	5—30	48″	Repeat

A *rugosa* hybrid. Upright, open. Vigorous grower. Large 3-inch (7.5-cm) flowers with a lavender tinge. Listed as deep pink to mauve crimson in catalogs. Golden stamens. Quite fragrant. Good repeat bloom. Dark green foliage on very prickly cane. Disease resistant. Occasionally, dies back in severe weather, but basically hardy (zone 4).

VARIETIES	COLOR	PETAL COUNT	HEIGHT	BLOOM
'Amiga Mia'***	Light to medium pink	25—35	48″	Repeat

A Buck rose. May be listed as a Grandiflora by some. Upright dense. Nice landscape plant. Clusters of almost red flowers just under 5 inches (12.5 cm) across and quite fragrant, and barely double. Good repeat bloom. Dark, leathery foliage. Somewhat susceptible to black spot. Consider covering the crown with potting soil and a thick layer of leaves in late fall (zone 4 to 5).

VARIETIES	COLOR	PETAL COUNT	HEIGHT	BLOOM
'Applejack'***	Medium pink/gold eye	5—30	48″	Repeat

A Buck rose. Vigorous, upright, spreading growth. Moderately fragrant blossoms just under 4 inches (10 cm) wide. Fairly good repeat bloom. Dark, leathery apple-scented grayish green foliage. Thorny cane. Nice scarlet hips. One of the tougher Bucks. Still wise to cover crown with potting soil and a thick layer of leaves for optimal winter protection (zone 4).

VARIETIES	COLOR	PETAL COUNT	HEIGHT	BLOOM
'Assiniboine'***	Deep pink	10—13	48″	Intermittent

A Parkland rose. A *suffulta* hybrid. Dense, upright. Lovely, 3-inch (7.5-cm) flowers with slight fragrance. Flowers often red to purplish. Attractive, glossy foliage. Hips in fall. Somewhat susceptible to disease, but very hardy and a good choice for the far north (zone 3).

VARIETIES	COLOR	PETAL COUNT	HEIGHT	BLOOM
Astrid Lindgren®***	Light pink	17—25	60″	Intermittent

A Danish rose bred by Poulson. Vigorous, bushy. Medium pink buds open into an amazing first bloom with clusters of nodding fragrant (lightly fruity) flowers about 4 inches (10 cm) wide. Some repeat bloom possible. Foliage is dark green, glossy, and resistant to black spot. May produce hips (not seen by us). Deserves to be more widely grown. Cover the crown with potting soil and and whole leaves to provide optimal winter protection (zone 5).

VARIETIES	COLOR	PETAL COUNT	HEIGHT	BLOOM
'Aunt Honey'***	Medium pink	35—40	48″	Repeat

A Buck rose. Carmine buds open into clusters of 4 to 5-inch (10 to 12.5 cm) flowers with a wonderful Old Garden rose fragrance. Foliage is medium to dark olive green and generally disease resistant. Somewhat thorny cane. Cover the crown with potting soil and a thick layer of whole leaves for optimal winter protection (zone 5).

VARIETIES	COLOR	PETAL COUNT	HEIGHT	BLOOM
'Belle Poitevine'***	Medium pink	12—16	40″	Repeat

A *rugosa* hybrid. Upright, dense, twiggy. Will sucker if grown on its own roots. Fairly vigorous and can form nice hedge in warmer areas. Large 3-inch (7.5 cm) flowers with purplish cast and flat form with cream-colored stamens. Noticeable fragrance. Marginal repeat bloom. Medium green foliage on thorny cane. Limited, but large orange to scarlet colored fruits. Disease resistant. Some dieback in severe weather but not much (zone 3 to 4).

'Black Boy'***	Dark red	13—17	48″	June

A *kordesii* hybrid with an offensive name, once meant as a tribute to Australian bushmen. Upright, arching. Will form a thicket in time. Wonderfully rich coloration and abundant bloom. Clusters of flowers up to 5 inches (12.5 cm) wide with apple fragrance. Stunning foliage and vigorous growth. Acts as a climber in more southerly areas, but remains a bush in cold climates. Susceptible to black spot. Cover the crown with potting soil and a thick layer of whole leaves in late fall for optimal winter protection. Expect severe dieback (zone 5 to 6).

'Blanc Double de Coubert'****	Snow white	30—40	48″	Repeat

A *rugosa* hybrid. Upright, mounded growth. Vigorous growth. Good as hedge or mass planting (suckers profusely). Lovely 3-inch (7.5-cm) flowers, large but loose in form with greenish eye. Extraordinary fragrance. Good repeat bloom. Thorny canes with wrinkly, dark green foliage. Nice yellow fall foliage. Few, but lovely scarlet to red-orange hips. Disease resistant. Some dieback in severe weather, but very hardy overall (zone 3 to 4).

Blushing Knock Out®***	Pink to pale pink	5	36″	Repeat

A Radler rose. Bushy, rounded. Slightly fragrant flowers are about 3 inches (7.5 cm) wide in shades of pink with a gold center. Green foliage with a bluish tint. Disease resistant. Cover the crown with potting soil and a thick layer of whole leaves for good winter protection (zone 5).

Bonica®***	Light pink	17—35+	60″	Repeat

A Meidiland rose. Pronounced "bow-KNEE-kuh." A somewhat gangly plant that can be used as a Climber. Nice light pink 3-inch (7.5-cm) slightly fragrant flowers. Color fades quickly to white. Good repeat bloom. Delicate, deep green, glossy foliage. Bright orange hips. Occasionally bothered by black spot. If used as a Climber all cane should be bent over and covered with potting soil and a thick layer of leaves in late fall (crown hardy to zone 5).

Brother Cadfael®****	Deep pink	25—45	60″	Repeat

An Austin rose. Upright, vigorous growth. Produces two good flushes of bloom only. Crimson buds open into cupped flowers up to 5 inches (12.5 cm) wide with a strong Old Garden Rose fragrance. Foliage is dark green and disease resistant. Cane is not thorny. Will perform best if protected with the Minnesota Tip or Hass Teepee method. Or cut cane back to 12 inches (30 cm) and cover with potting soil (zone 5 to 6).

Cape Diamond™***	Deep pink	38	48″	Repeat

A Canadian introduction. Upright, spreading. Clusters of medium to deep pink flowers roughly 4 inches (10 cm) wide have a lovely scent and bright gold center. Marginal repeat bloom. Foliage is glossy, light green, and disease resistant. Cover the crown with potting soil and a thick layer of whole leaves for optimal winter protection (zone 4).

'Captain Samuel Holland'***	Medium red	15—25	60″	Repeat

An Explorer rose. Lax, rambling. Pointed reddish buds open into clusters of loose, slightly fragrant pink red flowers up to 2 to 3 inches (5 to 7.5 cm) wide. Good repeat bloom. Glossy, dark green foliage. Susceptible to black spot. Crown hardy, but protect if using it as a Climber by bending cane over and covering with cardboard and soil (zone 3 to 4).

VARIETIES	COLOR	PETAL COUNT	HEIGHT	BLOOM
Carefree Beauty™*****	Medium pink	15–24	36″	Repeat

A Buck rose. Upright, spreading. Long buds burst into lightly fragrant, 2 to 4-inch (5 to 10-cm) coral pink blooms. Excellent landscape plant with superb repeat bloom. Delicious, orange hips. Occasionally bothered by black spot, but one of the more disease resistant shrub roses overall. Fairly hardy, but best if given good winter protection (zone 4 to 5).

Carefree Celebration™***	Orange red pink	15–20	48″	Repeat

A Radler rose. Loose, rounded form. Cupped mildly fragrant flowers are just under 3 inches (7.5 cm) wide and fade to pinkish coral. Lovely dark green foliage that is slightly susceptible to black spot. New cane has reddish tones. The plant will die back to the ground but is crown hardy. Cover the crown with potting soil and a thick layer of whole leaves for optimal protection, especially in the far north (zone 4 to 5).

Carefree Spirit™***	Red blend	5	48″	Repeat

A Meidiland rose. Mounded bush. Produces large clusters of slightly fragrant bright red 3-inch (7.5-cm) flowers with white eyes and yellow centers. Glossy, dark green foliage. Very disease resistant. Cut canes back to 12 inches (30 cm) and cover with potting soil as winter protection. Worth the extra effort (zone 5).

Carefree Wonder™**	Bright pink/white/yellow	20–30	30″	Repeat

A Meidiland rose. Neat, bushy. Will form thicket if grown on its own roots. Lovely, 4-inch (10-cm) flowers with slight fragrance. Good repeat bloom if deadheaded regularly. Small, medium green, semi-glossy leaves. May produce orange reddish hips. Very insect and disease resistant. Requires excellent protection to survive (cover the crown with potting soil and a thick layer of whole leaves in late fall) despite being listed as zone 4 hardy.

'Carmenetta'**	Light pink	5	72″	June

Rosa glauca × *rugosa* hybrid. Spreading, arching, vigorous. Can be used as Climber. Clusters of slightly fragrant, star-like, pinkish purple blossoms tinged green just under 2 inches (5 cm) wide with yellow center. Richer color in slight shade. Reddish or coppery gray-green foliage on red canes. Reddish-purple hips. Disease resistant. Very hardy (zone 2 to 3).

'Champlain'***	Dark red	30	36″	Repeat

An Explorer rose. A *kordesii* hybrid. Spreading. Clusters of slightly fragrant, velvety 3-inch (7.5-cm) blooms with yellow stamens. Good repeat bloom. Immature bronze reddish foliage turns pale green on prickly cane. Prone to black spot and powdery mildew, although it fights off mildew well. Orange hips in fall. Basically hardy, but susceptible to dieback. Winter protection does help (zone 4, but often listed zone 3 to 5).

'Charles Albanel'****	Medium red	15–20	24″	Repeat

An Explorer rose. A *rugosa* hybrid. Spreading, good ground cover. Clusters of nearly mauve flowers up to 3 inches (7.5 cm) across with slight to moderate fragrance and yellow stamens. Good repeat bloom. Light yellowish, green foliage. Thorny. Nice orange-red hips. Very disease resistant. Very hardy (zone 3 to 4).

'Chuckles'****	Rich deep pink/white	5	24″	Repeat

A Floribunda rose often listed as a Shrub because it is quite hardy. Compact, spreading. Produces fragrant 3-inch (7.5 cm) flowers with lots of rebloom throughout the season. Dark, leathery foliage. Benefits from spraying. Severe dieback common. Cover the crown with potting soil and a thick layer of whole leaves for optimal winter protection (crown hardy to zone 4).

VARIETIES	COLOR	PETAL COUNT	HEIGHT	BLOOM
Coral Drift™***	Orange salmon/yellow	15—35	18″	Repeat

A Meilland rose. Full, spreading and ideal as a ground cover plant. Showered with 1-inch (2.5-cm) blossoms from spring to fall. Glossy, dark green foliages is disease resistant. No hips. Cover the crown with potting soil and a layer of whole leaves for optimal winter protection (zone 5).

'Corylus'***	Medium pink	5	24″	June

A *nitida* hybrid. A tidy plant that hugs the ground. Suckers and will form a thicket. Loose, but lovely 3-inch (7.5-cm) pink flowers with bright, yellow stamens and strong fragrance. Medium to deep green, shiny foliage susceptible to powdery mildew. May produce scarlet hips in fall. A one-time bloomer. Mound potting soil over the crown and cover with a thick layer of leaves for winter protection (zone 4 to 5).

'Country Dancer'***	Deep pink	15—30	36″	Repeat

A Buck rose. Dense, compact. Vigorous grower. Fragrant, rose-red flowers up to 5 inches (12.5 cm) across. Papery petals. Profuse repeat bloom. Glossy, dark green foliage. Crown hardy (dies back to ground level each year). Will form beautiful plant even after severe dieback (zone 4 to 5).

Crown Princess Margareta®***	Apricot orange	40—100+	60″	Repeat

An Austin rose. Dense, arching. Produces large clusters of many-petalled flowers (up to 100 or more) with fruity fragrance. Cupped blossoms roughly 4 inches (10 cm) wide with gold stamens. The first bloom is very good, sporadic from then on. Semi-glossy, dark green foliage on nearly thornless cane. Will perform best if protected with the Minnesota Tip or Hass Teepee method. Or cut cane back to 12 inches (30 cm) and cover with potting soil and a thick layer of whole leaves (zone 5 to 6).

'Cuthbert Grant'***	Dark red	15—20	36″	Repeat

A Parkland rose. A *suffulta* hybrid. Spreading. A nice rose, but very slow growing. Lovely, 4-inch (10-cm) flowers with varying fragrance and purplish red coloration. Expect no more than two good flushes of bloom. Lush glossy, light green foliage, susceptible to black spot. Not as hardy as often claimed, but generally crown hardy. Cover the crown with potting soil and a thick layer of whole leaves in the far north (listed as zone 3 hardy, more likely zone 4).

Dagmar Hastrup® (see Frau Dagmar Hastrup®)

'Dart's Dash'***	Crimson pink to dark red	17—20	36″	Repeat

A *rugosa* hybrid. Dense, compact. Excellent ground cover or hedge. Crimson purple buds open into sweetly fragrant, 3-inch (7.5 cm) flowers. Ideal substitute for its larger version 'Hansa.' Fairly vigorous thorny canes with dark green foliage. Lovely, large orange-red hips contrast to nice fall foliage color (typically orange). Resistant to black spot, but prone to mildew (especially late in the season). Very hardy (zone 3).

'David Thompson'***	Medium red	25	48″	Repeat

An Explorer rose. A *rugosa* hybrid. Upright, rounded, dense. Moderately vigorous. Nicely scented flowers (deep pink with crimson) just under 3 inches (7.5 cm) across with yellow stamens. Excellent repeat bloom. Purplish pink prickly cane with small medium green leaves. Very disease resistant. No hips. Needs acidic soil. Very hardy (zone 3 to 4).

'De Montarville'***	Medium to dark pink	25	36″	Repeat

An Explorer rose. A *kordesii* hybrid. Compact, bushy. Makes a nice hedge. Crimson buds open into bright pink flowers with lighter undersides and deep golden centers. Lightly scented (fruity) blossoms about 3 inches (7.5 cm) wide. Good repeat bloom. Dark foliage tinged blue is quite disease resistant. Few thorns. Expect dieback, but crown hardy (zone 3).

VARIETIES	COLOR	PETAL COUNT	HEIGHT	BLOOM
'Delicata'****	Light pink mauve	18–24	48″	Repeat

A *rugosa* hybrid. Upright, dense. Blooms early in the season with papery, ruffled flowers slightly less than 3 inches (7.5 cm) across with a lilac hue and creamy yellow stamens. Good repeat bloom. Variable fragrance. Light green wrinkled leaves on prickly cane. Scarlet, orange hips. Lovely fall foliage. Disease resistant. Very hardy (zone 3 to 4).

'Diantheflora' or 'Dianthiflora' (see 'Fimbriata')				

'Distant Drums'****	Tannish mauve	40	48″	Repeat

A Buck rose. Upright, mildly spreading. Vigorous. Pale purple buds open into large clusters of cupped, reddish purple flowers up to 5 inches (12.5 cm) with myrrh-like scent. Often ruffled with bronze center. Dark reddish tinged, leathery foliage. Thorny. Disease resistant. Cover the crown with potting soil and a thick layer of whole leaves in late fall for optimal winter protection (zone 5).

Dortmund®***	Medium red/white	5	60″	Repeat

A *kordesii* hybrid. Short climber. Vigorous. Clusters of white-eyed flowers up to 5 inches (12.5 cm) wide with apple-like fragrance and yellow stamens. Deadhead regularly. Dark, glossy foliage. Lots of orange hips. Susceptible to black spot on occasion, but quite disease resistant overall. Any cane you want to save should be bent over, covered with soil, and a thick layer of whole leaves. This is particularly important when using this shrub as a replacement for a true Climber (zone 5).

Double Knock Out®***	Dark pink red	18–25	36″	Repeat

A Radler rose. Compact, mounded. Lovely cherry red buds open into dark pinkish 3-inch (7.5-cm) blooms with light spicy fragrance. Semi-glossy, medium green foliage with a purplish tint. Disease resistant. Cover the crown with potting soil and a thick layer of whole leaves for good winter protection (zone 5).

'Dwarf Pavement' ('Rosa Zwerg')***	Medium pink	12–16	30″	Repeat

A Pavement series *rugosa* hybrid. Compact, spreading. Clusters of fragrant light to medium pink flowers up to 3 inches (7.5 cm) wide. Foliage is glossy green. Many large orange red hips. Increasingly prone to black spot, but very hardy (zone 3).

'Earth Song'***	Ruby pink	25–30	60″	Repeat

A Buck rose. Technically, a Grandiflora. Nice landscape plant with upright, open growth. Small, narrow buds open into upped, fragrant flowers up to 5 inches (12.5 cm) wide. Good repeat bloom. Dark, leathery foliage tinged red when immature. Will die back to the ground each year. Cover the crown with potting soil and a thick layer of whole leaves for optimal winter protection (zone 5).

'El Catalá'***	Red silvery white	24–40	30″	Repeat

A Buck rose. Technically, a Grandiflora. Upright, bushy. Red to silvery pink buds form clusters of cupped flowers 4 inches (10 cm) or more wide. Blossoms are a red and white blend with slight fragrance. Nice, glossy dark green foliage. Thorny. Will die back to ground each year but crown hardy. Cover the crown with potting soil and a thick layer of whole leaves in late fall for optimal winter protection (zone 5).

Emily Carr™	Dark red	13–17	40″	Repeat

A Canadian Artist rose. Upright. Deep red buds. Clusters of mildly fragrant dark red blooms roughly 3 inches (7.5 cm) wide form at the ends of canes. Foliage is deep green tinged red and disease resistant. Consider covering the crown with soil and a thick layer of whole leaves in late fall for optimal winter protection. Not rated since we have not yet field-tested it. Reported to be crown hardy to zone 2.

VARIETIES	COLOR	PETAL COUNT	HEIGHT	BLOOM
Felix Leclerc™	Deep pink	15–25	Variable	Repeat

A Canadian Artist rose. Arching. Clusters of mildly fragrant flowers up to 4 inches (10 cm) across. The plant tends to spread out with canes that can be trained to a trellis for maximum effect. Foliage is dark green and disease resistant. Drought tolerant once mature. Consider covering the crown with soil and a thick layer of whole leaves in late fall for optimal winter protection. Not rated since we have not field-tested it yet. Reported to be crown hardy to zone 2.

'Fimbriata'***	Light pink to white	17–25	40″	Repeat

A *rugosa* hybrid. Upright, dense. A good hedge plant. Clusters of small, frilly carnation-like flowers just under 3 inches (7.5 cm) wide with a sweet fragrance and golden eye. Light to medium green foliage sometimes turning orange in fall. Very few hips. Disease resistant. Crown hardy (zone 4 to 5).

'F.J. Grootendorst' ('Grootendorst Red')***	Medium red	25–35	48″	Repeat

A *rugosa* hybrid. Upright vase. Good landscape or hedge plant. Will tolerate partial shade. Small 1-inch (2.5-cm), carnation-like flowers in large clusters. Prolific bloom varying from deep pink to light red depending upon soil. Slightly spicy to no fragrance. Very vigorous growth with deep green foliage. Thorny stems. Not grown for hips. Prone to mite and aphid attack. Susceptible to black spot. Cover crown with potting soil and a thick layer of whole leaves in late fall (zone 4).

'Folksinger'***	Yellow apricot	25–30	36″	Repeat

A Buck rose and one of the best. Upright, bushy. Lovely buds open into large clusters of cupped, sweetly fragrant flowers up to 5 inches (12.5 cm) wide. Glossy foliage tinged bronze. Susceptible to powdery mildew. Thorny. Often dies back to the ground. Cover the crown with potting soil and a thick layer of whole leaves in late fall for good winter protection (zone 4 to 5).

Foxi®***	Lavender pink	15–25	36″	Repeat

A Pavement series *rugosa* hybrid. Mounded, dense. Clusters of fragrant (spicy) flowers up to 3 inches (7.5 cm) wide with golden center. Excellent repeat bloom. Foliage is glossy light green and quite disease resistant. Abundant bright red hips against yellowish fall foliage. Salt tolerant. Very hardy (zone 3).

Frau Dagmar Hastrup® ('Frau Dagmar Hartopp')***	Light to medium pink	5	36″	Repeat

A *rugosa* hybrid. Dense, compact, mound to slightly spreading. Will form thicket if grown on its own roots. Clusters of large 3½-inch (9 cm) cupped flowers (somewhat like poppies) with slight to extreme fragrance depending upon growing conditions. Blossoms often have silvery tinge and delicate creamy yellow to light pink stamens. Wrinkled dark green foliage and dense growth. Good ground cover or hedge, but very thorny. Foliage turns from maroon to yellow in fall. Abundant large, deep red fruits (like cherry tomatoes). Disease resistant. Basically very hardy with occasional dieback in severe weather (zone 3 to 4).

'Frontenac'***	Deep pink	20–25	48″	Repeat

An Explorer rose. Upright, dense. Very free-flowering with clusters of bright pink to crimson 3-inch (7.5-cm) mildly fragrant flowers with rich golden centers and whitish base. Dark, glossy green foliage is quite disease resistant. Very hardy (zone 3).

'Fru Dagmar Hartopp' (see Frau Dagmar Hastrup®)

VARIETIES	COLOR	PETAL COUNT	HEIGHT	BLOOM
'Frülingsanfang'**	Ivory white	5	72″	Intermittent

A *spinosissima* hybrid. Upright, arching. Clusters of slightly fragrant 4-inch (10-cm) flowers shower the plant in one main flush in spring. Color sometimes light yellow. Leathery foliage on prickly cane. Big maroon hips possible in fall. This plant sets buds in fall—for good bloom it needs gradual cooling in the fall and gradual warming in spring. Provide it with good winter protection. Bend arching cane over and cover with potting soil and a thick layer of whole leaves (zone 5).

VARIETIES	COLOR	PETAL COUNT	HEIGHT	BLOOM
Frühlingsgold®**	Medium yellow	9–16	72″	June

A *spinosissima* hybrid. Upright, arching. Pointed reddish buds streaked orange burst into clusters of fragrant, golden yellow 3-inch (7.5 cm) flowers with dark gold stamens. Flowers vary from yellow to cream. Leaves are grayish green on yellowish canes with numerous thorns. Round black hips possible in fall. Susceptible to black spot. Order plants grown on their own roots. Budded stock often dies off in the winter (zone 5).

VARIETIES	COLOR	PETAL COUNT	HEIGHT	BLOOM
'Fürstin von Pless'**	White cream/yellow	41+	48″	Repeat

A *rugosa* hybrid. Upright, dense. Clusters of 3 inch (7.5 cm), moderately fragrant cupped blossoms that are lovely when first opening and loose later. Sometimes, tinged light pink. Foliage is deep green and disease resistant. Crown hardy. Winter protect by covering with potting soil and a thick layer of whole leaves (zone 4).

VARIETIES	COLOR	PETAL COUNT	HEIGHT	BLOOM
'George Vancouver'****	Medium red	24	36″	Repeat

An Explorer rose. A *kordesii* hybrid. Upright, open, arching. Deep crimson red buds burst into clusters of roughly 2.5-inch (7-cm) mildly fragrant medium red to dark pink cupped flowers with creamy pink yellow stamens. Good first and reliable repeat bloom in most years. Foliage is mid to deep glossy green on thorny cane. Red hips in fall. Susceptible to some types (races) of black spot but very hardy (zone 3).

VARIETIES	COLOR	PETAL COUNT	HEIGHT	BLOOM
'Geranium'****	Medium red	5	72″	June

A *moyesii* hybrid. Nice, upright, compact growth in colder areas, more arching to the south. Forms clusters of 2-inch (5-cm) scarlet mildly sweet fragrant flowers, often waxy and tinged orange. Deep red, flagon-like fruit. Protect as much cane as possible since the plant flowers only on old wood. the previous year's growth (zone 4 with protection).

VARIETIES	COLOR	PETAL COUNT	HEIGHT	BLOOM
'Golden Unicorn'***	Yellow/pink orange	20–30	30″	Repeat

A Buck rose. Rounded, spreading shrub. Orange buds tinged red open into fruity to lightly sweet fragrant clusters of cupped yellow to orange-pinkish 4-inch (10-cm) flowers. Dark, olive green leathery foliage on prickly cane. Susceptible to disease, but crown hardy. Cut back, cover with potting soil, and mound with whole leaves for winter protection (zone 4 to 5).

VARIETIES	COLOR	PETAL COUNT	HEIGHT	BLOOM
'Golden Wings'****	Light lemon yellow	5	36″	Repeat

A *spinosissima* hybrid. Bushy, dense, vigorous. Long golden buds burst into sulfur yellow flowers up to 5 inches (12.5) across with dark reddish amber stamens and slight fragrance. Best bloom in spring. Deadhead but prune as little as possible. Light green foliage. Round orange hips. Prone to black spot. Will tolerate light shade. Must have excellent winter protection to survive. Bend cane over and cover with soil and a thick layer of whole leaves. Such a beautiful plant that the extra work is worth it (zone 4 to 5).

VARIETIES	COLOR	PETAL COUNT	HEIGHT	BLOOM
Graham Thomas®***	Deep yellow	35–45+	30″	Repeat

An Austin rose. Bushy. Cupped, very fragrant flowers up to 4 inches (10 cm) across. Dull medium green foliage. One of the most prized and popular roses worldwide. Susceptible to black spot. Will not survive in colder climates without good winter protection (Minnesota Tip or the Hass Teepee methods are best), but a five-star gem in warmer areas (zone 5 to 6).

VARIETIES	COLOR	PETAL COUNT	HEIGHT	BLOOM
'Grootendorst Supreme'***	Crimson red	25—40	48″	Repeat

A *rugosa* hybrid. Upright, bushy, dense. Sometimes, bare at its base. Noted for almost constant bloom of small frilly, 1-inch (2.5-cm) carnation-like flowers in large clusters. Color varies from red to deep pink. Slight to strong fragrance depending upon location and growing conditions. Deadhead to increase repeat bloom. Small light green leaves. Prone to disease and insect infestations. Some dieback in severe weather, but basically very hardy (zone 3 to 4).

'Haidee'***	Light pink	25—35	96″	June

A cross between **Rosa laxa** and **Rosa spinosissima**. Arching, spreading. Suckers freely and may be wider than it is tall. Blooms early in the season. Cupped flowers with creamy centers up to 3 inches (7.5 cm) wide with strong fragrance. Dark green ferny foliage on reddish, thorny cane. Large red fruits. Hardy (zone 4).

'Hansa'**	Medium red	25—30+	72″	Repeat

A *rugosa* hybrid. Upright, vase shape. Will sucker. Excellent specimen plant, since it often grows as wide as it is tall. Lovely 4-inch (10-cm) or larger flowers with a violet tinge tending toward purple. Sweetly fragrant with a clove-like scent. Medium to dark green foliage on thorny cane. Nice orange-red hips like cherry tomatoes for tea or wildlife. Once disease resistant, now very susceptible to black spot. Very hardy (zone 3 to 4).

Hansaland®***	Dark red	9—16	40″	Repeat

A *rugosa* hybrid. Upright, bushy. Deep red buds open into clusters of mildly fragrant 3-inch (7.5 cm) flowers that are a lovely scarlet red with yellow stamens. Foliage is semi-glossy, medium green, and generally disease resistant. Hardy (zone 4).

'Harison's Yellow'****	Deep clear yellow	15—25	72″	June

A *foetida* hybrid. Known as 'The Yellow Rose of Texas.' Upright, dense, arching. Suckers freely. Magnificent display of up to 3-inch (7.5-cm) yellow blossoms with golden stamens. Very slight fruity fragrance. Luxuriant, dainty gray-green foliage and prickly cane. May form blackish red hips in fall. Responds beautifully to staking to increase canes (arch canes over and peg to ground). Susceptible to black spot. Try to buy plants on their own roots. If budded, plant the bud union well below the soil line. Very hardy (zone 3 to 4).

'Hawkeye Belle'****	White pink	35—40	36″	Repeat

A Buck rose. Erect, bushy. Very fragrant (sweet) flowers up to 4 inches (10 cm) across or larger. Turns from white to light pink to darker pink. Mild repeat bloom. Dark, leathery foliage with bronze tones when young. Often dies back to ground each year. Cover the crown with potting soil and a thick layer of whole leaves for winter protection (zone 4 to 5).

'Henry Hudson'****	White tinged pink	20—25	36″	Repeat

An Explorer rose. Seedling from **Rosa rugosa** 'Schneezwerg.' Bushy, dense, spreading. Suckers freely to form thicket. Round, light pink to crimson buds burst into white, almost flat, 3-inch (7.5 cm) papery flowers with yellow stamens. Abundant bloom, but individual flowers do not last long. Nice clove-like fragrance. Deadhead religiously. Deep green, crinkly foliage on thorny cane. Reddish orange hips in fall. Disease resistant. Very hardy (zone 4).

'Henry Kelsey'**	Medium red	25—28	Variable	Repeat

An Explorer rose. A **kordesii** hybrid. Very vigorous, trailing. Large clusters of 3-inch (7.5-cm) bright crimson flowers with golden stamens. Fairly good repeat bloom. Nice, spicy to fruity fragrance. Glossy, deep green foliage on thorny stems. Not grown for hips. Quite resistant to powdery mildew, although susceptible to black spot. Very hardy, but demands good winter protection if used to replace a true Climber. Lay cane down on ground, cover with soil, and then a thick layer of whole leaves (zone 4).

VARIETIES	COLOR	PETAL COUNT	HEIGHT	BLOOM
Heritage®****	Soft pink to cream	70+	48″	Repeat

An Austin rose. Upright, vigorous. An excellent first bloom. Clusters of cupped flowers are very fragrant (citrus), just over 3 inches (7.5 cm) wide, but do tend to shatter easily. Still, a winner. Dark, semi-glossy foliage on almost thornless cane. Susceptible to black spot. Will perform best if winter protected with the Minnesota Tip or Hass Teepee method. Or cut cane back to 12 inches (30 cm) and cover with potting soil and a thick layer of whole leaves in late fall (zone 5).

VARIETIES	COLOR	PETAL COUNT	HEIGHT	BLOOM
'Highdownensis'***	Cherry crimson	5	72″	May to June

A *moyesii* seedling. Upright, arching. Ideal for woodland plantings (a bird magnet). Clusters of bright flowers up to 3 inches (7.5 cm) across with a white center and slight sweet fragrance. Dark foliage tinged copper on thorny cane with crimson thorns. Profuse, large orange-red hips. Hardy (zone 4).

VARIETIES	COLOR	PETAL COUNT	HEIGHT	BLOOM
'High Voltage'****	Medium yellow	22	48″	Repeat

A Bailey rose in the Easy Elegance® series and considered by many as the best yellow in this group. Upright vase. Lovely clusters of 2 to 3-inch (5 to 7.5 cm) fragrant flowers. Good disease resistance. Crown hardy, but winter protect with potting soil and a thick mulch of whole leaves in the far north (zone 4).

VARIETIES	COLOR	PETAL COUNT	HEIGHT	BLOOM
Home Run®**	Dark red/yellow center	5	36″	Repeat

A Carruth rose. A relative of Knock Out® rose. Compact, mounded. Grown on its own roots. Clusters of bright red flowers (about 3 inches/7.5 cm across) with golden centers and little fragrance (some people detect a tea scent). Dark green foliage resistant to black spot but prone to other diseases. Demands lots of heat to grow well. Limited repeat bloom. Marginally hardy. Winter protect using the Minnesota Tip or Hass Teepee method or cut canes back to 12 inches (30 cm) and cover with potting soil and a thick layer of whole leaves (zone 5)

VARIETIES	COLOR	PETAL COUNT	HEIGHT	BLOOM
'Honeysweet'**	Orange pink	25—30	36″	Repeat

A Buck rose. Erect, bushy. Large clusters of very fragrant (honey), cupped flowers up to 5 inches (12.5 cm) wide. Flowers a mix of orange, red, and yellow. Dark, leathery foliage with hints of copper on thorny stems. Will often die back to ground. Cover the crown with potting soil and a thick layer of whole leaves for winter protection (zone 4 to 5).

VARIETIES	COLOR	PETAL COUNT	HEIGHT	BLOOM
'Hope for Humanity'****	Dark red	15—25	24″	Repeat

A Parkland rose. Upright, compact. Burgundy wine-colored buds open to clusters of lightly fragrant 2-inch (5-cm) blossoms with purplish tint. Light green, semi-glossy foliage is fairly disease resistant but susceptible to some black spot. Very hardy (zone 3).

VARIETIES	COLOR	PETAL COUNT	HEIGHT	BLOOM
'Jens Munk'***	Rich medium pink	25	60″	Repeat

An Explorer rose. A *rugosa* hybrid. Upright, dense. Clusters of spicy flowers up to 3 inches (7.5 cm) across with lovely, yellow stamens. Excellent repeat bloomer—almost continuous bloom under ideal conditions if deadheaded regularly. Small, medium green leaves on thorny cane. Limited, but lovely red hips. Disease resistant. Very hardy (zone 3).

VARIETIES	COLOR	PETAL COUNT	HEIGHT	BLOOM
'John Cabot'***	Deep pink to medium red	40	Variable	Repeat

An Explorer rose. *Rosa kordesii* 'Wulff' × 'Masquerade' × *Rosa laxa* hybrid. Arching canes make it suitable for a Climber. Clusters of slightly fragrant flowers just under 3 inches (7.5 cm) across develop a lovely orchid pink to purplish hue. Flowers most heavily in early summer but has strong second flush. Yellow green foliage on thorny cane. Disease resistant. Hardy. However, if used to replace a true Climber, lay canes on the ground before covering them with soil and a thick layer of whole leaves (zone 2 to 3).

VARIETIES	COLOR	PETAL COUNT	HEIGHT	BLOOM
'John Davis'*****	Light to medium pink	40	72"	Repeat

An Explorer rose. A *kordesii* hybrid. Large, arching. May be used as a climbing pillar rose. Clusters of fragrant (spicy), clear pink flowers up to 3 inches (7.5 cm) across with a yellow eye. Almost continuous bloom under ideal conditions. Leaves maturing from reddish to dark green glossy foliage on thorny cane. Somewhat susceptible to disease. Very hardy, but if used as a Climbing rose, lay cane on the ground before covering it with soil and a thick layer of whole leaves (zone 3).

'Kamchatica'***	Bright pink	5	60"	June

A *rugosa* hybrid. A chance seedling discovered in Siberia in 1770. Also known as *Rosa rugosa kamtchatica*. Dense, rounded. A shower of mildly fragrant 3-inch (7.5-cm) single blooms in clusters early in the season. Marginal repeat bloom if spent blossoms are removed religiously. Bright medium green crinkly foliage is quite disease resistant. Small, bright red hips. Hardy (zone 3 to 4).

'Karl Förster'***	White	25–35+	36"	Repeat

A *spinosissima* hybrid. A nice, if somewhat obscure, rose. Spreading, arching. Ivory white fragrant flowers just over 3 inches (7.5 cm) across with bright yellow stamens. Reddish stems with grayish to light green wrinkled foliage. Very vigorous grower, but dieback expected. Best to cover the crown with potting soil and a thick layer of whole leaves for winter protection (zone 5 to 6).

'Kashmir'****	Dark red	50	36"	Repeat

A Bailey rose in the Easy Elegance® series. Rounded. The 3-inch (7.5- cm) flowers have a soft, velvety appearance (cashmere). Foliage is medium green and disease resistant. A very healthy and hardy rose. Mound potting soil over the crown and cover with a thick layer of whole leaves in the far north for added winter protection (zone 4).

Knock Out®***	Cherry red to pink	5	36"	Repeat

A Radler rose. Bushy, rounded. Grown on its own roots. Clusters of flowers just under 3 inches (7.5 cm) wide and mildly fragrant (slightly spicy) with white center. Plants tend to bloom freely, but deadheading recommended. Medium green, semi-glossy foliage on somewhat thorny cane. Disease resistant with occasional black spot. Winter protect by covering the crown with potting soil and whole leaves. A five-star plant in warmer zones (zone 5).

Lady Elsie May™****	Orange pink	8–16	30"	Repeat

A Noack rose. Upright, dense. On own roots. Abundant 4-inch (10-cm) bright and moderately fragrant flowers in clusters. Best bloom early in the season with limited repeat bloom in summer. Foliage is a waxy dark green and offers good disease resistance. Expect die back to the snow line or cover the crown with potting soil and a thick layer of whole leaves for maximum winter protection (zone 4 to 5).

'Lambert Closse'**	Medium pink	50+	36"	Repeat

An Explorer rose. Deep pink buds open into light to medium pink lightly fragrant, ruffled flowers about 3 inches (7.5 cm) wide. Whitish pink on the outside, deeper pink inside. Foliage is glossy, light green and disease resistant. Moderate repeat bloom. Hips in fall. Very hardy (zone 3).

'L. D. Braithwaite'***	Crimson red	45+	48"	Repeat

An Austin rose. Upright, open. Red buds open into a 3-inch (7.5 cm) crimson red rose with a lovely fragrance. Foliage is medium green and susceptible to black spot. The plant is vigorous and bounces back well from any defoliation. It is crown hardy but best if given winter protection. Cut back the stems to 12 inches (30 cm), mound with potting soil, and cover with leaves or use the Minnesota Tip or Hass Teepee methods for even better winter protection (zone 4 to 5).

VARIETIES	COLOR	PETAL COUNT	HEIGHT	BLOOM
'Lillian Gibson'*****	Salmon pink	17–25+	96″	June

A *blanda* hybrid. Wonderful, cascading form. Looks lovely on pole or pillar. Flowers just under 3 inches (7.5 cm) across with strong fragrance. Nearly thornless. Often much hardier than listed, although mild winter protection is advised. Also, protect from rabbits which are particularly drawn to this rose (surround with wire). A shame that this rose is not more readily available. Hardy (zone 3 to 4).

VARIETIES	COLOR	PETAL COUNT	HEIGHT	BLOOM
'Louis Jolliet'**	Medium pink	25–30+	60″	Repeat

An Explorer rose. Use as a climbing pillar or ground cover rose. Clusters of flowers up to 3 inches (7.5 cm) across with slight spicy fragrance. Semi-glossy, medium green foliage. Susceptible to leaf spot. If used as a climbing pillar rose, bend cane over, cover with soil and a thick layer of whole leaves to prevent die back. Hardy when used as a Shrub (zone 3).

VARIETIES	COLOR	PETAL COUNT	HEIGHT	BLOOM
'Mabelle Stearns'***	Medium pink	55–65	24″	Repeat

Known as the 'Door Yard Rose.' Spreads out. Clusters of peachy pink to whitish pink flowers 1 inch (2.5 cm) across with strong fragrance. Small, glossy, dark green leaves. Severe dieback common. Cover the crown with potting soil and a thick layer of whole leaves for winter protection (zone 5 to 6).

VARIETIES	COLOR	PETAL COUNT	HEIGHT	BLOOM
'Magnifica'***	Dark red	20–25	40″	Repeat

There is more than one 'Magnifica.' This one is the *rugosa* hybrid. Upright, dense. Clusters of flowers larger than 3 inches (7.5 cm) across and very fragrant, almost spicy. Flowers open flat and combine purple and crimson hues. Good repeat bloom. Glossy, dark green foliage. Large, orange hips. Colorful fall foliage on occasion. Disease resistant. Very hardy (zone 3).

VARIETIES	COLOR	PETAL COUNT	HEIGHT	BLOOM
'Marguerite Hilling'***	Medium pink	8–10+	72″	Repeat

A *moyesii* hybrid. A sport of 'Nevada.' Arching. Flowers just over 3 inches (7.5 cm) across with noticeable, but light fragrance and bright gold stamens. Excellent first but marginal repeat bloom. Medium green foliage slightly prone to black spot. Will dieback. Cover crown with potting soil and a thick layer of whole leaves in late fall (zone 5).

VARIETIES	COLOR	PETAL COUNT	HEIGHT	BLOOM
'Marie Bugnet'****	Pure white	17–25	36″	Repeat

A *rugosa* hybrid. Compact. Vigorous. Prolific bloom with some blossoms medium pink. Clusters of flowers up to 3 inches (7.5 cm) across and very fragrant, but especially prone to rain damage. Dark olive green, elongated leaves on reddish canes. Not grown for hips. Basically disease resistant, although slightly susceptible to black spot. Very hardy (zone 3).

VARIETIES	COLOR	PETAL COUNT	HEIGHT	BLOOM
'Marie-Victorin'***	Pink apricot	26–40	48″	Repeat

A *kordesii* hybrid. An Explorer rose. Deep coral peach buds open into clusters of mildly fragrant flowers roughly 3 inches (7.5 cm) wide with yellowish undersides and a yellow eye once open. Foliage is dark green, glossy, and quite disease resistant. Expect severe dieback although crown hardy (zone 3).

VARIETIES	COLOR	PETAL COUNT	HEIGHT	BLOOM
'Martin Frobisher'**	Light pink	25–40	60″	Repeat

An Explorer rose. A *rugosa* seedling. Upright, good for hedges. Dainty flowers over 2 inches (5 cm) across with variable fragrance. Retains spent blossoms. Dense growth with light green foliage. Maroon canes with few thorns on upper portions. Not grown for hips. Once disease resistant, now very susceptible to black spot. Normally hardy, but some dieback possible (zone 2).

VARIETIES	COLOR	PETAL COUNT	HEIGHT	BLOOM
'Mary Manners'***	White blushed pink	11—17+	48″	Repeat

A *rugosa* hybrid. White sport of 'Sarah Van Fleet.' Upright. Flowers are 3 inches (7.5 cm) across with a yellow center and strong fragrance. Crown hardy. Cover with potting soil and a thick layer of whole leaves in late fall for optimal winter protection (zone 3).

'Mary Queen of Scots'****	Pink white	5	48″	June

A *spinosissima* hybrid. Suckers freely. An old but lovely, fragrant rose. Flowers are pinkish tinged white with purple markings and a bright gold center. Very open (about 3 inches/7.5 cm) and nearly flat. Foliage is medium to dark green and disease resistant. Forms many purplish black hips. Gets quite large where canes don't die back. Marginally hardy so cover the crown with potting soil and a thick layer of whole leaves for winter protection (zone 4).

'Métis'***	Medium pink	28—35	48″	June

A *nitida* hybrid. Pinkish lilac crimson flowers just over 2 inches (5 cm) across with mild fragrance and golden stamens. Glossy, deep green foliage. Red, bronze coloration in fall. Lovely, dark red hips. Some dieback possible, but basically hardy with adequate snowfall. Consider covering the crown with potting soil and a thick layer of leaves in late fall (zone 4 to 5).

'Mrs. Anthony Waterer'***	Vibrant red	25+	72″	Repeat

A *rugosa* hybrid. Vigorous growth. Good as landscape plant or hedge. Very fragrant, loose flowers up to 3 inches (7.5 cm) across. Limited repeat bloom. Dark green foliage on prickly cane. Not grown for its hips. Somewhat prone to disease, especially black spot. Hardy (zone 4).

'Mrs. Doreen Pike'**	Medium pink	80+	36″	Repeat

A *rugosa* hybrid (David Austin). Compact, spreading. Clusters of large 4-inch (10-cm) somewhat frilly, fragrant flowers in flushes throughout the summer. Light lime green, semi-glossy foliage on thorny cane. Quite disease resistant. Crown hardy. Strongly advise good winter protection. Mound potting soil over the crown and cover with a thick layer of leaves. Or, use the Minnesota Tip or Hass Teepee winter protection method (zone 5).

'Mrs. John McNabb'***	White tinged pink	15—20	60″	May to June

A *rugosa* hybrid. Spreading. Slight to moderately fragrant, 3-inch (7.5-cm) flowers tinged pink. Long, initial bloom period, but not a true repeat bloomer (although often listed that way). Lovely, dark green foliage on arching red canes with few thorns. Quite disease resistant, although slightly susceptible to black spot. Very hardy (zone 2 to 3).

Moje Hammarberg®***	Mauve	17—25	48″	Repeat

A *rugosa* hybrid. Upright, vigorous. Reddish, violet flowers, single or loosely double, over 3 inches (7.5 cm) across with strong fragrance. Dense growth with large, glossy medium to deeper green leaves. Forms few, but lovely, large scarlet hips. Fall foliage color is possible. Disease resistant. Very hardy (zone 3 to 4).

'Mont Blanc' ('Montblanc')***	White	9—16	36″	Intermittent

A *rugosa* hybrid. Upright, dense. Good early season bloom. Fragrant, 3-inch (7.5 cm) somewhat loose white flowers tinged pink. Limited repeat bloom. Medium to darker green foliage. Abundant orange-red hips. Quite disease resistant. Very hardy (zone 3).

'Monte Cassino'***	Red pink	17—25	36″	Repeat

A *rugosa* hybrid. One main flush of very fragrant flowers followed by limited repeat bloom. Flowers about 3 inches (7.5 cm) wide vary from deep red to red pink. Shiny light green disease resistant foliage. Large orange-red hips. Hardy (zone 3).

VARIETIES	COLOR	PETAL COUNT	HEIGHT	BLOOM
Moore's Striped Rugosa™***	Red white stripes	25—35+	48″	Repeat

A Moore rose. A *rugosa* hybrid. Upright, bushy. Clusters of large 4-inch (10-cm) flowers with unusual coloration, mild clove scent, and loose Hybrid Tea form. Dark green foliage with purplish overtones, especially when young. Moderately thorny. Disease resistant. Crown hardy but winter protection strongly advised. Mound potting soil over the crown and cover with a thick layer of whole leaves (zone 4 to 5).

VARIETIES	COLOR	PETAL COUNT	HEIGHT	BLOOM
'Morden Blush'****	White to light pink	51	24″	Repeat

A Parkland rose. Bushy, compact. Excellent bedding plant. Flowers with pink to white coloration up to 3 inches (7.5 cm) across with slight to moderate fragrance. One of the best shrubs for flower arrangements in the bud stage. Medium green, dullish foliage. Very susceptible to black spot. Basically hardy, but some dieback is common in severe winters (zone 3 to 4).

VARIETIES	COLOR	PETAL COUNT	HEIGHT	BLOOM
'Morden Centennial'****	Medium pink	40	36″	Repeat

A Parkland rose. Bushy, dense. Excellent bedding plant. Clusters of lovely, 4-inch (10-cm) flowers with slight to moderate fragrance. Pink often rose to lilac floral coloration. Good spring and fall bloom. Nice, semi-glossy, dark foliage. Susceptible to black spot. May die back in severe weather, but basically hardy. (zone 3 to 4).

VARIETIES	COLOR	PETAL COUNT	HEIGHT	BLOOM
'Morden Fireglow'***	Orange red	28	24″	Repeat

A Parkland rose. Bushy. Uniquely colored orange to scarlet flowers just over 3 inches (7.5 cm) across with slight fragrance. Medium green, dullish foliage. Red hips possible. Susceptible to black spot. Expect severe dieback although crown hardy (zone 3 to 4).

VARIETIES	COLOR	PETAL COUNT	HEIGHT	BLOOM
'Morden Ruby'***	Red pink	35—45	36″	Repeat

A Parkland rose. A *suffulta* hybrid. Upright, spreading, vigorous. Clusters of flowers up to 3 inches (7.5 cm) across with slight fragrance. Sparse growth habit may require frequent pruning. Dark green foliage turning yellow in fall. Orange hips. Susceptible to black spot. May die back in severe weather, but basically hardy (zone 3 to 4).

VARIETIES	COLOR	PETAL COUNT	HEIGHT	BLOOM
'Morden Sunrise'**	Yellow pink	12	36″	Repeat

A Parkland rose. Upright, open. Pointed orange-yellow buds open into flowers up to 4 inches (10 cm) wide varying greatly in color from light yellow to dark orange yellow with pink tones. Very informal appearance with differing colors on each plant. Glossy, dark green foliage susceptible to black spot. Crown hardy (zone 4).

VARIETIES	COLOR	PETAL COUNT	HEIGHT	BLOOM
'My Girl'****	Deep reddish pink	28—30	40″	Repeat

A Bailey rose in the Easy Elegance® series. Not well known yet. Upright, compact, twiggy. Round deep pink buds open into clusters of 2 to 3-inch (5 to 7.5-cm) ruffled flowers in a lovely contrast to the medium green foliage. Very disease resistant. Crown hardy, but protect with potting soil and a thick layer of leaves in the far north (zone 4).

VARIETIES	COLOR	PETAL COUNT	HEIGHT	BLOOM
'Nearly Wild'****	Medium pink	5	24″	Repeat

A Floribunda often listed as a Shrub in catalogs because of its overall hardiness. Upright, twiggy growth. Flowers up to 3 inches (7.5 cm) across with whitish petal bases and golden eyes. Slight scent. Excellent repeat bloom. Dark green foliage is disease resistant. No hips. Quite hardy, but benefits from having its crown covered with potting soil and a thick layer of whole leaves in late fall (zone 4).

VARIETIES	COLOR	PETAL COUNT	HEIGHT	BLOOM
'Nevada'****	White	8—10	84″	Repeat

A *moyesii* hybrid. Arching. Excellent first bloom with blossoms up to 4 inches (10 cm) across in clusters. White petals often tinged pink contrasting to bright yellow stamens. Mild scent. Light green foliage. Purplish cane. Vigorous grower used as Climber in warmer areas. Prone to black spot. Dieback is probable, but crown hardy (zone 4).

VARIETIES	COLOR	PETAL COUNT	HEIGHT	BLOOM
'Nova Zembla'***	Light pink	25—40	48″	Repeat

A *rugosa* hybrid. A sport of 'Conrad F. Meyer.' Upright, vigorous. Clusters of large (4 inch/10 cm), fragrant, creamy white, cupped flowers tinged pink. Hybrid Tea-like blooms as they open. Medium green foliage susceptible to disease. Not reliably hardy. Cover the crown with potting soil and a thick layer of whole leaves for winter protection (although commonly listed as zone 3).

VARIETIES	COLOR	PETAL COUNT	HEIGHT	BLOOM
'Nyveldt's White'****	Pure white	5	72″	Repeat

A *rugosa* hybrid. Upright. Very fragrant, open, almost flat flowers up to 4 inches (10 cm) wide with golden stamens. Attracts bees. Good repeat bloom. Deep green foliage susceptible to disease. Lovely orange red hips in fall. Very hardy (zone 3 to 4).

VARIETIES	COLOR	PETAL COUNT	HEIGHT	BLOOM
'Paint the Town'***	Medium red	20	30″	Repeat

A Bailey rose in the Easy Elegance® series. Mounded, spreading. Blooms freely with clusters of large 3-inch (7.5 cm) flowers. Glossy, dark green foliage that is typically disease resistant but occasionally infected with black spot. Does well in containers. Crown hardy, but benefits by covering with potting soil and thick layer of leaves in the far north (zone 4).

VARIETIES	COLOR	PETAL COUNT	HEIGHT	BLOOM
Parkdirektor Riggers®***	Dark red	9—16	72″	Repeat

A *kordesii* hybrid. Vigorous growth habit with spiny canes. Good as climber. Enormous clusters of gorgeous, velvety, crimson flowers up to 3 inches (7.5 cm) across with little scent and bright gold stamens. Dark, glossy foliage. Slight susceptibility to black spot. Expect severe dieback unless crown is covered with potting soil and a thick layer of whole leaves. When used as a Climber, winter protection is essential (zone 5).

VARIETIES	COLOR	PETAL COUNT	HEIGHT	BLOOM
'Paulii'****	White	5	24″	June

A *rugosa* hybrid. Crawls and sets roots where strong, thorny stems touch the ground. Good for ground cover. Clusters of large, clematis-like (4-inch/10-cm) blossoms with moderate clove-like fragrance and deep yellow stamens. Dark green foliage. Abundant, large orange red hips. Disease resistant. Specify plants growing on their own roots. Hardy (4 to 5).

VARIETIES	COLOR	PETAL COUNT	HEIGHT	BLOOM
Perdita®***	Creamy white	60—80+	48″	Repeat

An Austin rose. Upright, bushy. Clusters of 5-inch (12-5 cm) cupped and very fragrant (spicy) flowers several times a season. Flowers have creamy white edges and apricot centers. Medium green, semi-glossy foliage is quite disease resistant. Use the Minnesota Tip or Hass Teepee method for optimal winter protection. Or cut cane back to 12 inches (30 cm) and cover with potting soil and a thick layer of whole leaves in late fall (zone 5 to 6).

VARIETIES	COLOR	PETAL COUNT	HEIGHT	BLOOM
'Persian Yellow'*****	Medium yellow	40	72″	May to June

A *foetida* hybrid. Upright, arching, vigorous. Richly colored but unpleasantly scented flowers just under 3 inches (7.5 cm) across. Green foliage on thorny cane. Susceptible to black spot. Very hardy if grown on its own roots. If budded, plant the bud union 3 inches (7.5 cm) under the soil to increase winter hardiness (zone 4).

VARIETIES	COLOR	PETAL COUNT	HEIGHT	BLOOM
'Phoebe's Frilled Pink' (see 'Fimbriata')				
Pierette®***	Deep pink	17—25	36″	Repeat

A Pavement Series hybrid *rugosa*. Compact, spreading. Purplish buds open into loose, fragrant (clove) 3-inch (7.5-cm) flowers with a golden center. Dense, deep green foliage with good disease resistance. Large, red hips. Very hardy (zone 3).

VARIETIES	COLOR	PETAL COUNT	HEIGHT	BLOOM
Pink Double Knock Out®***	Bright pink	18–24	36″	Repeat

A Radler rose. Upright, compact. Pink buds open into clusters of gorgeous, slightly fragrant 3-inch (7.5-cm) flowers. Light green foliage on mildly thorny cane. Disease resistant. Cover the crown with potting soil and a thick layer of whole leaves for good winter protection (zone 5 to 6).

'Pink Grootendorst'***	Clear medium pink	25–30	48″	Repeat

A *rugosa* hybrid. Lovely, upright vase (pillar in South). Good hedge in cold climates. Covered in clusters of 1-inch (2.5-cm), carnation-like flowers (fringed petals). Slight to no fragrance, but exceptional repeat bloom. Foliage varies from glossy medium to dark green. Very thorny. Prone to both disease and insect infestation (mainly aphids). Occasional dieback in severe weather, but basically hardy (zone 3 to 4).

Pink Knock Out®***	Rich pink	7–11	36″	Repeat

A Radler rose. A sport of Knock Out®. Rounded, bushy. Small bright pink buds burst into lovely clusters of 3-inch (7.5-cm) flowers with little fragrance. Medium green foliage with a bluish tint. Disease resistant. Marginally hardy, so cover the crown with potting soil and a thick layer of whole leaves for good winter protection (zone 5 to 6).

'Pink Pavement'***	Mauve pink	9–16	30″	Repeat

A Pavement series hybrid *rugosa.* Compact, spreading. Red buds open into large fragrant blossoms roughly 3 inches (7.5 cm) wide on arching canes. Foliage is deep green and generally disease resistant although somewhat susceptible to black spot. Very hardy (zone 3).

Pink Robusta®***	Medium pink	9–15	60″	Repeat

A *kordesii* hybrid. Upright, spreading. A very vigorous grower. Rich pink flowers about 3 inches (7.5 cm) across with varying fragrance. Large, glossy, dark green leaves. Susceptible to black spot. Generally, crown hardy, but best if winter protected with potting soil and a thick layer of whole leaves over its crown (zone 3 to 4).

Polareis® ('Polar Ice')***	White tinged pink	25	40″	Repeat

A *rugosa* hybrid. Bushy. Will sucker freely. Greenish buds open into creamy white 2-inch (5-cm) blossoms with deep pink centers and edges. Flowers hang down and are lightly fragrant. The glossy, medium green foliage with gray accents has a scent as well (earthy). Thorny canes. No hips. Very hardy (zone 3 to 4).

Polarsonne®***	Medium pink	15–20	48″	Intermittent

A *rugosa* hybrid. Upright, bushy. Main bloom in June with limited bloom the rest of the season. Clusters of loose flowers about 2 inches (5 cm) across with nice fragrance. Limited repeat bloom after the first main flush in spring. Medium green wrinkled leaves. Cover the crown with potting soil and a thick layer of whole leaves in late fall for best winter protection (zone 5).

'Polstjärnan'****	Pure white/gold center	9–16	96″	June

A *beggeriana* hybrid. May be the 'White Rose of Finland.' Vigorous, sprawling. Can be grown as a Climber or ground cover. Large clusters of small 1-inch (2.5-cm) flowers with golden centers and little fragrance. Lime green foliage on arching cane with maroon hue. Prune as little as possible since it blooms on the previous year's growth (old cane). Very hardy. However, if used as a Climber, bend the canes over, cover them with soil and a thick layer of whole leaves to minimize dieback (zone 2).

'Prairie Dawn'***	Medium pink	25–35	60″	Repeat

A Parkland rose. Upright, lanky, arching. Nodding flowers up to 3 inches (7.5 cm) across and mildly fragrant with golden stamens. Varies in coloration by soil. Fairly good repeat bloom. Light to medium green, glossy foliage. Prone to black spot, but very hardy (zone 3).

VARIETIES	COLOR	PETAL COUNT	HEIGHT	BLOOM
'Prairie Princess'****	Salmon orange pink	17—25	60"	Repeat

A Buck rose. Upright, open. Plant will climb. Ruffled flowers up to 4 inches (10 cm) across with slight fragrance. Very good repeat bloom. Dark, leathery foliage. Very hardy for a Buck rose but expect severe dieback. Can't hurt to cover crown with potting soil and a thick layer of leaves for optimal winter protection (zone 4 to 5).

'Prairie Star'***	White	20—50+	36"	Repeat

A Buck rose. Erect, bushy. Orange red buds open into fragrant (fruity), large 4-inch (10-cm) flowers often tinged pink or yellow. Dark green, leathery foliage on thorny cane. Often dies back to the ground. Somewhat prone to black spot. Cover the crown with potting soil and a thick layer of whole leaves in late fall for winter protection (zone 5).

'Prairie Youth'***	Salmon pink	9—16	72"	Spring/fall

A *spinosissima* hybrid. Arching, vigorous. May be used as Climber. Clusters of salmon to orange pink flowers just over 2 inches (5 cm) across with mild fragrance and golden stamens. Blooms once in spring, once in fall—quite unusual. Somewhat susceptible to black spot. Very hardy (zone 3 to 4).

'Purple Pavement' ('Rotes Meer')***	Crimson red	12—16	36"	Repeat

A Pavement Series hybrid *rugosa*. Compact, mounded. Color close to a purple crimson. Flowers about 2 inches (5 cm) across with gold stamens in the center. Very fragrant. Dense, medium green, glossy foliage somewhat susceptible to black spot. Large scarlet hips in fall. Very hardy (zone 3).

'Quadra' ('J. F. Quadra')***	Dark red	65+	Variable	Repeat

An Explorer rose. A *kordesii* hybrid. Low, trailing. Dark red buds open into rich red 3-inch (7.5-cm) fragrant flowers that fade slightly to deep pink as they mature. Attractive red hue on newly forming leaves that turn a glossy, deep green. Thorny. Grows vigorously and may be used as a Climber. If used in this way, bend canes over and cover with soil and a thick layer of whole leaves (zone 3).

'Queen of Sweden'***	Pink	60—100+	36"	Repeat

An Austin rose. Upright. A very prolific bloomer with clusters of 3-inch (7.5-cm) or larger flowers that form wide shallow cups with a light fragrance. Foliage is medium green and quite disease resistant. Somewhat thorny. Best if winter protected using the Minnesota Tip or Hass Teepee method. Or cut cane back to 12 inches (30 cm) and cover with potting soil and a thick layer of whole leaves (zone 5 to 6).

'Quietness'***	Soft pink	60	40"	Repeat

A Buck rose. Upright, spreading. Clusters of flowers up to 4 inches (10 cm) across and very fragrant. Foliage is medium green, semi-glossy, and somewhat susceptible to black spot. Mildly thorny. Will die back to ground each winter. Cover the crown with a mound of potting soil and whole leaves in late fall (zone 5 to 6).

Rainbow Knock Out®****	Coral pink/yellow eye	5	36"	Repeat

A Radler rose. Rounded, bushy. Small pink buds open into clusters of stunning coral pink 2-inch (5-cm) flowers with slight scent. Best bloom early and late in the season. Petals tend to cling rather than falling off cleanly. New foliage purplish turning to medium to dark green on thorny cane. Generally disease resistant but sometimes susceptible to leaf spot. Hips possible. Cover the crown with potting soil and a thick layer of whole leaves for good winter protection (zone 5).

'Red Grootendoorst' (see 'F.J. Grootendoorst')

'Repens Alba' (see 'Paulii')

VARIETIES	COLOR	PETAL COUNT	HEIGHT	BLOOM
Rheinaupark®***	Medium red	20	60″	Repeat

A *kordesii* hybrid. Vigorous, upright growth. Flowers just under 3 inches (7.5) across with slight fragrance and yellow eye. Marginal repeat bloom. Glossy, dark green foliage. Susceptible to black spot. Expect some dieback. Best if winter protected by covering the crown with potting soil and a thick layer of whole leaves in late fall (zone 3 to 4).

VARIETIES	COLOR	PETAL COUNT	HEIGHT	BLOOM
Robusta®***	Medium red	5	72″	Repeat

A *rugosa* hybrid. Upright, dense, vigorous. Forms good hedge. Lovely, 2½-inch (6.25-cm) scarlet red flowers with yellow stamens. Slight fruity fragrance. Excellent repeat bloom. Glossy, deep green foliage contrasting to rich red cane (very thorny). Not grown for hips. Prone somewhat to black spot. Expect severe die back, but crown hardy (zone 3 to 4).

VARIETIES	COLOR	PETAL COUNT	HEIGHT	BLOOM
'Rose à Parfum de l'Haÿ'***	Medium red magenta	25—35+	48″	Repeat

A *rugosa* hybrid. Lovely cherry-colored, 3-inch (7.5-cm) blossoms with strong fragrance. Good initial but sporadic rebloom. Dense growth with deep green, but smooth foliage. Few hips. Prone to dieback in severe winters and susceptible to black spot. Expect severe dieback, but crown hardy (zone 4 to 5).

VARIETIES	COLOR	PETAL COUNT	HEIGHT	BLOOM
'Roseraie de l'Haÿ'***	Dark red	25—30+	60″	Repeat

A *rugosa* hybrid. Upright, arching. Vigorous, dense growth. Lovely buds open into clusters of large, purple crimson 4-inch (10-cm) globes (peony-like) with creamy yellow stamens and a strong fragrance. Medium green, wrinkly foliage. Thorny canes. Few hips, if any. Disease resistant. Quite hardy, but mild dieback is possible (zone 4).

'Rotes Meer' (see 'Purple Pavement')

VARIETIES	COLOR	PETAL COUNT	HEIGHT	BLOOM
'Royal Edward'**	Medium pink	12—18	18″	Repeat

An Explorer rose. A *kordesii* hybrid. Compact, spreading. Makes a good ground cover. Deep pink buds open into clusters of slightly fragrant soft pink blooms about 2 inches (5 cm) wide that fade to white. Lovely golden centers. Good first but sporadic rebloom. Glossy medium green foliage is disease resistant. Hardy (zone 3).

VARIETIES	COLOR	PETAL COUNT	HEIGHT	BLOOM
Rugelda®***	Yellow edged red orange	25—40	36″	Repeat

A Pavement series hybrid *rugosa*. Upright, bushy. Clusters of slightly fragrant frilly yellow orange pink flowers with red edge up to 4 inches (10 cm) wide. Shiny deep green foliage on thorny cane is susceptible to black spot. Cover crown with potting soil and a thick layer of leaves in late fall for optimal winter protection (zone 4).

'Rugosa Magnifica' (see 'Magnifica')

VARIETIES	COLOR	PETAL COUNT	HEIGHT	BLOOM
'Ruskin'***	Dark red	26—50	48″	Repeat

A *rugosa* hybrid. Upright, open, vigorous. Clusters of very fragrant cupped flowers just over 2 inches (5 cm) across turning almost purple as they age. Marginal repeat bloom. Dark green, leathery foliage slightly susceptible to black spot. Small red fruits possible, but uncommon. Hardy (zone 4 to 5).

VARIETIES	COLOR	PETAL COUNT	HEIGHT	BLOOM
'Sarah Van Fleet'****	Medium pink	9—16	48″	Repeat

A *rugosa* hybrid. Dense, erect to spreading growth. Large, fragrant 3-inch (7.5-cm) magenta flowers shaped like a cup with creamy stamens. Deep green, glossy foliage on thorny stems. Not grown for its hips. Somewhat susceptible to disease. Winter protect by covering the crown with potting soil and a thick layer of leaves in late fall (zone 4 to 5).

VARIETIES	COLOR	PETAL COUNT	HEIGHT	BLOOM
'Scabrosa'***	Deep pinkish mauve	5	72″	Repeat

A *rugosa* hybrid. Vigorous, dense bush will form thicket (suckers freely). Stunning, cerise-colored flowers up to 5 inches (12.5 cm) across with prominent creamy stamens. Rich, heavy scent. Medium, glossy green foliage. Excellent orange-red hips, sometimes the size of crabapples against yellow to maroon fall foliage. Disease resistant. Very hardy (zone 2 to 3).

VARIETIES	COLOR	PETAL COUNT	HEIGHT	BLOOM
'Scarlet Meidiland'***	Medium red	17–30	48"	Repeat

A Meidiland rose. Spreads beautifully and makes a good ground cover. Clusters of 1½-inch (4-cm) flowers bloom mainly in early summer. Some sporadic bloom until fall. No scent. Dark glossy, green foliage. Generally crown hardy, but benefits from good winter protection. Cover the crown with potting soil and a thick layer of whole leaves in late fall for optimal winter protection (zone 5).

VARIETIES	COLOR	PETAL COUNT	HEIGHT	BLOOM
'Scarlet Pavement'**	Medium pink red	17–25	36"	Repeat

A Pavement series *rugosa* hybrid. Compact, spreading. Deep lilac buds open into large 4-inch (10-cm) fragrant flowers with yellow centers. Glossy medium green crinkly leaves. Large dark red hips. Susceptible to black spot, but very hardy (zone 3).

VARIETIES	COLOR	PETAL COUNT	HEIGHT	BLOOM
'Scharlachglut'**	Medium to dark red	5	72"	June

A *kordesii* hybrid. Dense, upright, arching. Velvety, rich red flowers up to 5 inches (12.5 cm) across with yellow centers and slight fragrance. Dull green foliage, often tinged red. Large, red hips. Bush may be sparse and require pruning to encourage a full look. Severe dieback expected. Would get much higher rating if grown in warmer climates where it spreads freely (zone 4).

VARIETIES	COLOR	PETAL COUNT	HEIGHT	BLOOM
'Schneezwerg'**	White	9–16	60"	Repeat

A *rugosa* hybrid. Upright, spreading form. Suckers freely. Can make good, informal hedge. Slow growing, but often worth the wait. Clusters of attractive, 2-inch (5-cm) flowers with moderate to strong fragrance and lovely, golden stamens. Blossoms flatten out as they mature. Dark green, glossy foliage on thorny cane. Abundant, small, scarlet hips. Susceptible to black spot, but very hardy (zone 3).

VARIETIES	COLOR	PETAL COUNT	HEIGHT	BLOOM
Sea Foam®****	White to pinkish	25–30+	36"+	Repeat

Vigorous, rambling. Clusters of flowers just under 3 inches (7.5 cm) wide vary from white to pinkish. Excellent ground cover with slightly fragrant and profuse bloom. Nice, glossy medium to dark green foliage with good disease resistance. Orange red hips possible. Crown hardy, but benefits from good winter protection. If used as a replacement for a true Climber, remove canes from their support in late fall and lay them on the ground. Cover the cane with soil and a thick layer of whole leaves for optimal winter protection (zone 4 to 5).

VARIETIES	COLOR	PETAL COUNT	HEIGHT	BLOOM
'September Song'***	Apricot blend	30	36"	Repeat

A Buck rose. Technically, a Grandiflora. Nice, erect growth. Orange buds open in clusters of gorgeous pale peach flowers up to 5 inches (10 cm) wide. Fruity scent. Dark, leathery foliage tinged copper. Typically dies back to the ground each year. Cover the crown with potting soil and a thick layer of whole leaves for optimal winter protection (zone 5).

VARIETIES	COLOR	PETAL COUNT	HEIGHT	BLOOM
'Serendipity'***	Orange blend	20–25	36"	Repeat

A Buck rose. Upright, vigorous, to spreading. Orange to yellow flowers up to 5 inches (12.5 cm) wide with noticeable apple fragrance. Glossy, dark green, leathery foliage on thorny cane. Typically, canes die back to ground each year. Cover the crown with potting soil and a thick layer of whole leaves for optimal winter protection (zone 5).

VARIETIES	COLOR	PETAL COUNT	HEIGHT	BLOOM
'Showy Pavement'**	Medium pink	9–16	30"	Repeat

A Pavement series *rugosa* hybrid. Nice, compact plant. Deep purple buds emerge into clusters of large (3 inch/7.5 cm) fragrant flowers on arching canes. Light green foliage. Orange hips in the fall. Susceptible to black spot, but very hardy (zone 3).

VARIETIES	COLOR	PETAL COUNT	HEIGHT	BLOOM
'Simon Fraser'***	Light to medium pink	5–22	24″	Repeat

An Explorer rose. Nice ground cover. Forms clusters of mildly fragrant single 2-inch (5-cm) flowers early in the season and semi-double flowers later. Dark green, semi-glossy foliage has become susceptible to disease. Expect severe dieback, but very hardy (zone 3).

'Sir Thomas Lipton'***	White	30	48″	Repeat

A *rugosa* hybrid. Nice upright, V-shape, to arching. Quite fragrant, 3 inch (7.5 cm) flowers on extremely thorny cane. Lovely first bloom, but varying rebloom by season. Dark, leathery foliage. Susceptible to insect and disease problems, although often listed as disease resistant. Expect severe dieback, but crown hardy (zone 3 to 4).

'Snow Dwarf' (see 'Schneezwerg')

'Snow Owl' (see 'White Pavement')

'Sophia Renaissance'	Amber yellow	70–90	48″	Repeat

A Poulsen rose. Bushy, upright growth. Fairly vigorous. Large, long buds open into cupped to quartered flowers up to 4 inches (10 cm) wide. Long bloom period but darker color turns pale. Mild honey scent. Medium to dark green leathery foliage. Average disease resistance. Cover the crown with potting soil and a thick layer of whole leaves in late fall for optimal winter protection (zone 5). Not rated since we have not yet field-tested this rose.

'Souvenir de Philémon Cochet'****	White	26–40+	60″	Repeat

A *rugosa* hybrid. Sport of 'Blanc Double de Coubert.' Plant grows vigorously. Hollyhock-like flowers, up to 3 inches (7.5 cm) or larger, often have a pink center. Thick petals often knocked off by rain. Lovely scent. Foliage is a lush, medium to deep green. Not grown for hips. Disease resistant. Very hardy (zone 3 to 4).

'Spring Gold' (see 'Frühlingsgold')

'Stanwell Perpetual'***	White (pink)	45	40″	Repeat

A *spinosissima* hybrid. Upright, spreading. Will sucker. Cupped flowers up to 3 inches (7.5 cm) across with pink tinge and lovely fragrance. Marginal repeat bloom. Smallish deep green leaves with purplish sheen. Very thorny, cane with thin branches. Tolerates salt well, so good for roadsides. Severe dieback possible. *Select plants grown on their own roots only* (zone 3 to 4).

Starry Night™**	Pure white/gold	5	24″	Repeat

A French introduction by Pierre Orard. Compact, spreading. Stunning in bloom, particularly during the first flush. The plant produces many clusters of nearly 3-inch (7.5-cm) flowers with golden centers and no scent. Deadhead for limited repeat bloom. Medium to dark green, glossy foliage. Disease resistant. No hips. This lovely rose is quite tender and best protected using the Minnesota Tip or Hass Teepee method (zone 5).

'Summer Wind'**	Orange pink	5–8+	36″	Repeat

A Buck rose. Upright, dense, bushy. Moderately vigorous growth. Large, 4-inch (10-cm) flat flowers with mild clove-like scent. Excellent repeat bloom. Dark green, leathery foliage. Thorny. Quite susceptible to black spot. Crown hardy when mounded with potting soil and covered with whole leaves in late fall (zone 4 to 5).

Sunny Knock Out®***	Bright yellow to cream	5	36″	Repeat

A Radler rose. Upright. Clusters of delightful and abundant bright yellow 3-inch (7.5-cm) flowers fade to cream and contrast nicely to dark green, semi-glossy foliage with leaf stems exuding an interesting fragrance. Disease resistant. One of the hardier roses in this group, but still cover the crown with potting soil and a thick layer of whole leaves for winter protection (zone 5).

VARIETIES	COLOR	PETAL COUNT	HEIGHT	BLOOM
'Sunrise Sunset'*****	Pink to apricot	13—19	24″	Repeat

A Bailey introduction in the Easy Elegance® series. Very vigorous, dense, and spreading. Excellent ground cover rose. Long, consistent bloom. Clusters of pink flowers (just over 2 inches/5 cm wide) with apricot tones around a golden center. Deep green, semi-glossy foliage with a bluish tinge is somewhat susceptible to black spot, but fights it off fairly well. Crown hardy without protection although it benefits from being covered with potting soil and a thick layer of whole leaves in the far north (zone 4).

VARIETIES	COLOR	PETAL COUNT	HEIGHT	BLOOM
'Super Hero'***	Medium to darker red	35—40	48″	Repeat

A Bailey rose in the Easy Elegance® series and one of the group's finer reds. Dense, spreading. Clusters of Hybrid Tea-like blossoms are up to 3 inches (7.5 cm) wide and contrast nicely to the satiny medium to dark green foliage. Good disease resistance. Crown hardy, but best protected with potting soil and a thick layer of whole leaves in the far north (zone 4).

VARIETIES	COLOR	PETAL COUNT	HEIGHT	BLOOM
'Suzanne'***	Orange pink	30—40	60″	Repeat

A *spinosissima* cross. Upright, arching. Plant suckers freely often becoming twice as wide as it is tall. Cupped coral pink flowers just over 1 inch (2.5 cm) across with slight fragrance. Some repeat bloom. Dark green foliage. Excellent for erosion control or in naturalized settings. Lovely hips and possible reddish purple fall foliage. Hardy (zone 4).

'The Polar Star' (see 'Polstjärnan')

VARIETIES	COLOR	PETAL COUNT	HEIGHT	BLOOM
'Thérèse Bugnet'***	Medium pink	35—40	72″	Repeat

A *rugosa* hybrid. Upright, open, vase. Suckers freely. Clusters of deep pink buds burst into ruffled flowers 4 inches (10 cm) across with heavenly fragrance. Blooms best if pegged or trained on a fence with canes running laterally—produces lots of laterals and sub laterals (much more bloom). Few thorns close to blossoms make this one of the better shrub roses for cut flowers. Lovely gray-green to bluish foliage on attractive red canes. Nice orangish red fall coloration with remote chance of orange hips. Prune as little as possible. Resistant to black spot but often infected with powdery mildew and leaf spot diseases. Some dieback common, but basically very hardy (zone 3 to 4).

VARIETIES	COLOR	PETAL COUNT	HEIGHT	BLOOM
Topaz Jewel®****	Medium yellow	20—30	60″	Repeat

A *rugosa* hybrid (Ralph Moore). Upright, spreading to arching. Suckers freely. Clusters of lovely blossoms with fruity scent about 4 inches (10 cm) across with bright yellow stamens. Flowers fade to cream. Medium green foliage on thorny cane. Not grown for hips. Fairly disease resistant, but not reliably hardy. Best to winter protect for at least the first few years by covering the crown with potting soil and a thick layer of whole leaves in late fall. Worth the effort to get it fully established (zone 4).

VARIETIES	COLOR	PETAL COUNT	HEIGHT	BLOOM
Turbo™**	Medium pink	20—25	48″	Repeat

A *rugosa* hybrid (Meilland). Upright, bushy. Large 3-inch (7.5-cm) or larger mildly fragrant pinkish flowers with whitish bases. Light to medium green semi-glossy foliage. Thorny cane. Disease resistant. Crown hardy but best covered with potting soil and whole leaves in winter (zone 3)

VARIETIES	COLOR	PETAL COUNT	HEIGHT	BLOOM
'Wasagaming'***	Medium to deep pink	30	48″	Repeat

A *rugosa* hybrid. Vigorous. Plant suckers freely. Fragrant, rose-colored flowers just under 3 inches (7.5 cm) across. Deep green foliage turns reddish yellow in fall. Thorny. Slightly susceptible to black spot. Very hardy (zone 3 to 4).

VARIETIES	COLOR	PETAL COUNT	HEIGHT	BLOOM
'White Grootendoorst'***	White	25—35	48″	Repeat

A *rugosa* hybrid. A sport of 'Pink Grootendoorst.' Upright, dense. Often bare at its base (cut back regularly for best form). Clusters of small (1-inch/2.5-cm) carnation-like blossoms with little to moderate scent. Foliage tends to be light green. Slightly susceptible to black spot, and very commonly invaded by red spider mites. Not grown for hips. Occasional dieback in severe winter, but basically hardy (zone 4).

White Out™***	Creamy white	5	24″	Repeat

A Radler rose. A compact, vigorous plant with good branching. Very free flowering with white petals surrounding a golden center. Flowers just more than 3 inches (7.5 cm) wide. Mild fragrance (citrus). Excellent repeat bloom. Dark bluish green foliage with good disease resistance. Best if winter protected with potting soil and a thick layer of whole leaves over the crown (zone 5).

'White Pavement'***	White	12—16	36″	Repeat

A Pavement Series hybrid *rugosa*. Compact, spreading. Makes a good ground cover. Pure white, cup-shape blooms with nice fragrance and golden centers. Flowers roughly 3 inches (7.5 cm) wide. Deep green, glossy foliage. Red hips contrast to potentially colorful fall foliage. Susceptible to black spot. Very hardy (zone 3).

'Wildenfels Gelb'****	Light yellow	5	36″	Repeat

A *spinosissima* hybrid. Compact. Flowers more than 2 inches (5 cm) across with slight fragrance. Gorgeous early season bloom. Marginal repeat bloom. Medium green foliage on thorny stems. May have to be purchased from a European source. Hardy (zone 4).

'Will Alderman'***	Lilac pink	35—40	48″	Repeat

A *rugosa* hybrid. Upright, bushy. Deep lilac pink buds open into loose, very fragrant flowers (about 3 inches/7.5 cm wide) with golden centers. Good repeat bloom. Mid green foliage is disease resistant. Will form round red hips in fall as nice contrast to its fall foliage colors and reddish stems. Very hardy (zone 2).

'William Baffin'****	Deep pink to light red	15—20	Variable	Intermittent

An Explorer rose. A *kordesii* hybrid. Extremely vigorous. Large clusters of slightly fragrant flowers just under 3 inches (7.5 cm) across with yellow centers. More red than pink. One excellent first flush and some limited rebloom later. Glossy medium green foliage on reddish, very thorny stems. Generally disease resistant, but vulnerable to certain types (races) of black spot. Extremely hardy with minimal dieback in severe winters. An excellent choice to replace true Climbers in the far North (zone 2).

'William Booth'***	Light to medium red	5	72″	Repeat

An Explorer rose. A *kordesii* hybrid. Canes arch and can be trained to a trellis for best effect. Deep red buds open into clusters of small (2 inch/5 cm), lightly fragrant flowers with pink undersides and golden centers. Flowers fade to an awkward pink. Dark green, semi-glossy foliage generally disease resistant but prone to some leaf spot damage. Round orange red hips. Very hardy (zone 3).

'Winnipeg Parks'***	Deep pink to red	15—22	36″	Repeat

A Parkland rose. Upright, bushy. Reddish pink flowers just over 2 inches (5 cm) across with slight fragrance. Matte medium green foliage often tinged red in fall. Susceptible to black spot. Expect severe dieback, but crown hardy (zone 3 to 4).

'Yellow Dagmar Hastrup' (see Topaz Jewel®)

Rosa foetida bicolor

SPECIES (WILD) ROSES

Species roses are those that exist naturally in the wild. They tend to be vigorous and disease resistant. Many produce lovely hips that are excellent for jams and jellies. The hips and foiliage of some varieties are also highly decorative in floral arrangements. Species roses often spread freely through suckers or seedlings, so they are relatively easy to propagate. They are virtually maintenance free if fully hardy (not all are). Flowers vary from inconspicuous to stunning. These roses feed and protect bees, birds, and wildlife. Most bloom only once in the spring. If canes die back, they bloom only in the following season. They are more difficult to find than more popular roses, but for avid growers with enough space they are well worth seeking out. An interesting footnote is that no rose has ever been found growing wild south of the equator.

How Species Roses Grow

Species roses are the only ones that grow true from seed, producing plants identical to their parents. If you grow a number of Species roses close together, you may get crosses if you allow seed to drop to the ground and form seedlings. Species vary tremendously in their growth patterns. Plants available through catalogs are generally grown from seed or taken as suckers from parent plants. These are growing on their own roots. However, some growers bud them to produce more plants quickly. Look for plants grown on their own roots, so you do not have to worry about suckers forming from alien rootstock. Any suckers, little plants off to the side of the mother plant, from plants grown on their own roots will produce plants identical to the parent plant. Many of these roses become quite large and spread out over the years. Some will even form patches of roses if left unattended.

Where to Plant

Site and Light Plant Species roses in an open, sunny area away from trees or shrubs. They have an extensive root system and need lots of space to grow freely. Although some of the Species tolerate shade, full sun is still preferred.

Soil and Moisture Species roses are far less fussy about soil preparation than other roses. However, they do best in loose soil that drains freely. Replace clay or rock with loam purchased from a garden center or in bags as potting soil. Add lots of organic matter, such as compost, leaf mold, peat moss, or rotted manures. These keep the soil moist and cool during dry periods. They also help the soil drain freely, encourage the growth of beneficial soil microorganisms, and attract worms which keep the soil aerated and fertilized. Soil should be light, loose, and airy. This allows quick root growth and good drainage so that plenty of oxygen can get to the tender feeder roots. A handful of bonemeal (or superphosphate) mixed into the soil provides additional nutrients for the young plant. Proper soil preparation takes a little time, but it is critical to good growth.

Spacing Space according to the potential height of the plant. Most plants will grow almost as wide as they are tall (some even wider). These plants do well as accent or specimen plants (roses planted on their own in an open space) or in areas where they are allowed to roam freely, as along fences.

Planting

Most Species roses are sold as bare root plants. Order them through the mail.

Bare Root It is always preferable to buy Species roses growing on their own roots. As mentioned, a number of suppliers are now offering budded stock. These plants will produce unwanted suckers and are often less hardy than plants grown on their own roots.

Plant bare root Species roses directly in the garden as soon as the ground can be worked in spring. The exact steps for doing this are outlined in Part II.

Each plant should have healthy cane. The number of canes will vary by variety and supplier. Most will be 18 inches (45 cm) high for easy shipping. The length and width of canes also vary by variety.

If plants are budded, bury the bud union 3 to 4 inches (7.5 to 10 cm) below the soil. This will encourage the upper portion of stem to form roots and thus increase the chance for the plant to survive in cold climates.

Potted Plants It is rare to find wild roses for sale in garden centers. If you belong to a local rose club, you may be able to get suckers from other members. Wild roses may also be available from arboretums and public rose gardens.

Pick out plants with healthy canes and luxuriant foliage. Check plants carefully for any signs of disease or insect infestations to avoid buying infected plants. Avoid leaving the plant in the car while you do other errands, since the heat buildup can damage it. Also, if the plant will be exposed to wind, protect it well by wrapping it in plastic or paper. As soon as you get home, remove the cover, and water the soil if it's at all dry. Plant exactly as outlined in Part II.

How to Care for Species Roses

Water Species roses tolerate drought better than other roses. However, in the early stages of growth keep the soil evenly moist at all times. When plants mature, they can survive dry spells. Still, they prefer even moisture throughout the growing season. If there is a dry spell in late summer or early fall (common in cold climates), water them well until the first frost to avoid stress.

Mulch After the soil warms up to 60°F (15.6°C), apply a mulch around the base of the plant. The mulch should come close to but not touch the cane. The most commonly used mulch is shredded, not whole, leaves. Place at least a 3-inch (7.5-cm) layer over the soil. If you have enough leaves to make a thicker mulch, so much the better. Leaves keep the soil moist and cool which encourages rapid root growth. The mulch feeds microorganisms and worms which enrich and aerate the soil. The mulch also inhibits weed growth and makes pulling any weeds that do sprout much easier. Other good mulches include pine needles and grass clippings. The latter should be only 2 inches (5 cm) deep, or they will heat the soil and may smell. These mulches are all inexpensive and effective. If you use chipped wood or shredded bark, which are lovely around larger plants, apply additional fertilizer to the soil since these rob the soil temporarily of nitrogen as they decompose. Because mulch is eaten by soil microorganisms and worms, it needs to be replenished regularly throughout the growing season. Remove and compost all mulch in the fall.

Fertilizing Species roses are far less finicky about fertilizer than other roses. In fact, they thrive in less than ideal soil conditions. Add bonemeal or superphosphate to the soil at planting.

Using organic fertilizers for Species roses is preferred. Good organic fertilizers are alfalfa meal (rabbit pellets), blood meal, bonemeal, compost, cow manure, Milorganite, or rotted horse manure. Do not overfertilize. Doing so may result in dense foliage but little bloom. These plants are very different from Floribundas, Grandifloras, and Hybrid Teas which require lots of fertilizer to do well. Never add fertilizer late in the season (after late July). This may cause new growth, which will die back during cold winters.

Small doses of inorganic fertilizer are okay but second-best to organic ones. Fertilize once in spring every third or fourth year, but not more often, unless the plant is growing poorly. Match the amount of fertilizer to the size of the plant.

If a plant is growing vigorously but blooming poorly, don't fertilize at all.

Weeding Species roses compete well with weeds but thrive in weedless conditions. Kill off all perennial weeds before planting with an herbicide such as Roundup®. If you're an organic gardener, dig up the entire weed, including even the tiniest bit of root. Some perennial weeds (especially thistles) sprout from infinitesimal root pieces. Keep weeds at bay with a heavy application of mulch, as explained in the preceding "Mulch" section. Hand pull any weeds that appear through the mulch. Avoid digging deeply around the plant, since roots are near the surface.

Staking One of the nice things about Species roses is that they require little care compared with other roses. You really don't need to give them support.

Rosa hugonis, however, may look sparse unless staked in the following manner. Gently arch over longer canes. Avoid crimping the base. Tie the tip of the cane to a stake pounded into the ground. When lots of new branches begin to form at the base of the cane, cut off the bent cane at its base. This process may take a full year, but the plant will be much bushier and produce an abundance of blossoms.

Disbudding Some growers remove flower buds in the first year, directing all energy to the plant itself. Remove buds just as they begin to form, for best results. Let a few buds bloom late in the season to encourage the plant to go into dormancy for the winter.

There is no reason to remove buds on mature Species roses. Most Species bloom once in spring and then form hips for beautiful fall coloration. A few are repeat blooming, and there are people who would argue that the removal of some buds on these plants would encourage heavier bloom later in the season. We think that this is just a waste of time. However, we strongly agree that removing spent blossoms on repeat-blooming roses is an excellent idea if flowers rather than hips are your goal (as outlined next).

Deadheading Most Species roses bloom only once a season, so removing spent blossoms is a waste of time. However, as noted, a few will repeat bloom. If you remove spent blossoms on these plants, you will have intermittent bloom throughout the season. Remember that wild roses produce lovely hips, so you should stop removing blossoms by mid-summer on wild roses that bloom more than once.

Pruning Cut out dead wood each spring. Also, consider cutting out cane that is too spindly to produce bloom. (Such cane is usually less wide than the diameter of a pencil.) This keeps the center of the bush open and airy. It also prevents disease and promotes bushy growth from remaining canes. To encourage the growth of laterals, branches off the main cane, cut canes back by one-third. Cut laterals back to two or three eyes to encourage larger bloom. *Note that the last two steps are strictly optional.* Species roses require little pruning to remain healthy and hardy, which is one of their major advantages.

Winter Protection Not all Species roses are hardy in colder areas. Avoid the more tender varieties if burying or covering crowns is more work than you're willing to do. For the most tender varieties listed in the following table, mound soil around the base of the plant and hope for lots of snow. Remove the soil as soon as it warms up in spring. The hope is that part of the cane will survive. Prune off any dead portions. The plant will do the rest by growing vigorously if watered and fertilized properly. A few of the listed plants may not survive unless protected using the Minnesota Tip Method, outlined in Part II.

Problems

Insects Most common problems are aphids (spray off with a hose) and spider mites (spray leaves with a gentle mist in hot, dry weather). Use a miticide to kill mites if they persist.

Disease Powdery mildew and black spot are occasional problems. Control them with an appropriate fungicide. If they tend to be a problem each year, then spray plants every 7 to 10 days to prevent disease in the first place. Vary the fungicide used for best results.

Propagation

Division Some Species roses send out underground stems, or stolons, which in turn produce new plants. This growth pattern is familiar to many gardeners growing lilacs or raspberries. Early in the season, before plants begin to leaf out, begin dividing mature plants. With a sharp spade, dig straight down between the parent plant and the plantlet to the side. Your spade will hit the stolon. Step on the spade with as much force as necessary to sever the underground stem. Then dig around the small plant with your spade. Dig up the plant with as much soil around the roots as possible. Plant immediately as you would a bare root plant. Keep the plant consistently moist until it is growing well.

Cuttings You may want to experiment with growing Species roses from softwood cuttings. Take cuttings in late June or early July from the center of a firm cane. Firm canes are mature, not too green. Refer to Chapter 7 in Part II for specific tips on propagation with softwood cuttings.

Seed Species roses grow well from seed. Collect hips from the plant before they soften in the fall. Remove the seeds and place them in moist peat moss in a plastic bag. If possible, place the bag in an empty crisper of the refrigerator (fruit gives off ethylene gas which can kill seeds). In February or March, fill up shallow plastic dishes, such as margarine containers, with sterile growing medium—peat moss, vermiculite, or perlite (or a mixture of all three). Remove the peat moss containing the seeds from the plastic bag and press it onto the growing medium. Keep the growing medium moist and warm but not wet. Keep the lid on the container until the seeds start to germinate. Treat the young seedlings with a gentle misting of a solution containing a fungicide to prevent damping off (seedlings topple over as disease organisms attack the base of the plant—very common). Keep the growing medium just moist enough to encourage growth. When the seedlings form a second pair of leaves, gently dig them out of the growing medium and plant them in a small peat pot filled with potting soil. To do so, hold the seedlings gently by one leaf, and drop the roots into a hole made with a pencil. Firm the soil around the base. Keep the rooting medium moist until plantlets form. Then pot them up as necessary.

Note that some growers plant seeds immediately without chilling them at all. Sometimes this works; sometimes it doesn't. But, it's worth a try. The moist chilling (stratification) as outlined in the preceding paragraph may be necessary to overcome a seed's natural tendency to dormancy in the first season. A simple solution is to split the seeds up and try both methods for any particular Species. If you get immediate growth, great. If not, the other seeds will be moist chilled and will sprout in spring.

Note too that you can plant seeds in a bed outdoors and just let nature take its course. If you plant seed in rows, it's easier to distinguish them from small weeds in spring. Always be patient with seeds grown outdoors—some may sprout a year or 2 later.

Special Uses

Cut Flowers Cut canes just as buds begin to unfurl. They can be breathtaking in arrangements but are rarely long-lasting.

Dried Flowers Species roses are generally not grown for this purpose, because their flowers are rather small and fragile. However, their hips are often used in dried floral arrangements with eye-catching results.

Hips Many Species roses form beautiful, distinctive hips. These are stunning in the fall and add a special dimension to garden design. They are also attractive to birds, and many of them are excellent for jams and jellies. Never use hips for food, however, if plants have been sprayed to protect them from insects or disease. Since most Species roses are naturally disease and insect resistant, they rarely require spraying, making them highly desirable for hips.

Sources

Angel Gardens, P.O. Box 1106, Alachua, FL 32616, (352) 359-1133

Antique Rose Emporium, 9300 Lueckmeyer Rd., Brenham, TX 77833, (800) 441-0002

Blackfoot Native Plants Nursery, P.O. Box 761, Bonner, MT 59823, (406) 244-5800

Brushwood Nursery, 431 Hale Lane, Athens, GA 30607, (706) 548-1710

Burlington Rose Nursery, 24865 Rd. 164, Visalia, CA 93292, (559) 747-3624

Burnt Ridge Nursery, Inc., 432 Burnt Ridge Rd., Onalaska, WA 98570, (360) 985-2873

Corn Hill Nursery Ltd., 2700 Rte. 890, Corn Hill, New Brunswick E4Z 1M2, Canada, (506) 756-3635

Countryside Roses, 5016 Menge Ave., Pass Christian, MS 39571, (228) 452-2697

Eastern Plant Specialties, 660 A Berrys Mill Rd., West Bath, ME 04530, (207) 504-4405

Elk Mountain Nursery, P.O. Box 599, Asheville, NC 28802, (828) 683-9330

ForestFarm, 990 Tetherow Rd., Williams, OR 97544, (541) 846-7269

Fragrant Path (seed), P.O. Box 328, Fort Calhoun, NE 68023, (no phone by request)

Fritz Creek Gardens, P.O. Box 15226, Homer, AK 99603, (907) 235-4969

Gossler Farms Nursery, 1200 Weaver Rd., Springfield, OR 97478, (541) 746-3922

Greenmantle Nursery, 3010 Ettersburg Rd., Garberville, CA 95542, (707) 986-7504

Heirloom Roses, 24062 Riverside Dr. NE, Saint Paul, OR 97137, (503) 538-1576

High Country Roses, P.O. Box 148, Jensen, UT 84035, (800) 552-2082

Hortico, Inc., 723 Robson Rd., RR# 1, Waterdown, ON L0R 2H1 Canada, (905) 689-6984

Mary's Plant Farm & Landscaping, 2410 Lanes Mill Rd., Hamilton, OH 45013, (513) 894-0022

McKay Nursery Co., P.O. Box 185, Waterloo, WI 53594, (920) 478-2121

North Creek Farm, 24 Sebasco Rd., Phippsburg, ME 04562, (207) 389-1341

Oikos Tree Crops, P.O. Box 19425, Kalamazoo, MI 49019, (269) 624-6233

Out Back Nursery, Inc., 15280 - 110th St. S, Hastings, MN 55033, (651) 438-2771

Pickering Nurseries, 3043 County Rd. 2, RR #1, Port Hope, ON L1A 3V5 Canada, (905) 753-2155

Prairie Moon Nursery, 32115 Prairie Lane, Winona, MN 55987, (507) 452-1362

Raintree Nursery, 391 Butts Rd., Morton, WA 98356, (800) 391-8892

Rogue Valley Roses, P.O. Box 116, Phoenix, OR 97535, (541) 535-1307

Rose Fire, Ltd., 09394 State Rte. 34, Edon, OH 43518, (419) 272-2787

Rosemania, 4920 Trail Ridge Dr., Franklin, TN 37067, (888) 600-9665

Roses of Yesterday and Today, 803 Brown's Valley Rd., Watsonville, CA 95076, (831) 728-1901

Roses Unlimited, 363 N Deerwood Dr., Laurens, SC 29360, (864) 682-7673

St. Lawrence Nurseries, 325 State Hwy 345, Potsdam, NY 13676, (315) 265-6739

Select Plus International Lilac Nursery, 1510 Pine Rd., Mascouche, QC J7L 2M4 Canada, (450) 477-3797

S & W Greenhouse, Inc., P.O. Box 30, 533 Tyree Springs Rd., White House, TN 37188, (615) 672-0599

Spring Valley Roses, P.O. Box 7, Spring Valley, WI 54767, (715) 778-4481

Tripple Brook Farm, 37 Middle Rd., Southampton, MA 01073, (413) 527-4626

Vintage Gardens (custom propagation), 4130 Gravenstein Hwy N, Sebastopol, CA 95472, (707) 829-2035

VARIETIES

Following are some of the more popular Species roses grown in colder climates. Of course, there are many more Species, but these represent a good cross section of possible choices. Some of these are not truly hardy and will require winter protection. Most are very hardy, and a few will survive to temperatures of −50° F (−45°C). Some are rather timid bloomers but may have extremely attractive foliage, form, hips, or thorns. A few are showered in blossoms and have exquisite fragrance. Most wild roses bloom only once. Naturally, the bloom time could overlap or vary by year, depending upon the severity of the winter and the zone you live in. When the roses bloom more than once, you'll see the word "repeat" under the Bloom column. To our knowledge there are no known Species roses originating from the Southern Hemisphere. Some of these roses are highly prone to black spot, so keep this in mind if you want to avoid spraying. A number of them will grow well in partial shade.

VARIETIES	COLOR	PETAL COUNT	FRAGRANCE	HEIGHT	BLOOM
Rosa acicularis (ROW-suh ah-sick-you-LAY-riss)	Deep pink	5	Variable	36″	June

The 'Arctic Rose' or 'Prickly Rose.' Spreading. Grayish green leaves on wiry, thorny, russet stems may turn color in fall. Hips are over an inch (2.5 cm) long and look like coral, waxy pears. Very hardy (zone 2).

Rosa alpina var. *pendulina* (see *Rosa pendulina*) (ROW-suh al-PEE-nuh variety pen-doo-LEYE-nuh)					

Rosa altaica (see *Rosa spinosissima*) (ROW-suh al-TAY-i-kuh)					

Rosa × andersonii

Rosa multiflora inermis

Rosa primula

Rosa spinosissima plena

VARIETIES	COLOR	PETAL COUNT	FRAGRANCE	HEIGHT	BLOOM
Rosa × andersonii	Soft pink/gold	5	Slight sweet	36″	June

(ROW-suh an-der-SOHN-ee-eye)

Technically, a hybrid. Arching, dense, vigorous. Covered with a shower of blooms just under 3 inches (7.5 cm) across. Medium to deep green foliage on prickly stems. Medium sized roundish to oval red hips. Will die back or out unless winter protected from extreme cold by covering the crown with soil and a thick layer of whole leaves in late fall or by using the Minnesota Tip or Hass Teepee method (zone 5 to 6).

Rosa arkansana var. *suffulta* (see *Rosa suffulta*)					

(ROW-suh are-kan-SAY-nuh)

This name is often used in error to refer to the rose described later.

Rosa beggeriana	White	5	Variable	96″	June

(ROW-suh beg-er-ee-AY-nuh)

Bushy, arching, vigorous. Suckers. Forms clusters of white flowers just larger than 1 inch (2.5 cm) across with bright yellow centers. Subtly scented dainty green foliage with bluish tinge on canes with crimson thorns. Grown more as an oddity than for its flowers. Small orange to red to purplish round fruits. Very hardy (zone 3 to 4). (see 'Polstjarnan' p. 114).

Rosa blanda	Medium pink	5	Variable	60″	June

(ROW-suh BLAN-duh)

'Hudson Bay,' 'Labrador Rose,' or 'Smooth Rose.' Upright, arching. Spreads rapidly. Wonderfully fragrant with rosy, pink 2-inch (5-cm) flowers with yellow centers. Long, thin leaves on cane on nearly thornless cane (except at base). Medium round, red hips and potentially stunning fall foliage. Tolerates shade. Hardy (zone 2 to 4).

Rosa canina inermis	Pink/white	5	Variable	96″	June

(ROW-suh ka-NEYE-nuh in-UR-muhs)

The 'Dog Rose,' used as rootstock for many hybrids. Upright, arching, vigorous (almost a climber) with mildly fragrant, with almost 2-inch (5-cm) flowers with golden eyes. Light to medium green foliage on very thorny cane. Delicious elongated oval, orange-red hips. Grows extremely well in clay-based soils. Susceptible to black spot. Expect considerable dieback in colder climates unless winter protected. Cover the crown with soil and a thick layer of whole leaves in late fall and protect as much as possible with the Hass Teepee Method (zone 4 to 5).

Rosa carolina	Bright medium pink	5	Variable	36″	June

(ROW-suh care-oh-LINE-uh)

The 'Pasture Rose' or 'Carolina Rose.' A dense, prickly plant (when older) which suckers freely. Usually forms 3-inch (7.5-cm) pink flowers which occasionally are white and have golden centers. Shiny medium green leaves may turn golden in fall. Forms medium sized round orange red hips. Tolerates drought well. Hardy (zone 4 to 5).

Rosa eglanteria	Deep pink	5	Moderate	144″	June

(ROW-suh egg-lan-TEE-ree-uh)

'Sweet Brier' or 'Eglantine' rose. Good as a Climber. Fragrant blossoms roughly 1 inch (2.5 cm) across, rarely semi-double, with yellow centers. Apple-scented, dark green foliage. Thorny. Small, red, delicious hips. Hardy (zone 4 to 5).

VARIETIES	COLOR	PETAL COUNT	FRAGRANCE	HEIGHT	BLOOM
Rosa foetida (ROW-suh FEH-ti-duh)	Pure bright yellow	5	Unpleasant	72″	May to June

Commonly called 'Austrian Yellow.' A stunning upright, dense plant with many blooms just larger than 2 inches (5 cm) across. Attractive green toothed foliage on prickly cane. Attractive round hips reddish brown to almost maroon in color. Does exude a somewhat unpleasant odor. Highly prone to black spot and usually requires spraying. Very hardy (zone 3 to 4).

VARIETIES	COLOR	PETAL COUNT	FRAGRANCE	HEIGHT	BLOOM
Rosa foetida bicolor (ROW-suh FEH-ti-duh BICK-ku-lohr)	Red orange/yellow	5	Unusual	60″	June

'Austrian Copper.' Upright, dense. Blooms profusely along entire cane. Blossoms up to 3 inches (7.5 cm) across are coppery red on the outside, rich yellow underneath. Highly prone to black spot which requires spraying. Now often sold on budded cane. Depending on the rootstock, this once hardy plant may now need winter protection. Try to get plants grown on their own roots. Will revert to pure yellow on occasion. Keep growing it anyway, since new canes often form bicolor blossoms. Very hardy (zone 3 to 4).

VARIETIES	COLOR	PETAL COUNT	FRAGRANCE	HEIGHT	BLOOM
Rosa foetida persiana (ROW-suh FEH-ti-duh per-see-AY-nuh)	Egg yolk yellow	20—25	Unusual	72″	June

'Persian Yellow Rose.' Upright, arching. Produces flowers like globes with thin petals displaying wonderful bright coloration. Blooms affected badly by rain. Plant has sprawling habit and often thin, very thorny branches with bright green foliage. Some say the plant has an objectionable odor. Very susceptible to black spot. Plant it deep if you get a budded plant. Very hardy if grown on its own roots (zone 4).

VARIETIES	COLOR	PETAL COUNT	FRAGRANCE	HEIGHT	BLOOM
Rosa glauca (ROW-suh GLOWE-kuh)	Medium pink	5	Variable	72″	May to June

The 'Red Leaf' rose. Upright, mildly arching. Favorite of flower arrangers for its blue- to red-tinted foliage (varies by light exposure and soil) on reddish stems (thorny with age). Produces 1½-inch (3.75-cm) blossoms, which are star-shaped and fragrant (sometimes) with pale yellow centers. Its small, red (occasionally tinted purple) to bright orange fruits point out from the foliage and make wonderful jams and jellies. Very hardy (zone 2 to 3).

VARIETIES	COLOR	PETAL COUNT	FRAGRANCE	HEIGHT	BLOOM
Rosa hugonis (ROW-suh hu-GOH-niss)	Primrose yellow	5	Variable	48″	June

'Father Hugo Rose.' Upright, arching. Suckers freely. Cup-like, 3-inch (7.5 cm) or smaller soft yellow flowers, with mostly mild fragrance. Small, gray green leaves are almost ferny on mildly prickly brownish cane. Very small round deep scarlet to black hips (but few). Drought resistant. Hardy (zone 4 to 5).

VARIETIES	COLOR	PETAL COUNT	FRAGRANCE	HEIGHT	BLOOM
Rosa laxa (ROW-suh LACKS-uh)	White blushed pink	5	None	96″	June

Upright, arching. Becomes as wide as it is tall. A shower of 2-inch (5-cm) or wider blossoms in spring with bright yellow centers. Occasional marginal rebloom in fall. Light bluish green foliage. Pear-shaped hips change color as they mature, going from orange to deep scarlet. Extremely hardy (zone 2).

VARIETIES	COLOR	PETAL COUNT	FRAGRANCE	HEIGHT	BLOOM
Rosa macounii (ROW-suh muh-COW-nee-eye)	Light to medium pink	5	Variable	84″	June

Upright, spreading (will sucker). Attractive, 1½-inch (4-cm) blossoms with yellow centers. Normally, mildly fragrant. Pale green to bluish foliage. Small, red hips. After covering the crown with soil and a thick layer of whole leaves, protect as much cane as possible using the Hass Teepee Method (said to be hardy to zone 3; still, winter protect).

VARIETIES	COLOR	PETAL COUNT	FRAGRANCE	HEIGHT	BLOOM
Rosa mollis (ROW-suh MOL-liss)	Deep pink	5	Variable	96″	June

The 'Soft Downy Rose.' Upright, arching (suckers). Nice 1-inch (2.5-cm) or larger blossoms (sometimes white) with bright yellow centers. Grayish green foliage. Red hips. Plants seem to vary greatly in hardiness (zone 4 to 5).

VARIETIES	COLOR	PETAL COUNT	FRAGRANCE	HEIGHT	BLOOM
Rosa moschata (ROW-suh moss-KAY-tuh)	White	5	Moderate	96″	August

The 'Garland Rose.' Can climb. Clusters of flowers less than 2 inches (5 cm) across with a strong musk fragrance. Tolerates low light conditions. Light green foliage. Clusters of small red hips. Use the Minnesota Tip or Hass Teepee method for optimal winter protection (zone 6).

VARIETIES	COLOR	PETAL COUNT	FRAGRANCE	HEIGHT	BLOOM
Rosa moyesii (ROW-suh moy-EE-see-eye)	Deep pink to red	5	Slight	72″	June–July

'Moyes Rose.' Arching form. Will sucker. Parent to many Shrub roses. Large, rich pink, occasionally almost maroon flowers just under 3 inches (7.5 cm) with yellow red stamens. Dark foliage (small, rounded, leaflets) susceptible to leaf spots. Known especially for its small, lovely, elongated, orange-red hips, often referred to as "bottle-shaped." Cover the crown with soil and leaves before protecting canes with the Hass Teepee Method. Hardiness varies (zone 5).

VARIETIES	COLOR	PETAL COUNT	FRAGRANCE	HEIGHT	BLOOM
Rosa multiflora inermis (ROW-suh mull-ti-FLOH-ruh in-UR-muhs)	White	5	Variable	120″	June

Upright, arching. Used as rootstock for budding many other roses. Clusters of 1-inch (2.5 cm), blackberry-like blossoms (but only blooms on wood that comes through the winter—old wood). Dull, light green foliage. Nearly thornless cane. Small, red hips. Winter protect with the Hass Teepee Method. Hardiness varies (zone 4 to 5).

VARIETIES	COLOR	PETAL COUNT	FRAGRANCE	HEIGHT	BLOOM
Rosa nitida (ROW-suh nih-TEA-duh)	Bright pink	5	Variable	24″	June

The 'Shiny Rose.' Nice rounded shape. Suckers freely. Flowers are bright pink and about 2 inches (5 cm) across with a gold center. Mild fragrance (usually). Possible repeat bloom. Glossy medium green foliage often turning golden red purple in fall on reddish stems. Round medium sized orange red hips. Hardy (zone 4).

VARIETIES	COLOR	PETAL COUNT	FRAGRANCE	HEIGHT	BLOOM
Rosa pendulina (ROW-suh pen-doo-LEYE-nuh)	Deep pink	5	None to slight	48″	June

Upright, mildly arching. One variation of the 'Alpine Rose.' Mauvish pink flowers about 2 inches (5 cm) wide with yellow centers. Arching, reddish tinged cane with light green foliage. Forms blood red, pear to bottle-shaped hips over an inch (2.5 cm) long. Cover the crowns with potting soil and whole leaves before protecting exposed canes with the Hass Teepee Method (zone 4).

Rosa pimpinellifolia (see *Rosa spinosissima*)
(ROW-suh pim-pin-ell-i-FOH-lee-uh)

Rosa pomifera (see *Rosa villosa*)
(ROW-suh pom-IFF-er-uh)

VARIETIES	COLOR	PETAL COUNT	FRAGRANCE	HEIGHT	BLOOM
Rosa primula	Light yellow	5	Moderate	48″	May

(ROW-suh PRIM-you-luh)

The 'Incense Rose.' Upright, arching. Nice, 2-inch (5-cm) blossoms with deep yellow stamens. Early bloom for about 3 weeks in May. Nicely scented, feathery leaves on reddish brown stems. Occasionally, may bloom lightly in fall, but don't count on it. Round red purple to black hips somewhat flattened on the sides. Occasionally dies out in unusual way, one stem at a time from a canker in summer. Cut out dead canes immediately. Tolerates drought, but best to keep well watered and fed. Usually hardy, but for optimal winter protection in the far north, cover the crown with potting spoil and a thick layer of whole leaves before protecting exposed canes with the Hass Teepee Method (zone 4).

VARIETIES	COLOR	PETAL COUNT	FRAGRANCE	HEIGHT	BLOOM
Rosa roxburghii	Medium pink	5	Variable	36″	June

(ROW-suh rocks-BURG-ee-eye)

'The Chestnut Rose' or 'Burr Rose.' Mossy buds open into lightly fragrant flowers up to 3 inches (7.5 cm) across. Pale brown bark on prickly cane contrasts to many-leaflet leaves. Unusual, prickly, orange yellow hips up to 1½ inches (4 cm) across. Note there is a double form called *Rosa roxburghii plena* with many more petals (40+). These plants need good winter protection. The Minnesota Tip Method or the simpler Hass Teepee Method are preferred (zone 6).

Rosa rubiginosa (see *Rosa eglanteria*)
(ROW-suh roo-big-ghin-NOH-suh)

Rosa rubrifolia (see *Rosa glauca*)
(ROW-suh roo-bri-FOH-lee-uh)

VARIETIES	COLOR	PETAL COUNT	FRAGRANCE	HEIGHT	BLOOM
Rosa rugosa	Deep pink to red	5	Strong	60″	Repeat

(ROW-suh roo-GOH-suh)

The 'Japanese Rose' or 'Turkestan Rose.' Vigorous, shrubby plants. Forms suckers freely. Good for hedges and windbreaks. Tolerates salt. Very fragrant, 3-inch (7.5-cm) flowers vary widely in coloration and produce 1-inch (2.5-cm) red hips for jams, jellies, and teas (but you need two different plants in the Rugosa group for good pollination). Cane often very prickly. Leaves deep green and wrinkled (rugose). This plant is the parent to a wide number of hybrids and is extremely disease resistant. Lovely orange to red fall foliage. Extremely hardy (zone 2).

VARIETIES	COLOR	PETAL COUNT	FRAGRANCE	HEIGHT	BLOOM
Rosa rugosa alba	White	5	Strong	72″	Repeat

(ROW-suh roo-GOH-suh Al-buh)

Vigorous, shrubby plant producing silky, very fragrant 4-inch (10-cm) blossoms with showy yellow stamens. Foliage and growth similar to its parent. Lovely autumn colorations and large rounded orange red hips. Extremely hardy (zone 2).

VARIETIES	COLOR	PETAL COUNT	FRAGRANCE	HEIGHT	BLOOM
Rosa rugosa rubra	Reddish purple	5	Strong	72″	Repeat

(ROW-suh roo-GOH-suh ROO-bruh)

Dense, shrubby plant. Very fragrant flowers up to 4 inches (10 cm) across form delicious, large red hips. Coloration of blossoms varies from lavender to deep red to purple. Foliage and growth similar to its parent. Tolerates shade. Hardy (zone 4).

VARIETIES	COLOR	PETAL COUNT	FRAGRANCE	HEIGHT	BLOOM
Rosa sericea					
var. *pteracantha*	White	5	Slight	72″	May to June
(ROW-suh sair-REE-see-uh)					

The 'Wing Thorn Rose.' Not grown for its white, fleeting flowers with golden centers. Grown instead for the beauty of its large red thorns. These can be up to 2 inches (5 cm) long and stick out from the cane, almost looking like transparent mini-sails from a distance. Foliage is delicate. This rose is not reliably hardy, although it grows well even if it dies back to the ground. Consider covering the crown of this tender rose with soil and leaves before protecting exposed canes with the Hass Teepee Method of winter protection, to increase chances of limited bloom (zone 5 to 6).

Rosa setigera	Deep pink	5	None to slight	72″	July
(ROW-suh seh-TIJ-er-uh)					

The 'Blackberry Rose.' Long, rambling canes often arch over and root where tip hits ground, just like wild blackberries. Used in breeding Climbers. Use as Climber, ground cover, or rambler. Clusters of 2-inch (5-cm) or larger flowers bloom later in the season than most Species roses. Wild blackberry-like foliage. Lots of small brownish green hips on female plants (a rarity in the rose world in having both male and female plants—dioecious). Buy plants known to come from northern grown stock. Hardy (zone 4).

Rosa soulieana	Creamy white	5	Variable	72″	June
(ROW-suh soo-lee-AY-nuh)					

Tall, arching. Acts as a Climber in warmer areas. Vigorous. Pale yellow buds open into clusters of small 1½-inch (3.75-cm) blossoms with bright yellow centers. Rounded light green leaves with blue to gray tones on thorny cane. Lots of round orange-red hips. Play it safe by using the Hass Teepee Method of winter protection to preserve as much cane as possible (zone 4).

Rosa spinosissima	White cream	5	Variable	36″	June
(ROW-suh speye-noh-SISS-i-muh)					

The 'Scotch Briar Rose.' Upright, rounded, arching. Suckers freely. Good ground cover. Parent to a number of fine hybrids. Attractive pointed buds open into large, cream-colored blossoms with a bright yellow center. Very thorny, reddish cane with ferny foliage (small leaves with tiny leaflets). Noted primarily for its fall coloration. Tolerates drought. Forms small, black or purple hips. Very hardy (zone 3 to 4).

Rosa spinosissima					
var. *altaica*	White	5	Variable	60″	May to June
(ROW-suh speye-noh-SISS-i-muh al-TAY-i-kuh)					

Dense, vigorous, spreading. Blooms in clusters of blossoms up to 2 inches (5 cm) across. Open faintly yellow, turning to cream with prominent golden stamens. Ferny foliage. Extremely spiny canes. Lovely and delicious round medium sized black hips. Hardy (zone 4).

Rosa spinosissima plena	White cream	26—40+	Moderate	72″	June
(ROW-suh speye-noh-SISS-i-muh PLAY-nuh)					

Commonly known as the 'White Rose of Finland.' The plant spreads freely forming a large clump. The true variety is often hard to find because it is difficult to propagate. It has long, arching canes covered with small, 2-inch (5-cm) fragrant blossoms in early summer, each with bountiful, bright yellow stamens contrasting nicely with rich green foliage. A good replacement is Polstjärnan under Shrub roses (it too lays claim to the title 'White Rose of Finland'). Protect as much cane as possible using the Hass Teepee Method around scattered clumps of cane (zone 5 to 6).

VARIETIES	COLOR	PETAL COUNT	FRAGRANCE	HEIGHT	BLOOM
Rosa suffulta (ROW-suh suf-FULL-tuh)	Medium pink	5	Slight	24″	June

Upright, spreading. The 'Prairie Rose' or 'Wild Prairie Rose.' Flowers, occasionally white with pink tones, up to 2 inches (5 cm) across with prominent yellow stamens. Limited rebloom is possible. Produces small red hips. Withstands drought and high heat. Hardy (zone 3).

Rosa villosa (ROW-suh vill-LOH-suh)	Medium pink	5	Variable	72″	June

The 'Apple Rose.' Not grown for its 1½-inch (4-cm) medium pink flowers with bright golden centers. Grayish green foliage with a mild evergreen scent. Grown specifically for its wonderful, bristly, large, red orange hips that can be more than 1 inch (2.5 cm) wide. Disease resistant. Cover the crown with potting soil and whole leaves before protecting as much cane as possible with the Hass Teepee Method (zone 4).

Rosa virginiana (ROW-suh vir-gin-ee-AY-nuh)	Medium pink	5	Variable	48″	June

The 'Virginia Rose.' Upright, arching. Suckers freely. Flowers about 2 inches (5 cm) wide with golden centers. Glossy, deep green foliage sometimes turning golden to reddish in fall. Produces long-lasting, deep red orange hips, which are quite small and round. Reddish canes stand out in winter. Cover the crown with potting soil and whole leaves before protecting as much cane as possible with the Hass Teepee Method (zone 5).

Rosa wichuraiana (ROW-suh witch-oor-ee-AY-nuh)	White	5	Strong	12″	August–September

The 'Memorial Rose.' Fast-growing, trailing plant with 1-inch (2.5-cm) blossoms with yellow centers. Small, deep green, glossy leaves on pliable, very thorny cane. Excellent ground cover (really a rambler). Very small, orange to red, waxy hips and good fall foliage coloration. Cover the crown with potting soil and a thick layer of whole leaves before protecting as much cane as possible with the Hass Teepee Method (zone 5).

Rosa woodsii (ROW-suh WOOD-zee-eye)	Medium pink	5	Slight	24″	June

The 'Mountain Rose.' Upright, spreading. Fleeting flowers occasionally white, but usually lilac pink about 1½ inches (4 cm) across. Repeat blooms in warmer areas. Light green foliage, susceptible to black spot and other diseases. Produces large, red to orange, waxy almost heart-shaped fruits. Very hardy (zone 2 to 4).

Rosa xanthina hugonis (see *Rosa hugonis*)
ROW-suh zan-THEEN-uh hu-GOH-niss

'Showbiz'

TREE ROSES (STANDARDS)

Tree roses are popular with some growers. They are aristocratic and elegant in appearance. Some are bushy and rounded on top while others tend to arch down or weep. With proper care they bloom almost continuously throughout the season. Tree roses do well either in pots or planted directly in the garden. If planted in pots, they are easy to move around and look lovely on decks. In the garden, they make stunning focal points. Planted by themselves, they act as eye-catching specimen plants. The market for larger Tree roses is somewhat limited in colder climates. Your choice is often determined by growers. Specific plants may have to be specially budded and are extremely expensive. Others, while still expensive, are available through catalogs or in retail outlets. Miniature Tree roses are more commonly available and gaining in popularity.

How Tree Roses Grow

Miniature Tree roses are created by budding a specific named variety on the top of a tall cane of another variety (rootstock). The union is high up on the plant. The larger Tree roses used to consist of rootstock, a second rose budded to the rootstock for the standard (main stem), and a third named variety budded on top of the standard. Larger Tree roses are now grown in the same manner as Miniatures. In the United States the term *standard* generally refers to the stem or main cane, while the top of the plant is called the *crown* or *head*. In Great Britain the term *standard* refers to a Tree rose, not just the main cane. Some growers refer to large Tree roses as "standards" in the United States as well.

The rootstock most commonly used today in the United States is either 'Dr. Huey' or 'RW.' 'Dr. Huey' grows well for commercial growers who are able to get even extremely long stems from this rootstock. It is not hardy (winter protect well). The tendency of 'Dr. Huey' to get mildew is not passed on to the named variety budded on top of it.

Where to Plant

Site and Light Place in a sunny location. The best blooms occur if plants have at least 6 hours of direct sunlight a day. Sunlight also helps prevent many foliar diseases. Avoid planting under shade trees. Trees also compete for water and nutrients (if the rose is planted in the garden). Avoid planting under drip lines, because water running off the roof can damage the plants. Also, keep the plant out of wind if possible.

Soil and Moisture Tree roses need rich soil that retains moisture but still drains freely. Replace clay or rock with loam purchased from a garden center or in bags as potting soil. Add lots of organic matter, such as compost, leaf mold, peat moss, or rotted manures. These keep the soil moist and cool during dry periods. They also help the soil drain freely, encourage the growth of beneficial soil microorganisms, and attract worms which keep the soil aerated and fertilized. Soil should be light, loose, and airy. This allows quick root growth and good drainage so that plenty of oxygen can get to the tender feeder roots. A handful of bonemeal (or superphosphate) mixed into the soil provides additional nutrients for the young plant. Proper soil preparation takes a little time, but it is critical to good growth. If plants are kept in pots, repot each year. Remove and replace half of the old soil.

Spacing Space according to the expected height of the plant and width of the head (crown). Give enough space so that there is good air circulation around the upper canes to avoid disease. If you grow a number of Tree roses close together in pots, move them farther apart as the season progresses for proper exposure to sun.

Planting

You can buy either bare root (dormant) or already potted Tree roses. Either are fine. You often will have a greater selection if you order through the mail.

Bare Root You can plant bare root Tree roses directly in the garden or in a pot. Follow the steps outlined in Part II for best results.

Following are some special tips that apply to Tree roses:

The lower bud union should be one-half below and one-half above the soil on large Tree roses. There is no lower bud union on miniatures.

If a Tree rose arrives with long white sprouts already growing from the branches, simply snap these off with your fingers. This is not the way they are supposed to arrive. Buds should not be growing on properly shipped plants, but this does not mean you won't have such growth on occasion. The plants will form new buds. The long white growth usually dies off anyway—so, get rid of it.

Push a stake gently into the soil about 1 inch (2.5 cm) from the stem of the plant. The stake should be higher than the treetop to allow for future growth. Support the plant as outlined later under "Staking."

If you're growing a Tree rose in a pot, you may also want to anchor the pot to the ground with additional stakes to stop it from blowing over in strong winds.

Place moist leaves into the area between the upper canes. If canes are spread out, pull them together gently, and tie them into place with polyester twine. Just wrap the twine around the canes from the bottom up and gently pull them together. Fill all the space between the canes with leaves. These moist leaves keep canes from drying out and induce growth bud formation. Then moisten them thoroughly with a gentle spray of water.

Some growers use sphagnum moss—the long, stringy moss sold in nurseries to line baskets instead of leaves. Sphagnum moss can be difficult to find, is quite expensive, and has been linked with a rare fungal disease (see the Glossary). However, it is excellent used for this purpose. If you use it, wear gloves and a mask to protect yourself from potential infection.

Cover the upper head of the Tree rose with a piece of burlap. Burlap is sold in many discount lumber stores, garden centers, and county cooperatives. The upper portion of the burlap will rest on the support. Tie the lower part of the burlap around the stem. Mist the bag. Keep the fabric moist at all times.

Some growers use plastic instead of burlap, but you must be careful that heat does not build up on hot, sunny days to the point that growth buds are destroyed.

Check the plant every other day by lifting up the burlap. When buds begin to appear, remove it. Wearing gloves, carefully remove most of the leaves with your fingers. Wash off any bits of leaves that stick to the canes with a gentle spray from your hose.

Potted Plants Most nurseries and garden centers stock a few of the most popular Tree roses. Look for plants with as many healthy upper branches as possible and luxuriant foliage. A good plant will have a healthy, straight main stem (standard) to support the upper, branch-filled area. The stem should have no scars or damaged bark and be growing vigorously. Check the plant for any signs of disease or insect infestations to avoid buying a diseased plant. Take the plant directly home. Don't let it sit in the car while you run errands. If you have to expose the plant to wind, protect it well with plastic. Once home, water the soil immediately if it's at all dry. If the plant is in too small a pot, pot it up to a larger one. If you want to plant the tree directly in the garden, follow the steps for planting potted roses as outlined in Part II. Remember to support all Tree roses to avoid wind and storm damage.

How to Care for Tree Roses

Water Keep the soil evenly moist at all times throughout the season for best bloom. Saturate the soil thoroughly with each watering. If plants are kept in a pot, water them more frequently, since soil dries out rapidly in summer heat waves. Use plastic pots to cut down on watering. Although attractive, clay pots demand more watering. Water potted plants until water begins to pour out the drain holes. Then water again. This saturates the soil completely. If the soil moves away from the rim of the pot, it indicates that you need to water the plant more often. If this happens, push the soil back against the rim so that there is no gap between the soil and the pot. Saturate the soil immediately. Avoid letting this happen again because lack of water stresses plants badly.

Mulch After the soil in the garden warms up to 60°F (15.6°C), apply a mulch around the base of the plant. The mulch should come close to but not touch the cane. The most commonly used mulch is shredded (not whole) leaves. Place at least a 3-inch (7.5-cm) layer over the soil. If plants are growing in pots, simply cover the soil surface with pulverized leaves. Leaves keep the soil moist and cool which encourages rapid root growth. The mulch feeds microorganisms and worms which enrich and aerate the soil. The mulch also inhibits weed growth and makes pulling any weeds that do sprout much easier. Other good mulches include pine needles and grass clippings; the latter should be only 2 inches (5 cm) deep, or they will heat the soil and may smell. These mulches are all inexpensive and effective. If you use chipped wood or shredded bark, apply additional fertilizer to the soil, since these rob the soil of nitrogen as they decompose. Because mulch is eaten by soil microorganisms and worms, it needs to be replaced regularly throughout the growing season. Remove and compost all mulch in the fall.

Fertilizing If combining use of inorganic and organic fertilizers, be careful during the first year. Use inorganic fertilizer only after the plant has leafed out and is growing vigorously.

In subsequent years, as early as possible in spring, sprinkle 10-10-10 (or 5-10-5) granular fertilizer around the base of each plant. Use common sense in judging the amount, more for bigger plants, less for smaller. Hand sprinkle, never placing fertilizer against the stem of the plant. Water immediately to dissolve the granules and carry nutrients to the roots.

In both the first and subsequent years do the following: One to two weeks after the first feeding with 10-10-10 granules soak the base of each plant with a liquid containing 20-20-20 water soluble fertilizer following directions on the label. Use common sense, giving larger plants more, smaller plants less. This feeding will stimulate sensational bloom as the plant matures.

One week later soak the base of each plant with a liquid containing fish emulsion. Follow the directions on the label. Feed according to the size of the plant.

One week later give all plants the 20-20-20 treatment again. One week later, give all plants the fish emulsion soaking. Keep this rotation going until the middle of August. At this time, stop all use of fertilizers containing nitrogen, since nitrogen stimulates new growth which dies off in most winters. Use 0-10-10 fertilizer after mid-August if desired.

Additional fertilizing with Epsom salts (magnesium sulfate) is highly recommended. Give ½ cup (115 g) for larger plants, 1 tablespoon (14 g) for smaller, two times a year: in late May and in early July. Magnesium sulfate neutralizes the soil and makes it easier for the plant to take in nutrients. This may encourage the growth of new cane in the top portion (head) of the plant.

Many growers like to keep their Tree roses in pots throughout the plant's lifetime. Adding iron to the soil may be necessary in such cases to keep foliage healthy. Sprint is one of the best products for this. Follow the directions on the label carefully, matching the amount used to the size of the pot.

Good fertilizers for organic growers are alfalfa meal (rabbit pellets), blood meal, bonemeal, compost, cow manure, fish emulsion, fish meal, rotted horse manure, and Milorganite. Add bonemeal to the soil before planting. The others are effective added to the soil at planting time and as additional feedings on the surface of the soil throughout the season.

Weeding Destroy all perennial weeds before planting. Use a herbicide such as Roundup®. You can inhibit annual weeds with the use of mulch. Pull by hand any weeds that appear through the mulch. Avoid hoeing, since this can harm the shallow root system. Weeds compete with roses for nutrients and water. They also act as hosts for harmful insects. Get rid of all weeds.

If you're growing Trees in pots, pull out all weeds. Also, clean up all dead leaves and fallen petals from the soil, since they act as breeding areas for disease spores and insect eggs.

Staking Tree roses, whether grown in pots or directly in the garden, require support. You do not want the plant to jiggle or tip in heavy wind. This can damage the tender feeder roots and prevent nutrients and water from flowing freely throughout the plant. You also want to support upper branches to stop them from snapping in storms.

The type of support should match the potential size of the plant. For miniature Tree roses, a typical support would be a piece of ¼-inch (6-mm) stainless steel rod. For larger Tree roses, electric conduit is excellent and can be cut to size with a hacksaw. Green plastic supports of varying sizes are also a good alternative.

Push the support firmly into the ground or soil in the pot. The support should extend well above the base of the uppermost branches.

The simplest way to attach the main stem to a support is with electrical tape. It is easy to use and relatively inexpensive, doesn't restrict stem growth, and is very strong, providing just the kind of support needed.

A method that takes a little more time but which is preferred by many growers is to cut out a small piece of insulation or foam and wrap it around the stem. Use electrical tape or a plastic tie to hold the insulation in place. This keeps the plant firmly anchored, without any chance of the supports rubbing against the stem.

Casual growers can even use strips of cloth tied around the stem to the support in a figure-eight pattern. The knot should be firm. The cloth stops the stem from rubbing against the support.

Attach the plant to its support just under the upper cane and 4 inches (10 cm) or so from the base of the plant.

If the plant is exposed to heavy winds, tie individual upper canes to the support with polyester twine. This will often prevent canes from snapping in wind gusts. In effect, it can save your plant.

If a Tree rose is grown in a pot, it's important to support both the main stem and the pot itself. As mentioned earlier, lightweight pots (not heavy clay) can blow over in gusts, causing damage to the rose. So, stake the main stem in the pot and then the pot itself to another stake in the ground.

Disbudding Removing flower buds is rarely done with Tree roses. The mass effect of bloom atop the cane is the whole purpose of the plant.

Deadheading When flowers fade, remove them immediately by snipping off the blossoms to a point just above the nearest five- or seven-leaflet leaf below. This encourages the plant to form more blossoms. It also keeps the area around the plant tidy and prevents disease and insect infestations. You can also snip off growth to keep the plant at exactly the desired shape and size. This could fall under the category of pruning or pinching back. Don't overdo it, though, or you will harm the plant.

Pruning During the first year, do no pruning at all, other than to remove diseased or damaged cane with sharp shears. Cut off all buds to the nearest leaf below until late summer. This creates a healthier and more vigorous plant. Let the plant flower late in the season to prepare it for winter dormancy.

During the second year, remove any dead or damaged cane as you lift the plant from the ground (see "Winter Protection"). Prune only to shape the plant.

During the third and subsequent years, remove dead canes. Also remove canes that cross so that there is good air circulation in the upper area of the plant. Rub off inward-facing buds to get growth on outer portions of the plant. And prune to shape the plant for aesthetic reasons. Pruning encourages bushiness and fuller bloom.

Remove any growth from the base or stem of the plant. If a sucker forms, it is best to cut it out. Some growers cover the wound with commercial sealants; others don't.

Remove cane heads (see p. 249) by cutting back to an outward-facing bud below it. This produces healthier and more free-flowering cane.

Keep the top of the rose bushy and symmetrical by snipping off longer canes. Much of the beauty of Tree roses is in their formal symmetry.

Warning: Never cut the plant back to a point below the upper bud union. This would be the equivalent of destroying your plant. All Tree roses have a named variety budded onto the top of a cane from a different type of rose. You'll end up growing the wrong rose.

Winter Protection Potted Tree roses may be either buried using the Minnesota Tip Method or brought indoors during the winter. In colder climates, it's probably best to bury the plants unless you're willing to pay a local greenhouse to store them in ideal climatic conditions (lots of sun and adequate winter temperatures). Very few places are willing to do this anymore.

Some growers place perforated plastic bags in the pots when they plant their Tree roses. At the end of the season, they lift the bags from the pot and bury the plant (bag and all) in the garden. The following spring, they use a new bag and add fresh soil around the base of the tree when repotting (if they want to continue to grow the Tree rose in a pot).

If you're growing Tree roses in the ground, also protect them in winter with the Minnesota Tip Method as outlined in Part II.

A few additional tips for winter protecting Tree roses are important:

Remove the metal support. Take off any tape carefully. Avoid damaging the cane. Wounds are susceptible to infection.

Larger Tree roses may topple over once support has been removed and soil loosened around the base of the plants. If you want to lean something against the Tree to support it, that's fine; just make sure it's soft or padded so as not to scrape the bark.

If the plant has been budded at its base (some Tree roses are), *tip the plant over the bud union*. If you don't do this, the cane may snap off. Be sure also to take this into consideration when planting the rose directly in the garden in the first place.

Also, during the tipping process, grab the base of the plant around the union with your hand as an added way of preventing the plant from breaking at this point. If it's possible, have two people working together to tip really large Tree roses.

When raising the plant in the spring, get your support up right away. Otherwise, you could lose your rose in a single gust of wind.

Problems

Insects Follow a regular routine of spraying every 7 to 10 days to ward off insect infestations. During hot, dry weather watch out for spider mites. Begin using a miticide immediately if any appear.

Disease Follow a 7- to 10-day spraying routine to prevent common rose diseases. These diseases are much easier to prevent than to control.

Propagation

It is possible to create Tree roses on your own, but most home gardeners let professionals handle this.

Special Uses

Cut Flowers Tree roses are made from many classes of roses. Certainly, these produce some lovely cut flowers. However, the symmetry of the plant is the essence of its beauty. Therefore, it's common to cut only to keep the plant aesthetically pleasing and use the resulting cuttings, if desired, in floral arrangements. When taking cuttings, cut back to an outward-facing growth bud to stimulate new growth.

Dried Flowers The same general advice for cut flowers applies here. Remove flowers only to shape the plant. Use the flowers in potpourris or for dried wreaths or floral arrangements.

Sources

The list of Tree roses includes the ones you are most likely to find in retail outlets throughout the cold-climate region. Most of them are available through major wholesalers. So retailers will special order them for you if you ask months in advance. If you combine all of the offerings of the following companies, you'll have an interesting choice of varieties. Some companies only offer a few, while others offer many. Hortico will do custom budding. Call them to see whether the rose you want could be budded for you. Ask months in advance.

David Austin Roses Ltd., 15059 Hwy 64 W, Tyler, TX 75704, (800) 328-8893

Edmunds' Roses, 335 S High St., Randolph, WI 53956, (888) 481-7673

Garden Valley Ranch, 498 Pepper Rd., Petaluma, CA 94952, (707) 795-0919

Hortico, Inc., 723 Robson Rd., RR# 1 Waterdown, ON LoR 2H1 Canada, (905) 689-6984

Inter-State Nurseries, 1800 E Hamilton Rd., Bloomington, IL 61704, (309) 663-6797

Jackson & Perkins, 2 Floral Ave., Hodges, SC 29653, (800) 872-7673

Liferoses.com, 32235 SE Pipeline Rd., Gresham, OR 97080, (503) 757-0670

Regan Nursery, 4268 Decoto Rd., Fremont, CA 94555, (800) 249-4680

Witherspoon Rose Culture, 3312 Watkins Rd., Durham, NC 27707, (800) 643-0315

VARIETIES

Following are some excellent choices for Tree roses in cold climates. The top portions—the head, or crown—of Tree roses may be taken from a variety of rose groups as indicated under the column "Type." The best of these have vigorous, bushy or spreading growth which suits a Tree rose. The weeping varieties generally are budded with roses such as Sea Foam® (Shrub) or 'The Fairy' (Polyantha).

Tree Roses are generally classified by the height of the stem: standards—stems about 36 inches (90 cm), patio Tree roses—24 inches (60 cm), and miniature Tree roses—18 inches (45 cm) or less. Whenever ordering Tree roses, knowing the stem height is very important. Each type of rose has its place. Technically, all Tree roses could be called standards, as in Great Britain, since they are grown on one stem (a standard).

The ratings on Tree roses may differ from those of bush roses of the same name: a bush rose that does quite poorly may do extremely well when grown as a Tree (standard). The reason is that the upper budded portion may react better to the longer stem and produce exceptional growth. That is part of the reason why a few of these roses are not included elsewhere in the book, although the majority are.

The availability of Tree roses does vary quite a bit each year. However, a number of these are usually in stock. If a specific rose is not available, you can have it budded for you at an additional cost. That would include roses not in this list if the grower has access to the specific variety you want.

VARIETY	TYPE	COLOR	PETAL COUNT
'About Face'***	Grandiflora	Bronzy red gold orange	35
'Angel Face'****	Floribunda	Bluish purple	30
'Barbra Streisand'***	Hybrid Tea	Lavender pink	32—35
Betty Boop™****	Floribunda	Ivory white edged red	6—12
Black Magic™***	Hybrid Tea	Dark red	30—40
Brandy®***	Hybrid Tea	Rich apricot	28—30
Brilliant Pink Iceberg™***	Floribunda	Deep pink/cream white	18—24
Bronze Star™***	Hybrid Tea	Deep apricot and yellow	26—40
Burgundy Iceberg™***	Floribunda	Light purple/cream	20—25
Chihuly®****	Floribunda	Red yellow	25+
Chris Evert™****	Hybrid Tea	Orange yellow	30
'Chrysler Imperial'***	Hybrid Tea	Dark red	45—50
Cinco de Mayo™****	Floribunda	Rusty red orange	20—25
Double Delight®****	Hybrid tea	Red white	32
Dream Come True™***	Grandiflora	Yellow edged red	40
Easy Does It™****	Floribunda	Peachy orange apricot	25—30
Easy Going™***	Floribunda	Peach golden yellow	25—30
Ebb Tide™***	Floribunda	Plum purple	35+
Europeana®*****	Floribunda	Dark red	25—30
Falling in Love™****	Hybrid Tea	Rich medium pink	28
Fame!™***	Grandiflora	Dark pink	30—35
Firefighter®***	Hybrid Tea	Dark red	42
'First Prize'***	Hybrid Tea	Rose to light pink	28
'Fragrant Cloud'***	Hybrid Tea	Orange red	32
'French Perfume'***	Hybrid Tea	Cream yellow edged rose pink	26—40
'Full Sail'***	Hybrid Tea	White with pink tones	32
'Gourmet Popcorn'***	Miniature	White	15—20
Grace™****	Shrub	Deep apricot orange white	60—80+

'Gourmet Popcorn' (Miniature)

Just Joey® (Hybrid Tea)

'Peace' (Hybrid Tea)

Scentimental™ (Floribunda)

VARIETY	TYPE	COLOR	PETAL COUNT
Hot Cocoa™****	Floribunda	Russet red	25–30
'Iceberg'*****	Floribunda	White	35–40
Ingrid Bergman®***	Hybrid Tea	Dark red	35
In the Mood™****	Hybrid Tea	Rich medium red	25–30
Intrigue®***	Floribunda	Light purple	25–30
'Julia Child'****	Floribunda	Golden yellow	26–40+
Just Joey®***	Hybrid Tea	Coppery apricot	30
Knock Out®	Shrub	Cherry red to pink	5
Lasting Love™*****	Hybrid Tea	Dark red	25
Lavaglut® ('Lava Flow')****	Floribunda	Dark red	25–30
Legends™***	Hybrid Tea	Medium ruby red	30
Let Freedom Ring™***	Hybrid Tea	Dark red	17–25
Livin' Easy™****	Floribunda	Apricot orange	25–30
Mister Lincoln®***	Hybrid Tea	Dark red	35
Molineux®*****	Shrub	Deep yellow	120
Moonstone™****	Hybrid Tea	White edged pink	32
Outta the Blue™***	Shrub	Magenta tinged lavender	25–30
'Over the Moon'****	Hybrid Tea	Apricot caramel	30+
Pacesetter®****	Miniature	Creamy white	25+
'Peace'*****	Hybrid Tea	Yellow pink white	43
'Pink Promise'***	Hybrid Tea	Light to dark pink	28
Playboy®***	Floribunda	Red orange yellow/gold center	7–10
'Pope John Paul II'***	Hybrid Tea	Bright white	50
Rainbow Knock Out®	Shrub	Coral pink/yellow eye	5
Rainbow's End™****	Miniature	Yellow edged pink red	30–35
Rio Samba™***	Hybrid Tea	Yellow orange red	25
Rock & Roll™***	Grandiflora	Red white splashed burgundy	41+
Saint Patrick™***	Hybrid Tea	Deep yellow tinted green	32
Scentimental™***	Floribunda	Red white swirls	25–30
Sea Foam®****	Shrub	White to pinkish	30–35
Secret™****	Hybrid Tea	Pink edged deep pink	32
Sexy Rexy®****	Floribunda	Medium pink	40
Shockwave™***	Floribunda	Deep yellow pink blush	25+
'Showbiz'*****	Floribunda	Bright medium red	28–30
Strike It Rich™****	Grandiflora	Gold yellow edged pink	30
Sun Flare®****	Floribunda	Medium yellow	25–30
Sunset Celebration™***	Hybrid Tea	Apricot cream orange	36
'Sunsprite'****	Floribunda	Deep yellow	25–30
Sunstruck™***	Hybrid Tea	Apricot/yellow peach	26–40
'Tahitian Sunset'****	Hybrid Tea	Apricot pink	28
'The Fairy'****	Polyantha	Light to deep pink	35–40
Vavoom™***	Foribunda	Hot orange yellow	35+
Veterans' Honor®****	Hybrid Tea	Dark red	28
Whisper™***	Hybrid Tea	Cream white with peach tones	32
Wild Blue Yonder™***	Grandiflora	Bluish wine purple	25–30
Winsome®****	Miniature	Pinkish purple	25–40

PART II

The Basics of Growing Roses

The following chapters are filled with essential and unusual information about growing roses. Whether you're an expert or a beginning rose grower, you will find this information helpful and easy to understand. Our only concern is that you get the very most from your plants with a minimal amount of wasted money, time, and energy. If you study the following chapters, you will know more about roses than most master gardeners.

CHAPTER 2

UNDERSTANDING ROSES

Fossils of roses date back 35 million years. Roses grow wild only in the northern hemisphere—not south of the equator. In the United States alone, nearly 23 million people grow roses each year.

There are just under 60 classes of roses, based on somewhat artificial guidelines. The different classes vary in the amount of effort needed to grow them well in cold climates. Some are easy to grow, others much more difficult. Some are not even included in this guide because they do so poorly in all but warmer areas.

You can have roses blooming from spring until fall, cascading over trellises or blooming brightly in a small pot on a sunny porch. It's all up to you. This book shares critical secrets about growing roses—many kinds of roses—in cold climates.

Just how many roses you can grow successfully in cold areas will surprise you. There is an extreme range in color, flower form, plant size, foliage color and sheen, stem color, fragrance, fruit (rose hips have 20 times as much vitamin C as oranges), and use.

Roses are among the most versatile garden plants. Use them in beds or borders for a splash of color; in rock gardens to add new shape and form; as accent or specimen plants for bold punctuation in the landscape; over fences, pergolas, screens, trellises, or walls for stunning vertical displays; in containers or tubs; as indoor or greenhouse plants; for cut flowers or boutonnieres; for making jams, jellies, rose water, and perfumes; or even as ground covers and informal hedges.

How Roses Grow

Roses are woody plants in the same family as raspberries. Unlike trees which keep expanding in size with age, older canes (the main stems) of roses eventually die off and are replaced by new canes. The new growth, or cane, which originates from buds at the base of the plant is known more technically as a *basal break*. It can also be called a *shoot*. Each cane produces a bud at its tip (terminal) and a series of buds at the base of leaves where a leaf joins the stem (leaf axil). These buds grow into branches (laterals). On some roses, these branches form even smaller branches (sublaterals). New growth starts from buds. The uppermost bud always grows first. Flowering occurs on the tips of cane and on the laterals or sublaterals.

Leaves also form on canes, on laterals, and on sublaterals. On most roses, the number of leaflets on each leaf varies or may alternate. Many roses have three to five leaflets per leaf. Some roses that normally have five leaflets may have more under ideal growing conditions. A few roses have as many as 9 to 11 tiny leaflets on a leaf. Foliage is extremely important to

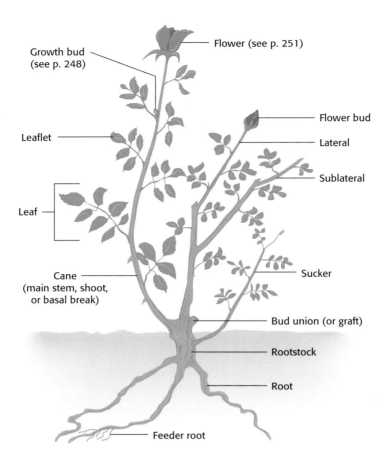

Growth bud
(see p. 248)

Flower (see p. 251)

Leaflet

Flower bud

Lateral

Sublateral

Leaf

Cane
(main stem, shoot,
or basal break)

Sucker

Bud union (or graft)

Rootstock

Root

Feeder root

The main parts of a growing budded rose. After the plant has flowered, cut the cane back to a bud above a leaf with five leaflets. Note the sucker growing from the rootstock: it should be removed as soon as it's noticed, as it depletes the plant's food reserves and forms different flowers than those on the plant's upper portion.

roses because it creates food necessary to produce good blooms. Healthy roses have lots of leaves.

The plant's root system is extensive. Its depth varies with the type of rose grown. Often, the roots are quite shallow, getting nutrients and water from the top portion of the soil. The roots themselves also vary by type of rose. Some are quite thick, while others are more fibrous and similar to those of perennials. Tiny hairs off these larger roots are called *feeder roots*. These feeder roots regenerate each year. The purpose of roots is to hold the plant firmly in place and to take in as well as store nutrients.

Some roses grow into bushes varying in size from tiny, 8 inches (20 cm), to extremely large, more than 96 inches (240 cm). Others form long canes, which make them good Climbers. However, roses do not produce tendrils to attach and wind themselves around supports. They must be attached to supports if you want to grow them vertically.

Some roses grow on their own roots. All of the canes on own-root roses produce identical flowers. Other roses do poorly when growing on their own roots. From centuries of experimentation, rose growers have learned techniques to attach one part of a rose to the rooted portion (rootstock, or understock) of a completely different rose, one that is much more vigorous and cold resistant. The techniques are known as *budding* or *grafting*. Roses that are not growing on their own roots are therefore known as budded or grafted roses.

The place where the desirable rose has been budded to the rootstock is called the *bud union*. The desirable canes produced by the budded or grafted upper portion of the plant are different from those produced from the rootstock, or lower portion. Undesirable canes that come up from the rootstock portion are referred to as *suckers* and must be removed. Suckers usually have different stem and leaf coloration—and they also may have different or fewer thorns—so, it is often quite easy to tell a sucker from a desirable cane.

Each rose plant is unique. Even rosebushes of the same name may grow differently depending on how they were propagated (see selective budding, p. 216). Each rosebush degenerates over the years. An analogy is to a person having a series of small strokes. In the case of the rose, the plant slowly goes downhill and should eventually be replaced. The time period is unpredictable. Some Species (wild) or Shrub roses may take decades to decline. Other types of roses may deteriorate more rapidly.

In colder climates, canes may be injured by cold weather which results in the death of the tip of a cane or an entire cane. This is known as *dieback*. How to deal with this problem is an important consideration and is covered both in this section and in the individual listings in Part I.

Buying Roses

Of the many classes of roses, each has advantages and disadvantages. The good and bad points of each group are covered in detail in the various sections in Part I. Note that individual roses may have more than one name; this can be confusing, but throughout the guide you'll find cross references in these instances. Depending on the source, roses also may be listed, or may have been listed in the past, under different categories. The Index will help you to quickly locate any rose included in this book.

Following is an overview "checklist" to use in buying roses.

Hardiness

Roses that survive unprotected in cold climates are called *fully hardy*. Roses that survive but may suffer some dieback are called *hardy*. The canes on a number of roses die back to the ground, but the rose sends up new growth to produce a lovely plant in one season; these roses are called *crown hardy*. (They act like a perennial.) Roses that would die out without winter protection are not hardy and are referred to as *tender*. However, a number of tender roses listed in this book are fully hardy with the proper winter protection. Even with winter protection, many roses will not survive in colder climates. You will find only a few of these roses listed in this book, with adequate warning that buying them entails a great risk of loss.

Winter Protection

If a plant requires winter protection, this means additional work for you. Some plants require less effort to protect than others. So, if you want to avoid lots of work in early fall, you might want to avoid certain types of roses altogether. Information on what is and isn't required for each type of rose or for a specific variety within a class of roses is given in detail throughout the book.

Overall Use

A certain rose may be excellent for cut flowers, as a specimen plant, as a plant for bedding in a perennial garden, for mass effect, as ground cover, as a hedge, and so on. Information on the best use for specific roses is included throughout the book.

Color

There is an infinite variety of coloration, but roses are *artificially* classified into a set range of colors, which, in our opinion, often bare faint resemblance to reality. There is even a green rose, but it's not the petals that are green (see p. 74). The breeder of a rose has

The Colors of Roses

BREEDERS CHOOSE COLORS FROM the following official rose color classifications, which may be artificial but are better than no system at all:

Apricot/apricot blend (ab)	Red blend (rb)
Mauve/mauve blend (m)	Medium red (mr)
Orange/orange blend (ob)	Dark red (dr)
Orange pink (op)	Russet (r)
Orange red (or)	White/white blend (w)
Pink blend (pb)	Yellow blend (yb)
Light pink (lp)	Light yellow (ly)
Medium pink (mp)	Medium yellow (my)
Deep pink (dp)	Deep yellow (dy)

You'll also see the following terms used to describe rose coloration:

Bicolor The color on the outside of the petal is different from the one on the inside.

Blend Two or more colors meld together on the inside of each petal.

Hand Painted One color blotched or feathered with another.

Multicolor Petals change color with age, giving a group of flowers a multicolored look.

Single All petals are one color. The color may fade or deepen with age.

Striped Petals of one color are banded with a different color.

the right to choose its official color. Since we may disagree, we give our own color descriptions in the tables throughout the book. This is a chief reason why you should see a rose growing before choosing it, whenever possible. Bear in mind also that color varies by weather, rainfall or artificial watering, and soil conditions.

Flower Size and Number of Petals

Flower size and number of petals can be an important consideration. Miniatures have extremely small blossoms (as their name implies), while other roses may produce blooms up to 5 inches (12.5 cm) wide. Throughout the book we give petal counts for individual blossoms. The official way of stating petal counts is single (4 to 8 petals), semi-double (9 to 16 petals), double (17 to 25 petals), full (26 to 40 petals), and very full (41+ petals).

Flower and Petal Form

Some people decide which roses to buy based on form alone, since shape makes such a distinct impression on the plant itself or in floral arrangements. Rose growers use specific terms to describe the shape and condition of a blossom and its petals:

Blown A flower past its prime, with the petals about to fall off.

Decorative A flower in which petals form a loose, rather than tight, bloom.

Exhibition Classic form. High-centered. Long petals form a lovely central cone.

Frilled A petal that is jagged at the tip like the end of a feather.

Garden Variety See "Decorative."

Globular A ball-like flower with a closed center.

Open-Cupped A cuplike flower with many petals and an open center.

Plain A typical rose petal, flat and somewhat rounded on the end.

Pompon A rounded flower, looking somewhat like a dahlia.

Quartered A many-petaled flower that appears to be divided into four sections.

Reflexed A petal with its tip bending down and tucking under at the end.

Rosette A flower looking somewhat like a flattened pompon.

Ruffled A petal with a wavy tip.

Split-Centered A flower with an irregularly formed central cone, appearing to have two centers.

Bloom Time

The length of the bloom period and the frequency of bloom each season is one of the most important features of any rose. A number of roses bloom only once a season. Repeat (remontant) or intermittent (off and on) bloom is a desirable quality found in a number of rose classes or in individual varieties within a larger class.

Foliage and Stem Color

Foliage and stem coloration, scent, shape, and texture are added elements often neglected by rose buyers but very important in determining the total value of any rose both in the garden and in arrangements.

The scented leaves and stems of *Rosa centifolia muscosa* (evergreen), *Rosa eglanteria* (apples), *Rosa gallica* (perfume), and *Rosa primula* (incense) are noteworthy.

Foliage shape and texture vary by group and individual variety. Some leaves are smooth, others wrinkled. Some are dull, others glossy. Some have many leaflets, others fewer. The size of the leaflets may give the rose a dainty or a bold appearance. The contrast of foliage to flowers is often one of the plant's most attractive features.

Fragrance

Some groups of roses are fragrant, while others aren't. The scent of a rose comes from evaporating oils emitted from cells (aiglets) at the base of petals. Scent varies by humidity, light, and temperature. High temperatures and high humidity increase fragrance. During cool to cold periods, many fragrant roses have little scent. Roses are often most fragrant just as dew begins to evaporate or later in the evening, because rose oils producing scent evaporate quickly in hot sun or drying winds. Fragrance also depends on the stage of maturity: some roses give off a scent only as they open, while others exude perfume when fully mature. Scents may be berrylike, fruity (apple, lemon, lime), heavy, honeylike, myrrhlike, nutty (almond), spicy

Unusual Foliage Coloration

SOME ROSES ARE ESPECIALLY desirable because of unusual foliage coloration such as *Rosa glauca*, often used in cut-flower arrangements. Rose foliage varies from green to bronze, gray, red, or purplish. Throughout the book you'll find information on foliage coloration to help you decide whether a plant may fit a special need. Foliage color is also an important indicator of plant health.

Bella Rosa®

Rosa glauca

'Lillian Gibson'

'Suffolk'

Rose Hips

HERE IS A SAMPLING of varieties worth growing for their hips alone.

VARIETY	TYPE
'Alba Semi-plena' (*Rosa alba semi-plena*)	Old Garden
Rosa canina inermis	Species
Carefree Beauty™	Shrub
Rosa eglanteria	Species
Frau Dagmar Hastrup®	Shrub
Rosa gallica 'Complicata'	Old Garden
Rosa gallica 'Alika'	Old Garden
'Geranium'	Shrub
Rosa glauca	Species
'Golden Wings'	Shrub
'Highdownensis'	Shrub
Rosa moyesii	Species
Rosa pomifera	Species
Rosa roxburghii	Species
Rosa rugosa alba	Species
Rosa rugosa rubra	Species

VARIETY	TYPE
'Scabrosa'	Shrub
Rosa spinosissima altaica	Species
Rosa villosa (see *Rosa pomifera*)	

Rosa glauca

'Alba Semi-plena'
(*Rosa alba semi-plena*)

(cinnamon, clove, licorice, mint, vanilla), sweet (perfumed), or tealike. People vary in ability to detect fragrance. Some people simply cannot smell scents.

Potpourris (Sachets)

Many rose growers use the petals of fragrant varieties for potpourris or sachets. Especially noteworthy are *Rosa centifolia* and *Rosa officinalis* ('Apothecary's Rose') both covered in the section on Old Garden Roses. (See Chapter 8 for instructions on making and storing potpourris.)

Food

Hips, the fruit of roses, are important for their beautiful fall color and as food for wildlife and humans. Hips vary in shape (elongated, round, prickly), size (pea to crab apple), color (black, green, orange, red), and taste. Hips not only are beautiful on plants, but also look lovely in ornamental wreaths or arrangements. Tips on using roses for food are included as applicable in Part I and in Chapter 8 (p. 225).

Resistance

A rose's resistance to disease and insects is important to know about ahead of time if you want to grow roses without using chemicals. You'll find quite a bit of information on resistance throughout this book, but remember that resistance is a quality that varies greatly by location. We strongly recommend joining a local rose club or society to take advantage of the knowledge gleaned by other local rose growers over the years.

CHAPTER 3

SELECTING AND PREPARING A SITE

Nothing is more important in growing roses than placing them in the correct spot and preparing the soil properly. The following information will help you do both in a way that ensures you'll get the best growth and bloom from every rose you grow.

Site and Light

Roses thrive in areas with good light, lots of air circulation, and protection from strong, drying winds.

Sunlight

Almost all roses need lots of sun, at least 5 to 6 hours of direct light per day (preferably morning light), to do well in colder climates. In general, sun is critical to vigorous growth, lush foliage, abundant bloom, and resistance to disease. Exceptions: *Rosa alba* (an Old Garden Rose), Hybrid Musks, and a few Species (wild) roses will grow well in partial shade.

Protection

Avoid planting roses that don't need winter protection under drip lines (the edge of roof eaves). During the winter and early spring, falling snow and ice will often damage roses planted under them.

Roses are prone to wind damage. Although they need air circulation, protection afforded by fences, hedges (not too close), and other barriers is helpful as long as they don't block out available sunlight.

Avoid planting roses in frost pockets, low-lying areas at the base of hills. They'll do much better placed higher up in a sunny, warmer area.

Keep roses away from hedges and trees to prevent competition for nutrients, light, and moisture. Rose roots spread as far out as the plant is tall, and often much farther. Furthermore, hedges and trees may need to be sprayed with chemicals that are potentially harmful to roses.

Also keep roses away from the bases of buildings, at least 18 to 24 inches (45 to 60 cm) away from any foundation. This allows for good air circulation and reduces the chance of infestations by spider mites. Foundations may leach harmful chemicals into the soil, make soil more alkaline, and draw out available moisture. They also heat up quickly in spring which may cause premature sprouting in nearby rosebushes. Such new growth is easily damaged by late cold snaps.

Soil and Moisture

While many conditions contribute to healthy roses, the beginning point is always the quality of your soil.

Providing roses with the right kind of soil is the starting point to success with these lovely plants. Ideally, you want to provide them with good soil to a depth of 15 to 18 inches (37.5 to 45 cm). Following is critical information on good soil.

Purpose of Good Soil

Good soil has certain characteristics. It is firm enough to hold your plants in place, yet loose enough for easy penetration of water and oxygen to a plant's root system. Loose soil is particularly important to roses that need to be covered during the winter for protection. Good soil also drains freely while having the ability to retain moisture during drought and heat waves. Good soil locks in essential nutrients as well and makes them available to the plant over a long period of time. Good soil is alive, filled with billions of microorganisms. These microscopic creatures benefit the plant by providing and helping the rose take in nutrients. Good soil attracts worms, which tunnel through the ground to keep it loose and also fill it with nitrogen through their droppings (castings). If weeds or grasses are thriving in an area, you have some good soil, at least on the surface. On the other hand, soil contaminated by pollution, herbicides, oil, or salt is often bare, a sign that the soil should be replaced.

Composition of Good Soil

Good soil is composed of both inorganic and organic materials. *Inorganic* means that a material does not come from plants or animals. *Organic* means that a material comes from the decomposition of anything that was once alive, whether plant or animal.

The inorganic materials are clay, silt, and sand. Clay is made up of minuscule particles that cling together when wet. When wet clay dries, it turns almost as hard as rock and tends to crack apart. Clay is usually a light to grayish tan. It sticks to your shovel and is very hard to work. However, nutrients cling to clay. So, some clay in your soil is beneficial because it locks in nutrients. Silt is made up of larger particles than clay. When wet, it feels somewhat slippery. Sand is made up of very large particles. When wet, sand has a grainy feel. Water slides through sand

quickly. Having silt and sand in the soil helps keep it loose.

The organic material found in good soils is the result of the decomposition (rotting) of anything once alive. Everything alive eventually dies. When it does, it rots. Actually, it is being eaten by billions of different creatures, many of them microscopic. These creatures fall into different categories: some are plants (fungi), others are animals (insects), and some have characteristics of both plants and animals. Organic material is especially attractive to worms. When worms die, their nitrogen-rich bodies decompose to give plants even more valuable food. The wide variety of unseen creatures digest organic matter into a light brown, fluffy material called *humus*. The benefits of humus to the soil are incredible. It keeps the soil loose and airy, holds moisture during drought, contains essential nutrients, provides a home for helpful soil microorganisms (many of which help roses take in food), and maintains soil at just the right pH.

The term *pH* refers to the activity of hydrogen ions in the soil. In simple language, it is a gauge of how acidic or alkaline soil is. The pH scale runs from 0 (totally acidic) to 14 (totally alkaline). Neutral soil has a pH of 7. Roses do best in a slightly acidic soil with a pH of 6 to 6.5. The foliage of plants will often appear diseased or distorted if roses are grown in overly alkaline soil. One of the most common conditions is called *iron chlorosis,* which results in yellowing foliage with distinctly green veins (add citric acid or Sequestrene to soil). The right pH determines the availability of nutrients to plants. If the pH is too high or too low, many essential plant foods will be locked into the soil and not be absorbed by the roots. If you're adding lots of organic matter to the soil on a regular basis, all of the major and minor nutrients necessary for good rose growth will be taken in by the plant; so, don't worry about pH. Just keep adding organic matter to the soil, and everything will be fine.

Organic Matter as a Soil Amendment

The organic matter you add to soil to improve its texture (structure) is called a *soil amendment*. Leaves

Excellent Soil Amendments

Here are the four most commonly used soil amendments in cold climates.

Leaf mold

Peat moss

Rotted horse manure

Compost

are generally readily available and are excellent for this purpose. Let them rot in a thick pile until they form what is called *leaf mold*—nothing more than soggy, digested leaves. Mix this material into the soil each year. Add as much leaf mold as possible throughout the growing season. The term *mold* implies disease, which is not the case. Leaf mold is like gold in the garden. All leaves are excellent. Since oak are slightly acidic, they're preferred. But, roses are not *that* fussy.

Peat moss, sold in most nurseries in bags, bales, or bulk, is a superb soil amendment. It is disease and weed free, and it's also slightly acidic—perfect for roses. Since peat moss tends to shed water when dry, moisten it before use. Peat absorbs warm to hot water much more readily than cold water. Slash open bags or bales, and pour hot water onto the peat. If purchased in bulk, spray it with a hose. Stir and moisten it over a period of several days. You can also just expose peat to a few rains. Turn it a couple of times until it's dark rather than light brown. Work with it once it is evenly moist (but not soggy).

Another fine soil amendment is rotted horse or other manures. Animal manures sometimes harbor disease organisms or contain weed seed, but if they're available, they are worth the gamble to use.

Finally, compost is another option. Composting is more popular in rural than suburban areas, but it can be done discreetly in large containers or attractive bins. However, while compost is a wonderful soil amendment, there is rarely enough of it after the pile breaks down over a period of weeks, months, or years.

Think of soil as a living creature, one that needs to be fed and cared for to be healthy. Every year, you must add lots of organic matter to it. The inorganic substances break down very slowly, but the organic matter disappears rapidly, eaten by all of the billions of living creatures in the soil itself.

Loam

When inorganic and organic materials are mixed together in just the right proportion, the soil is called *loam*. About one-third of the soil should be organic matter. The rest is best if a combination of clay, silt, and sand in nearly equal parts. Naturally, no soil will meet these standards exactly, but when soil gets close to these proportions, you'll know it. Loam is usually very dark, almost black. This coloration is very good because it attracts sunlight in spring and warms up quickly. It also retains heat in the fall. Loam usually has a loose feel to it. Superb loam is so loose you can almost dig into it with your bare hand. It does not get compacted easily. This means that water drains quickly through it. Water gets to the roots easily but does not pool up around the roots, shutting off oxygen. It drains away, carrying with it toxic salts. The looseness of the soil induces rapid root growth. Roots spread out in all directions and form an intricate system to support the plant and to provide it with nutrients essential to vigorous growth. Often, the root system is much larger than the above-ground portion of the plant. Loam, when dry, crumbles in your hand. If it gets wet, it is mildly sticky but doesn't form a solid, sticky ball; it will still crumble through your fingers. Loam has all of the good characteristics of the individual components without the negative ones. For example, the small amount of clay retains nutrients in the soil but doesn't make the soil hard or compacted when dry.

Getting Good Soil into Your Garden

You may already have good soil where you intend to plant your roses. Most people don't, however. The first thing to do is to dig into the ground with a spade. Dig down at least 18 inches (45 cm). Usually, the top portion of the soil will be dark brown or black. This may be 2 to 3 inches (5 to 7.5 cm), a half foot (15 cm), or even a full foot (30 cm) if you're lucky.

In most areas you will find very little loam, or topsoil, because it is expensive. When builders construct homes, they often cover the surface of the soil with loam, also known as "black dirt." Usually, it's enough for a good lawn. Underneath the topsoil is often clay, sand, rock, or less desirable soil.

In most instances, the area where you'll be growing roses will be covered with lawn or weeds. You must kill or remove these. Roundup® is one of the

Composting Made Easy

COMPOST IS NOTHING MORE than rotted organic debris. Many home gardeners build simple structures to produce compost. Here are the basic tips:

- Enclosures should be at least 4 feet (120 cm) square.
- Organic debris breaks down best if material is exposed to air. Use wire or slats for the side of the bin.
- Anything organic will break down in time. However, if you shred it before putting it into the pile, it will break down faster.
- Kitchen scraps are excellent in compost piles, but cover them well if you add them to the pile to avoid attracting animals.
- Keep the pile evenly moist at all times.
- Add a little 10-10-10 fertilizer to the pile every few weeks. The added nitrogen helps break down the organic matter.
- Turn the pile with a pitchfork or spade to speed up decomposition. If you don't have the energy, just let the material decompose slowly.
- Expect the pile to shrink to a small fraction of its original size. If you let the material decompose long enough, it will turn into a light brown, fluffy material called humus.

most effective herbicides; it will kill lawn and most types of weeds. Ideally, do this in late summer. The vegetation will die and turn brown. If you're an organic gardener, remove the lawn and dig up the weeds. One of the easiest ways to remove lawn is in stages with a pickax. Shake off all the topsoil, place the sod in a wheelbarrow, and compost it in another area.

Once you've removed or killed all weeds and lawn, dig up the good topsoil (black dirt or loam) and place it on a tarp or in a wheelbarrow to the side of the garden area. Naturally, if your soil is good to a depth of 18 inches (45 cm), you don't have to remove it; simply loosen it by digging or rototilling it. Whenever using a rototiller, stir up just the topsoil. If you have a limited amount of topsoil, you can just set it off to the side of the planting hole or flower bed. **Note:** Never work in the soil when it is wet, just when it is damp at most. Working wet soil will cause it to compact.

Dig out the poor soil underneath the topsoil to a depth of 18 inches (45 cm). Cart this poor soil off to another area of your property. This can be back-breaking work. If you're making a large garden in an accessible location, hire a sled steer (commonly referred to as a *bobcat*) to do this for you. A bobcat can do a job like this easily.

Then place the topsoil back in the garden. Your problem is now obvious: you haven't filled the hole. You may have to buy soil from a garden center or nursery. If you've rented a bobcat, have the appropriate amount of soil delivered ahead of time. The operator can dig the hole and fill it with good soil at the same time. Mix in lots of organic matter, so that it makes up about one-third of the soil.

You'll buy garden soil by the cubic yard (cubic meter). The price varies by quality. You want peat-based loam. It should come from a source not contaminated by herbicides or chemicals. Reputable garden centers will sell good garden soil, but a few try to get rid of problem soil by selling it to unsuspecting customers. Check the soil before buying it. It should be black and have a nice, loose feel to it. If you have a small garden or are just digging a few planting holes, you may choose instead to buy potting soil. The latter is often sold in large bags. Good potting soil contains loam, peat, and perlite. The value of loam has already been explained. Peat is an organic substance found in bogs that is slightly acidic and moisture retentive. It is also disease, insect, and weed free. Perlite looks like little, white, round balls mixed into the potting soil. It acts like a spacer in the soil to keep it loose and airy.

Raised-Bed Gardens

Even if your soil is solid clay, rock hard, or literally made of rock, you can still have a garden. The type of garden you'll be constructing is called a raised-bed garden. Mounding soil or enclosing soil with stones, timbers (no copper or creosote), or rot-resistant wood is a method that works fine for roses. Ideally, you'll want 18 inches (45 cm) of soil above the hard surface to work with. Some people get by with less. You'll have to buy the soil and have it placed where you want it. Growing plants in mounded soil has been done for thousands of years in China and for centuries in France. Again, when making a raised-bed garden, mix in lots of organic matter. The big advantage of a raised bed is that it warms up quickly in spring; its major disadvantage is that it tends to dry out faster in hot or windy weather.

Note: If you're extremely patient, you can make most soils (not rock) into loam over a period of years. Clay is one of the worst soils to work with. Some writers suggest adding sand to it. *Do not!* Instead, add lots of organic matter. Also spread gypsum (calcium sulphate) at the rate of 1 pound (454 grams) per 100 square feet (10 square meters) yearly. This helps break down clay. But the key is adding organic matter consistently. You *will* eventually get good soil, but most gardeners don't have that much patience.

Sandy soil can be a problem, since water drains through it quickly, carrying valuable nutrients away from a plant's root zone. Again, some writers suggest adding clay to it. And again, *do not!* Add lots of organic matter over a period of years, and your sand

Starting a Rose Garden from Scratch

No matter where you make your garden, you must kill all perennial weeds with Roundup®. Otherwise, they will spring up again from tiny portions of root left in the ground. **Tip:** Although the label recommends killing weeds when they are mature and close to flowering, you can often do this early in the season with just as much success and far less herbicide.

Remove trees that might be in or shade the bed. Dig up all roots. Dig around the base, exposing the roots, and cut them with loppers. For larger trees, have stumps removed professionally with a machine that chips the stump into usable mulch. If the area is inaccessible, you'll have to dig out the stump and cut roots by hand or with a chainsaw.

Dig into your soil with a spade to see how deep the topsoil is. If you have very little topsoil, you may want to make a raised bed. If you have quite a bit of topsoil, *loosen the topsoil only* with a rototiller or by hand with a spading fork. As you loosen the topsoil, you'll notice that it takes up much more space than when compacted.

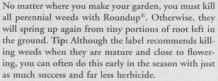

In this instance, we've decided to make a raised bed. The topsoil has already been loosened with a rototiller. When adding soil, work through it with your hands and remove any pieces of root, rock, or debris. Now start filling in the area with good soil. You want a total of 15 to 18 inches (37.5 to 45 cm) of good soil, which includes the loosened topsoil underneath the mounded area. Keep beds about 5 to 6 feet (150 to 180 cm) wide for easy access from both sides.

Add lots of organic matter to the soil. Peat moss is one of the best soil amendments and readily available in large bales, but compost, rotted manures, and leaf mold also are all highly recommended. Grass clippings, more commonly used as a mulch, are fine if mixed into the soil in the fall. A combination of these is ideal. The organic matter begins to break down right away. By spring this will be an ideal bed for planting bare root or potted roses. You want your soil to be easy to work, especially if you plan to bury (tip) your roses in the fall.

You can enclose your fresh beds with rocks or timbers, or you can just leave them open as illustrated. Smooth the upper surface with a garden rake, and slant the sides slightly. You can place any kind of mulch, such as leaves, on the slanted portion to stop soil from running off in rain. The wood chips at the base of the bed in this case make an attractive pathway between the beds.

will become good garden soil. Since sand doesn't compact the way clay does, it's an easier soil to work with from the start. As when working with clay, getting good soil will take years.

An Easier (but Inferior) Method

Not everyone has the energy, time, or money to create a garden as we outlined. As with everything, compromises are possible. Nevertheless, gardening does require some energy, some time, and some money—no matter what.

An easier alternative is to dig a hole at least 18 inches (45 cm) deep and 24 inches (60 cm) wide for each plant. Remove all grass, weeds, rocks, and debris from the soil. Save only the topsoil, adding soil purchased in bulk or in bags as necessary to get enough to fill the hole. Mix in the appropriate amount of organic matter before planting the rose in the prepared hole.

A drawback is that in heavy clay or rock, such a hole will gather water which may kill plants. In soils of this type, we therefore suggest building a raised bed instead. But in most soils this method works fine.

Making Your Own Potting Soil

You can make your own potting soil by mixing equal parts of loam, peat, and perlite. You can also buy it in bulk from garden centers. To make, place equal amounts of loam, peat, and perlite in a wheelbarrow, and mix these together with your hands or with a spade. While doing this, remove any bits of debris. You now have the ideal soil for planting individual roses.

Special Considerations

It is best not to work in soil when it is wet. If you do, it may get compacted. It is also best not to work the soil too often, as any farmer will tell you. When we say *work* the soil, we're talking about deep digging, not just scratching the surface to remove weeds or mix in fertilizer.

Once roses have been planted, try to avoid walking in the rose beds. Compacted soil stops oxygen from getting to the feeder roots. If the leaves of your plants have yellow veins, this indicates a lack of oxygen to the plant. The most common cause is compacted soil. Keep it loose and airy.

Of course, you will have to walk on soil to prune, remove spent blossoms, and keep the bed clean. After you do, loosen the soil with a pronged cultivator for good aeration. All cultivation should be just on the surface to avoid damage to shallow root systems. A summer mulch also helps avoid soil compaction.

Soil Poisoning

Soil poisoning is a highly disputed concept. Regardless, if you are planting a rose where another rose has grown recently, our recommendation is to remove all soil from the planting hole and replace it with fresh soil. Old, partially decomposed rose roots may give off a toxic substance. This is believed by some to harm the feeder roots of new roses. If the old rose roots have decomposed completely, toxicity is not a problem. Since it is hard to judge the stage of decomposition, you should replace soil on a routine basis to avoid soil poisoning. Some highly respected professional growers insist that soil poisoning is a myth. Since soil reverts to its original state in roughly three years, adding fresh soil to the planting hole is beneficial no matter what your belief.

Plants for Poor Soils

Some roses do well in poor soils without many nutrients. Most Species (wild) and some Shrub roses are good choices if you can't improve the soil as outlined in this section.

Salt Tolerance

A few roses tolerate salt (from the seashore or when used on roads in the winter). These include the *Rosa rugosas* (one group of Shrub roses) and *Rosa spinosissimas. Rosa spinosissima* is listed in Part I under Species Roses while its hybrids are listed in the Shrub Roses section. Other specific roses that are salt tolerant are pointed out in the Varieties tables for Old Garden Roses and Shrub Roses. Spreading gypsum around roses may help remove salt from the soil.

Moisture

Since roses need good drainage, you should avoid planting in low-lying or swampy areas. Most roses will die in such places. There are exceptions, but none are included in this guide because they are inferior plants. Even the so-called 'Swamp Rose' does well outside of bogs. This particular Species rose is not included in the guide.

Spacing

Roses are susceptible to many diseases, some of which are caused by poor air circulation or high heat. Air must flow freely around the plants. So, give roses plenty of space. Never crowd them. Proper spacing reduces the need for pruning and also makes it much easier to winter protect roses requiring the Minnesota Tip Method. Exact spacing suggestions are included in each section in Part I.

A number of rose growers do plant their roses close together and get good results, but these growers are meticulous in using sprays to prevent disease and insect infestations. Close growing has the advantage of shading out weeds and making it easier to spray and fertilize. Plants close together also shade the ground to keep it cool and moist. They offer wind protection as well. However, if you're opposed to the use of chemical sprays, spacing plants farther apart is recommended.

The effects of spacing are controversial. The debate about proper spacing has always gone on and will continue. This book gives you advice that has worked for decades. We suggest you let our experience in cold climates be your guide.

Designing with Roses

Designing with roses is far less complex than with perennials. The reason for this is that roses demand lots of sunlight. Thus, you can put them only in certain places. Nevertheless, as long as it's sunny, you can use them as hedges, in island beds, in perennial and rock gardens, or as isolated accent or specimen plants. A few spread out widely and make excellent ground covers on sunny slopes or over extended open spaces. Many roses also grow well in pots or containers, which you can move at will to add color to different places throughout the summer. Or, they can be placed against walls or supported vertically as Climbers if they grow in that manner.

The use of the plant will determine your design. Still, bear in mind that roses tend to be rather informal plants. For instance, if you want a tidy hedge, use a different plant. A rose hedge looks casual and tends

A delightful combination of 'Love' and perennials

to have petals falling on the ground, but it has a charm all its own.

Considerations to keep in mind include the type of rose you're growing (e.g., bush or Climber), whether or not it will have to be buried in winter for protection, the potential height and width of the plant, the color and form of blossoms, fragrance (especially if close to an open window or pathway), and purpose for the rose (varying from a bold statement to a tiny pot on a ledge).

For mass effect, plant a number of roses of the same variety together. This makes a bold statement and is extremely attractive.

Climbers or large shrubs covered in clusters of blooms soften horizontal lines, such as those of a split-rail fence. They often stand out from a distance and are good accent plants.

Light colors, particularly white, are most visible at night. The soft reflection of moonlight is just enough to enhance this subtle effect.

Bold colors often clash. In larger gardens this is usually inevitable, as it also is in cutting gardens where a number of varieties are grown for different reasons—some for their long stems, others for exquisite scent.

As with perennials, roses look best in groups of three to five of the same variety. Another advantage of planting this number is that you'll have a better chance of finding out whether this particular variety does well in your area. If you buy just one plant, you may be unlucky and get one propagated from a bad bud (see the discussion of selective budding in Chapter 7, p. 217).

Tree roses are quite formal and can be kept highly symmetrical with attentive pruning. They are an exception to the general rule that roses have a casual feel, although some growers let Tree roses do what they will and end up with free-flowing forms. Even though such Tree roses are casual, they still are quite formal compared with other roses.

Potted roses are charming on patios, on decks, or in sunny corners where just a splash of color adds zest to your exterior design. All they need is sun and constant attention to proper watering to bloom abundantly throughout the summer.

CHAPTER 4

PLANTING ROSES

This chapter will help you select healthy plants and get them off to a good start by following the correct planting steps. Many of these steps are simple, but not well known. If you follow them you will be surprised how much difference they make. Your plants will be much healthier and produce far more bloom.

Buying Bare Root Plants

If you do a lot of rose growing, you'll eventually be planting bare root roses.

A bare root rose is a dormant plant sold without any soil around its roots. Some bare root roses are packaged in plastic and sealed in cardboard boxes for sale in retail stores. Others are sent through the mail and may have far less packaging. Following are tips related to buying this kind of rose.

The Name

In catalogs, roses are often listed with a name and then a code name behind it in extremely small print. For instance, "'Snow Owl' var. UHLensch." The code name tells you the first three letters of the name of the company or the person who created or found the rose (in this case, UHL stands for a German hybridizer named Uhl). The lower-case letters in the code stand for the name given the rose by the company or per-

son who created it ("ensch" is the letter code for 'White Pavement'). However, distributors often change names in an effort to sell a plant that didn't sell well under its original name or was an inferior plant to begin with. It would be helpful to the consumer if catalogs would be upfront about this: "'Snow Owl,' formerly sold as 'White Pavement'" would be a lot easier for us to understand. But, catalogs rarely do this in an effort to disguise the original name so as to increase sales. Fortunately, the *Combined Rose List* addresses this problem by giving you the various names by which an identical rose may be sold (see Chapter 1, page 4).

Canes

Bare root roses should have healthy canes. Usually, healthy canes have a greenish, maroon, or light tan tinge. If you bend an individual cane slightly, it's firm and pliable. If you cut into it, the inside looks whitish green.

Unhealthy cane looks brown, black, or gray. If you bend it, it often breaks because it's so dry and

A healthy bare root rose

brittle. If you cut into it, it's brown because it's dead. It may also have a wrinkled appearance caused by dehydration. These canes will not grow.

Ideally, canes of bare root roses should be dormant, showing no white growth from the buds. Growth before planting may reduce the rose's overall vigor in its first year of growth. Unfortunately, it's common for cane to have sprouting buds. If these buds are long and white, break them off with your finger; if they are barely beginning to grow, just let them be. Longer buds generally just dry up and die anyway. New buds will form along the stem, so it's not a problem. Still, our advice is to buy plants that are not yet sprouting. Also, ask mail-order houses to ship your plants at the appropriate time, and ask that they be completely dormant when shipped.

To be considered a quality plant, each type of rose should have a specific number of canes. The number and size of canes vary by group. Information on this is included in each section in Part I. For example, a fine Hybrid Tea will generally have three thick, strong stems, while a fine Polyantha might have 16 thin, spindly canes.

Tip: Companies may sell young plants as *liners* or *bands*. Liners are young bare-root plants consisting of a single stem with roots attached. Bands are rooted cuttings grown and shipped in small pots. Young plants that are budded (not grown on their own roots) are known as *maidens* if they are in their first year of growth. These young plants may be sold at a reduced price.

Remove sprouted growth on bare root roses.

Roots

Healthy bare-root plants have firm, pliable roots that are not broken, diseased, or mashed. These roots should be protected well from drying by being wrapped in damp packing material—usually shredded newspaper or some similar substance. Some companies sell plants without wrapping the roots. In fact, more and more companies are doing this as a way of saving time and money. It is a poor practice.

Never expose the roots to sun or wind. As soon as you get these plants, get the roots into water for at least 12 hours, but never more than 24, or they may begin to rot or suffocate.

If some of the roots are broken, snip them off to a healthy section. Snipping off the ends of roots does not hurt the plant. In fact, it may stimulate growth.

Check the roots and canes carefully for signs of disease. The most common problem is the presence of white mold. Simply wash it off and disinfect the plant as outlined on the next page.

Return any bare root rose with galls (swollen areas that look like little balls) on its roots or canes, since these are signs of a disease over which you have little control. In fact, if you plant these, they will infect your soil.

Most bare root plants are sold free of disease. When in doubt, call the supplier for advice. Don't take chances with diseased plants, since they can infect your soil and cause problems for years to come.

Bare root plants sold in stores should have the same characteristics as those sold through mail order. Call ahead and ask when bare root plants are coming in. Then buy them *as soon as they arrive*, since they will begin to sprout in heated stores. Most plants sold in this way are enclosed in plastic. Remove them quickly from packaging, and pot or plant them as outlined later in this section.

When to Buy

Most people buy bare root roses in the spring. At this time of year, they are available in a wide number of stores as well as through the mail. However, a few rose growers buy bare root roses in the fall, for two main reasons: you have a better chance of getting a

Snip off broken, crimped, or damaged roots.

rare or newly introduced rose in the fall, and the roses are usually extremely high-quality plants, since the supplier can ship you the best of that year's crop just after they've been dug. What to do with these roses is covered later under "Planting Bare Root Roses in the Fall."

Preparing Bare Root Roses for Planting

This chapter includes step-by-step instructions on two methods of planting bare root roses, either in pots or directly in the garden. The following steps apply to either method you choose to plant your bare root roses in spring.

As soon as you get bare root roses, through the mail or from a retail store, remove them from the packaging. This packaging may be next to nothing, a plastic bag in a cardboard container or something similar.

Inspect each plant as outlined earlier to make sure that it is healthy and disease free.

Keep the plant out of direct sun.

If there is a metal tag around the stem, remove it. The wire will restrict growth and can kill the cane. Identify the plant in some other way, as with a metal, plastic, or wooden marker at its base.

If any portion of the stem is broken or brittle (dry), cut it back to a healthy section of cane. Cut back to a spot ¼ inch (6 mm) above a growth bud, a tiny bump on the cane.

If the plant has more than three canes and one of them is extremely large, cut it off. Contrary to what you might think, the large cane is unlikely to produce good bloom. If the plant has three canes or fewer and one of them is extremely large, do not cut off the larger cane until the following year. During the first year, it will help the young plant produce food to create new growth from the base of the plant. Whenever possible, avoid buying plants with this type of cane. Sometimes, you'll get stuck with a plant like this through mail order.

Plants sold in stores often are tied together tightly with string to take up less space. In most cases this is not a problem, but, occasionally, a cane can be crimped at its base. If the plant has several healthy canes and one that's crimped, go ahead and buy it. However, cut off the crimped cane at the base, since it will not do well.

If any roots are broken or crushed, cut them off to a firm section with pruning shears.

Many bare root plants sold in large retail outlets have canes covered in wax. This will inhibit new growth, especially in the second year. Cut all canes back to 1 to 2 inches (2.5 to 5 cm). It's counter-intuitive—but do it.

Soak the entire plant overnight in water, with the roots down and the cane up. This softens the cane and gets water into the root system which may have dried out somewhat in shipping. The easiest way to do this is in a plastic garbage can. You can soak a number of plants in one can. Although plants tend to float, all of the roots and most of the stem will be submerged. Soaking them overnight is long enough. You can soak them for up to 24 hours, but not longer, or they may begin to rot.

Just before planting, dip the entire plant in a solution of 5 gallons (19 liters) water and 1 cup (¼ liter) bleach. Do this for about 30 seconds. This kills many disease-causing organisms, especially those related to downy mildew and gall.

If you cannot plant roses right away, remove them from any packaging, and place them in a trench. If the ground is frozen or hard to dig, place the bare root plants horizontally on the ground and cover them completely with moist potting soil. If the ground is covered with snow, first shovel it off and pour hot water on it until the ground is completely exposed. Cover the plants completely with moist peat or purchased potting soil. Then shovel snow over the mound. If you have only a few plants, place them in a large container (garbage can, box, black plastic bag) filled with potting soil, sterile sand, or peat moss in your garage. Cover the plants completely. The protecting material should be moist. This method of burying bare root plants briefly until they can be planted is called *heeling in*. Get the roses into pots or directly

Preparing Bare Root Roses for Planting

BARE ROOT PLANTS NEED special preparation to do well whether planted in pots or directly in the garden. Here are the basic steps.

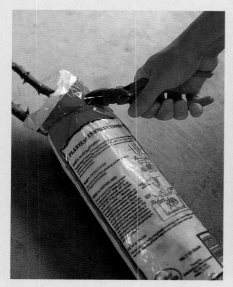

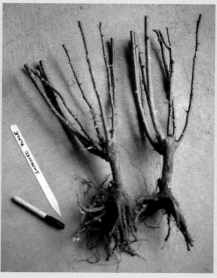

There are many types of packaging. Remove any boxes, plastic, paper, and metal binders or name tags as soon as the plant arrives in the mail or shortly after purchasing it at a store.

Most bare root roses are shipped with some sort of moist packing material around their roots. Hold the rose over a bag, and gently remove the material until the roots are fully exposed. Toss the packing material into the compost pile.

Label the plant in any way that suits you. Using permanent ink on plastic markers is an easy way to identify plants in the first year.

Remove any dead cane from the plant with sharp pruning shears. This includes the tips of the cane which may be brittle and dry. Cut back to live cane just above a bud. If the rose is covered in wax, cut all stems back to 1 to 2 inches (2.5 to 5 cm). A waxed plant may live through the first year, but often dies ("suffocates") in the second year.

Cut off damaged, crimped, or oversized cane as outlined in the text. Also, snip off damaged roots.

Place your bare root roses with their roots down in a large container filled with water. Although the plants tend to float, you can get almost the entire plant submerged. You can place a number of these plants together for simultaneous soaking.

After soaking the plants overnight, dip the roots and cane into a chlorine solution as explained in the chapter. Do this just before planting to kill off disease organisms.

into the garden as soon as possible after following this procedure. Heeling in is strictly a temporary measure which may be dictated by extremely cold weather outdoors.

Although many rose books suggest placing roses in the crisper of your refrigerator if they arrive too early, this is not practical unless you have an empty refrigerator set aside for this purpose. Some serious growers do. The temperature should be set at between 32 and 36°F (0 and 2.2°C). Remove the roses from all packaging. Clean them well, getting rid of any mold or damaged cane and roots immediately. Mist them daily to keep them moist. Plant them as soon as possible.

Planting Bare Root Roses in Pots

Some people prefer to grow roses in pots temporarily or permanently.

There are several advantages to growing roses in plastic pots on a short-term basis (1 year). You may get your roses way too early to plant them outdoors; you may not know where outdoors you will plant the rose; you may not have enough space for the rose at the moment; or you may believe, as some growers do, that growing a rose in a pot for a year gets it off to a better start than planting it directly in the garden.

Other growers simply prefer growing roses in pots. Miniatures, for example, do extremely well in pots. So do Tree roses. The advantage of potted roses is that they can be moved around at will to create whatever effect you might want. They can be buried with other plants in the fall for winter protection or, in limited quantities, moved into a garage and protected there. You can even transfer Miniatures indoors for the winter.

Jerry Olson's Method

The following method of planting bare root roses in pots was developed by Jerry Olson over a period of years. He has had tremendous success with it but emphasizes that many other methods also do well. Here are the steps:

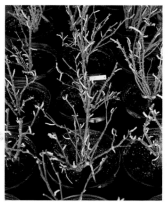

Newly potted bare root plants in early stages of growth

If you intend to remove the rose from the pot after a year or at the end of the season, you may want to first place a perforated plastic bag in the pot. The size and type of pot you use will vary with the type of rose grown. (Pot sizes are given in individual sections in Part I.) The plastic bag must have holes in the bottom for good drainage. The kind of plastic bags with handles on them used in most grocery stores work well. The only reason to do this is to be able to lift the rose out of the pot when you're ready to plant it directly in the garden or bury it for winter protection.

Pour 3 to 4 inches (7.5 to 10 cm) of potting soil into the bottom of the plastic pot, with or without the plastic bag in it.

Set the bare root plant in the pot with the roots spread out as best you can. Dump Pro-Mix® or a similar product around and into the gaps in the root system. Firm the mixture in place with your fingers. Cover the entire root system. Each class of rose must be planted at a slightly different depth, as indicated in the appropriate listings in Part I.

Sprinkle ½ cup (115 g) of Milorganite on top of the Pro-Mix. Avoid the use of any chemical fertilizers; these can damage the tender feeder roots.

Fill in the rest of the pot with potting soil. Press down on the soil with your fingers. This removes any air pockets and forces the Pro-Mix® and soil into contact with the plant. The soil should end up an inch or 2 (2.5 to 5 cm) from the top of the pot.

Fill the pot to the brim with water. Let the water soak in. Fill it as many times as necessary to get water draining slowly out of the bottom drain holes. Water again the following day so that the Pro-Mix® and soil are thoroughly saturated.

Keep the plant in your garage. Keep the canes moist by misting them. They will bud out over a period of days. Keep the soil moist but not soggy at all times. Check soil moisture by feel, not by sight. Just push your finger into the soil to feel whether it's drying out. Water as necessary.

Once the plant begins to form leaves, you don't have to mist it anymore. Just keep the soil in the pot moist.

Once the plant has budded out or even formed leaves, place it outside after all danger of frost. Place the rose first in partial shade. Gradually increase the amount of sun over 10 to 14 days until the plant is in full sun. This is called *hardening off,* a gardening term for allowing the plant to get used to increased light, varying temperatures, and drying winds. If you move the plant directly into sunlight, leaves often turn pale or gray and may die from a condition called sunscald. The plant will form new leaves, but your impatience will have slowed its growth.

Some growers spray these young plants with an antitranspirant during the hardening off period. These sprays are available in most nurseries. They coat leaves and stems with a substance that stops the plant from drying out. They have an added advantage of coating insect eggs and disease spores so that they cannot hatch or grow.

During the growing season, care for the plant as recommended in the individual sections of Part I which include detailed information on water and fertilizer, spray programs to prevent insects and disease, and exact recommendations for pruning.

At the end of the season, protect the rose as suggested. Mist it with dormant spray. You have to bury it, sometimes pot and all, in the rose bed alongside other roses being protected with the Minnesota Tip Method covered in Chapter 5.

The following spring, raise the plant up at the appropriate time. Plant it directly in the garden as you would a potted rose purchased at a garden center or nursery. Steps on how to do this are outlined later in the chapter. Or, if you prefer to raise roses in pots alone, you can pot it once again, adding some fresh potting soil, or pot it up to a larger pot.

Planting Bare Root Roses Directly in the Garden

Plant bare root plants as soon as the ground can be worked in spring—much earlier than you would potted plants that already have leafed out. If possible, plant on an overcast day or in late afternoon or early evening. This keeps the plant from drying out in high heat or hot sun. You want the canes to stay moist.

In extremely cold springs, it may be necessary to plant bare root roses in pots just to keep them healthy. You do not want to plant dormant roses when there is snow on the ground or when the ground is still frozen. You could also heel them in until the ground is ready for planting as outlined earlier.

Dig a hole 18 inches (45 cm) deep and 24 inches (60 cm) wide. Place the soil on a tarp to the side of the planting hole. If the soil is poor, replace it with potting soil sold at most nurseries in bags or by the cubic yard (cubic meter).

Mix 3 tablespoons (40 g) of superphosphate or a cup (about 225 g) of bonemeal into the soil at the base of the hole. Cover this with an inch or 2 (2.5 to 5 cm) of soil. Then pour some more potting soil into the hole, forming a small mound. Spread the roots of the plant out over the mound of soil. Check the level of the bud union, if there is one. Place it at the correct depth as specified in the guide for each category of rose. Spread out the roots in all directions. Never wind them around or cramp them into a tight space. If they are longer than the hole, widen the hole.

Some growers lay the roots out in one direction. This is common in Great Britain but rarely done in the United States. However, some U.S. growers are beginning to follow the British method because they believe that it makes tipping and burying roses in the fall easier.

Cover the roots with Pro-Mix® or similar light, fluffy material. Firm the light, fluffy material with your fingers.

Sprinkle a handful of Milorganite over the Pro-Mix®. Never add inorganic fertilizer which burns tender feeder roots.

Add good garden or potting soil on top of the Pro-Mix® to fill in the hole. Firm it with your hands. *Do not use your feet.* Soil should be firm but not heavily compacted.

Next, saturate the soil with water. Soak the soil for several minutes until water stands on the surface. Let the water soak in. Often the soil will settle at this stage; add more soil if necessary, and firm again. Water once more.

Traditionally, rose growers now mound the entire plant with soil. Cover all the canes completely. You may have to buy additional potting soil or bring in loam from another part of the garden to do this.

Mist the soil with water until it's evenly moist. Spraying it too hard will cause the soil to fall off the cane.

Keep the soil moist every day with a fine mist of water. It may take several days or longer for the growth buds on the plant to begin to sprout under the soil. The new growths will crack the soil open. When this happens, remove the soil by spraying it lightly with water over a period of days, not all at once. This allows the uncovered buds to get acclimated to sunlight and drying winds.

When new growth is exposed in many places, spray off all the soil on a cool, overcast day. Wash the plant down to remove any trace of soil. Some growers leave a shallow basin around the base of the plant to collect water in the plant's initial stages of growth. Doing so is strictly optional.

Keep the plant consistently moist. Mist it twice a day if necessary. Do this until the plant has leafed out fully.

Variations

A variation of this method is used by some growers: After saturating, mound all but the top 6 inches (15 cm) of the plant with soil. Mist the soil and then envelop it with black tar paper. Spray the exposed

cane with Wilt-Pruf® or a similar product to keep the cane from drying out. Keep the soil consistently moist, and check on growth every day. Once buds begin to break through the soil, remove the tar paper immediately. The advantage is that the tar paper traps moisture and keeps the soil in place, even in heavy rains.

The first method, however, is most highly recommended for the average gardener. The alternative just described requires more care because of potential heat buildup which can kill buds. It's also more expensive and more work.

The newest and perhaps easiest method of getting buds to grow (break) from dormant cane is becoming extremely popular among experienced growers. They have found that in removing soil from mounded plants newly formed buds may break off by accident. Instead of covering plants with soil, use moist leaves. You then cover the plant and leaves with a 5-gallon (19-liter) plastic pail pushed firmly into the ground. Place a rock or something else that's heavy on top to keep the pail in place. This creates a "sweatbox" with high humidity inside. Growers using this method say that it causes buds to grow within a few days, with less chance of breaking buds than with the traditional method. The moist leaves work well as long as you're careful to remove them gently from the cane. Wear gloves! And be sure to collect plenty of large pails in the winter so that you'll be prepared if you plan to plant a number of bare root roses in spring (see p. 229).

Planting Bare Root Roses in the Fall

This heading is a little bit misleading. When you order roses for fall delivery through a mail-order distributor, you're not actually going to plant the rose in the fall. You're just going to protect it over the winter and then plant it as a bare root rose the following spring. Here's what you do:

Far in advance of when the plants are likely to arrive, dig a hole wide and deep enough to accommodate the number of plants you've ordered.

Place a plastic bag or several bags filled with leaves in the hole, and cover the hole with plywood. Cover the plywood with the soil removed from the hole. Then cover the soil with a tarp. This will often stop the soil from freezing hard while you wait for your plants to arrive.

When your plants arrive, mist them, then wrap them in newspaper which you can hold in place with polyester twine. Moisten the paper. You don't have to wrap plants individually; wrap in bunches of five or so if you've ordered a number of plants. Use enough paper—perhaps a full section—around each bunch so that the plants are not breaking through.

Remove the tarp over the hole. The weather may be extremely cold, or it may even have snowed, by the time you do this. Shake any snow off the tarp. Now transfer the soil from the hole onto the tarp with a flat shovel. Remove the plywood and the bagged leaves.

Place the plants, still wrapped in moist newspaper, in the bottom of the hole. Also put poison bait in the hole to ward off rodents. Then open the bags and pour in the whole, loose leaves, filling up the entire hole. Cover the hole again with the plywood. Then cover the plywood with soil, preferably no less than 6 inches (15 cm) deep. This is like creating an outdoor root cellar.

In early spring, reverse the process. Get the plants out of the ground before they begin to sprout. Prepare and plant them as outlined earlier for bare root plants.

Planting Bare Root Tree Roses

You can plant Tree roses (standards) either in pots or directly in the garden as you would other bare root bush roses. The procedure for preparation and planting is the same as outlined in this chapter, except you must follow these extra steps carefully:

Place a stake next to the Tree rose as you plant it, whether in a pot or directly in the garden. Many growers place stakes up to the point where branches shoot out from the stem (the bud union). This is

wrong. The stake should be higher than the top canes (head or crown) to allow for additional growth. A good rule of thumb is to place the stake so that its top is 12 to 24 inches (30 to 60 cm) above the bud union, depending on the potential size of the upper branches. The top branches often must be attached to that higher portion of stake with polyester twine to protect the plant from storm damage. This advice can save your plant.

Use electrical tape to tie the plant to the stake. Secure the plant just under the upper cane, just below the bud union, and 3 to 4 inches (7.5 to 10 cm) from the base of the plant. You do not want the plant to jiggle or tip in heavy wind. This can damage the feeder roots. Electrical conduit cut to an appropriate length is strong, light, and easy to work with. You can find it in most hardware stores. Plastic stakes also work well. If you use materials that can rot (such as wood stakes), you'll have to replace them regularly. Metal or plastic supports are preferred.

An alternative method is to envelop the stem at the points where you want to secure it to the stake with a foam pad, then wrap tape around it or tighten it in place with a plastic tie, the kind you slip through a hole and tighten by pulling. You can buy this type of insulation at any hardware store. It's commonly sold as pipe insulation or just called foam rubber.

Place sphagnum moss (the long, stringy moss sold in nurseries to line baskets) in the area between the upper canes. If the plants have arrived with long white sprouts, knock these off with your finger. They'll just dry up and die anyway. If canes are spread out, pull them together gently and tie them into place with polyester twine. Fill all the space between the canes with moss. Moisten it with water. If you don't want to buy sphagnum moss, use wet leaves instead. **A word of caution:** Working with sphagnum moss has been linked to a somewhat rare fungal disease (*Cutaneous sporotrichosis*). Wear gloves and a long-sleeved shirt when working with it. You do not want to let this moss touch any open sore. Note, however, that there is no disease linked to peat moss—the loose, not stringy, moss sold in large bales or smaller bags as an organic soil amendment.

Cover the upper canes with a piece of burlap.

Planting Bare Root Roses in the Garden

AFTER PREPARING BARE ROOT roses for planting, most growers plant them immediately in the garden.

Dig a hole much wider and deeper than you would think. Rose roots spread out and are often more extensive than the upper portion of the plant. Loosen the bottom of the hole with a spade, but don't step into the hole once you've done this.

Place the recommended amount of superphosphate or bonemeal in the bottom of the hole. Mix it into the loose soil at the base of the hole. Phosphorus moves poorly through soil and must be placed underneath the root system for maximum growth; superphosphate, which contains phosphorus, has proven more effective. Now nearly fill the hole with potting soil.

Once the hole is nearly full, set the plant in place over the potting soil. Check to see that the bud union, if there is one, is at the correct level as indicated in the appropriate section in Part I. Add or remove soil to get it at the right level. Spread the roots out evenly in all directions.

Cover the roots with potting soil or Pro-Mix®, and gently firm it in place. Then sprinkle Milorganite over the mixture. This mild fertilizer will give the rose a quick boost for good growth.

Now add more soil over the Milorganite, firm it in place with your fingers, and then saturate the soil thoroughly. Let the water soak into the ground before saturating the soil a second time. You want to get the soil around the roots completely wet.

Traditionally, rose growers now bury the exposed cane with soil. Moisten this with a gentle mist from a hose. This moist soil helps the cane produce buds. As they begin to grow, or break, the buds crack the soil open. As this happens, wash the soil off in stages over a period of days.

Planting Bare Root Roses in the Garden (continued)

Other growers take a little more time and go to greater expense to keep the soil in place. They surround it with tar paper or a similar product. This keeps the soil from washing off in a heavy rain and creates both moisture and high heat in the enclosed soil. The top of the cane is often left exposed. Many growers mist the exposed cane with Wilt-Pruf®.

One of the newer methods of getting canes to break is the use of a "sweatbox." Although you cannot see it in the photo, the bare root rose has been covered with moist leaves before being placed under a black pot with a stone on top to hold it in place. This is a very simple and effective way to get canes to produce buds quickly. However, you should prepare well in advance by gathering appropriate containers of varying sizes during the winter if you intend to start a number of roses in this way.

Some people ignore all of the foregoing methods for protecting canes and simply spray exposed canes with Wilt-Pruf®. This product works only when exposed to sun. It stops cane from drying out. Even if you choose this shortcut, we still recommend that you mist canes frequently and keep the soil around the base of the plant consistently moist for good results.

Burlap is sold in many discount lumber stores, garden centers, and county cooperatives. If you know someone raising animals, ask if you may have the empty feed bags, and cut these up for burlap for free. The upper portion of the burlap will rest on the metal support. Tie the lower part loosely around the stem. Keep the fabric moist at all times by misting it with water whenever it starts to dry out.

Experienced growers use plastic instead of burlap. Usually, they puncture the plastic in several spots to avoid heat buildup. The plastic works well and keeps humidity high, but it requires constant attention to avoid burning the tender buds inside, which could mean opening the plastic every day or even more often on hot days.

Water the base of the plant as you would any other growing rose. Once the roots take, they will send nutrients up to the canes, where buds will develop and begin to form shoots.

Loosen the covering every other day to check on bud growth. When buds begin to appear, remove the burlap or plastic entirely. Untie the top canes. Remove the sphagnum moss (or moist leaves) by gently pulling it out with your fingers (wear leather gloves). Some moss or bits of leaves will stick to the thorns on the canes. Remove these by gently spraying the canes with water.

In the following days, keep the canes and main stem (standard) consistently moist with frequent mistings. It is just as important to keep the soil consis-

Tree Roses

BARE ROOT TREE ROSES require special attention, since they can dry out quickly or snap off in strong gusts of wind.

You need to place a stake next to a Tree rose when you plant it, either in a pot or in the garden. Note that the stake goes up into the crown or upper branches of the rose.

Just as with bush roses, it's important to keep cane moist to induce buds to grow (break). Tie the upper cane together with polyester twine. Place moist sphagnum moss in the space between the canes.

Some growers avoid the expense of sphagnum moss and place wet leaves in the space instead.

To keep the moss or leaves moist and the humidity high around the cane, cover the entire crown with burlap or similar moisture-retentive material. Tie the burlap in place around the cane.

Skilled growers who have been working with roses extensively use plastic instead. This requires extreme care, since heat can build up inside the plastic and harm the emerging buds. Nevertheless, it's an effective way to get canes to break. No matter what method you use, open the covering each day to make sure that the material stays moist and to check on the temperature inside. You want the cane to stay moist and the humidity to stay high, but you don't want the temperature to get so hot that buds are destroyed.

Once the buds begin to leaf out, remove the cover and slowly move the plant into increasingly bright light over a period of 10 to 14 days. Remove the moss or leaves gently to avoid breaking any buds. A gentle washing with a hose removes any moss or leaves that might cling to the thorns.

tently moist. If you do not keep both the canes and soil moist, the buds may begin to shrivel or turn brown from water loss.

If frequent misting is not possible, spray the entire plant with an antitranspirant such as Wilt-Pruf®. This helps prevent loss of new buds.

Buying Potted Roses

You'll find many of the roses listed in this guide in pots at local garden centers and nurseries. However, the number of roses in these stores is naturally limited. You may have to buy certain varieties through the mail as bare root plants.

Good potted roses have several or more healthy stems evenly spaced apart. Pull on the base of the plant *gently*. If the plant starts to dislodge from the soil, then it was recently planted. A plant that has been potted for some time will resist your gentle tug. It will have formed a solid rootball, and that's what you want from a potted rose!

Roses are often placed extremely close to one another in retail outlets, and people inadvertently damage or break canes as they select plants. Check canes carefully for any damage. Avoid plants with broken, cracked, or scraped canes.

These potted plants will have leaves. Foliage should be healthy. Check it carefully for any sign of disease or insect infestation to avoid buying a plant that is diseased or infested with bugs. The most common signs of disease are yellowing leaves, leaves covered with a whitish powder, or leaves with spots or lesions. Since potted roses are often sprayed at a nursery, they may look a little funny but be fine. If you notice any film on the leaves, ask what it is.

Check the soil. If the soil has moved away from the edge of the pot, this indicates inconsistent watering which stresses potted plants. If the soil is extremely dry, it also indicates careless handling. If such a plant otherwise looks extremely healthy, you can remedy both situations by pushing the soil back against the side of the pot and by watering immediately. However, it would be better if you could find a plant properly cared for from the start.

Whenever you buy a potted plant, get it home quickly. Avoid leaving it in a hot car while you run errands or do additional shopping. This overheating can stress a plant badly.

Also, if the plant will be exposed to wind on the way home, have the garden center or nursery wrap plastic or paper around the entire upper portion of the rose. Have them tie the protective covering in place with polyester twine around the base of the stem. Get the plant home quickly, remove the plastic at once, and soak the soil until water runs out of the bottom drain holes.

Slowly move the plant from light shade into full sun over a period of 10 to 14 days. This hardening off helps the plant adjust to bright light, varying temperatures, and drying winds. If a plant has been outdoors for a few weeks before you buy it, this process is not necessary. When buying at a nursery or garden center ask where the plant has been for the last 7 to 10 days to see whether it has already been exposed to varying light and temperature for the correct period of time.

Planting Potted Roses in the Garden

Here is a step-by-step planting method guaranteed to get your rose off to a perfect start. If you follow these steps carefully, you won't go wrong.

Plant a potted rose in the garden only after all danger of frost in spring. Otherwise, keep it in your garage or in a sheltered location if necessary until it's safe to plant it outside.

Move potted plants slowly into increasing light over a period of 10 days as temperatures warm up. This prevents sunscald (too much sun too quickly). This step has already been mentioned but is extremely important. Many people buy potted roses and return them with sunscald because leaves turn a grayish color or pale white; they mistake their improper handling for disease.

Dig a hole much larger than the size of the pot itself. A hole should be at least 18 inches (45 cm) deep and 24 inches (60 cm) wide for larger roses, a little less for smaller ones. Set the soil on a tarp. If it's poor soil, buy good potting soil to replace it.

At the base of the hole, sprinkle superphosphate according to the directions on the package. This product contains phosphorus, essential to good rose growth. Organic growers substitute 1 cup (about 225 g) of bonemeal, which is also a good source of phosphorus. Although it works (over a long period of time), it is slow acting.

Cover the superphosphate or bonemeal with an inch or 2 (2.5 to 5 cm) of soil. Potting soil is highly recommended. Most potting soils contain loam, perlite, and peat. Perlite keeps the soil loose and well aerated. Peat is acidic, keeps soil loose, and retains moisture. The loam locks in nutrients and provides support for the plant. The combination is ideal for roses.

Check the plant to make sure that the bud union (if there is one) is at the right depth. Not all growers who sell roses plant them properly. The correct depth of the bud union is outlined in each section in Part I. If the bud union has been left too high above the soil or buried too deeply, compensate for this when you place the rose in the planting hole.

Assuming that the bud level was originally planted at the correct level, set the pot temporarily in the hole. Check to see if the surface of the soil in the pot is almost level with the surrounding soil. Add additional soil to the hole as necessary until the pot rests just a half inch (12 mm) below the surrounding soil. Then remove the pot so you can cut it.

Potted roses come in an assortment of containers. Most nurseries use plastic pots, but a few sell roses in biodegradable material. In either case, cut out the bottom of the pot with a utility knife. Then, cut up along one side of the pot to within an inch (2.5 cm) of the upper edge.

Some growers put a thick rubber band around the severed pot to keep any soil from spilling out or to keep it from breaking off until it is in place. If you hold the pot firmly, this is not necessary. However, you do not want the root ball to break at all.

Set the pot back in the hole. Check the level again. Fill some of the space around the pot with soil. Firm it slightly. Now finish the cut with your utility knife. Then grasp the sides of the pot and gently pull the severed pot up and out of the hole. The plant remains snugly in place.

This method of cutting the pot causes no root disturbance whatsoever and is much better than tapping a plant out of a pot, which is fine for most plants, but not for roses. Often, when you tap a rose out of a pot, the root ball crumbles or breaks apart. This retards the plant's growth. So, follow the extra steps of cutting the pot to avoid this.

Sprinkle Pro-Mix® or a similar soilless product around the plant. Then add enough soil to keep the plant in place, but do not fill up the entire hole.

Sprinkle a handful of Milorganite around the plant. This is an organic fertilizer. Never add an inorganic fertilizer, since it will burn the tender feeder roots of the young plant. Press the soil firmly with your fingers around the plant.

Now add more soil until the hole is full. Press the soil firmly into place, again making sure that the bud union, if there is one, is at the correct level.

Soak the plant thoroughly with a hose. Let the water soak into the ground. The soil may sink slightly. If it does, add more soil, and press it lightly into place. Then water again. Saturate the soil.

The soil around the plant should be approximately a half inch (12 mm) below the surrounding soil to create a basin which traps water. This makes successive watering easier and more effective. However, having this slight indentation in the soil is strictly optional.

Growing Roses in Pots

The most common pots used for growing roses as container plants are ceramic (unglazed), clay, plastic, and wood (containing no toxic preservatives).

All pots should be large enough to accommodate the potential size of the mature rose. Deep pots are preferred to give plenty of room for roots and enough body to stop the plant from blowing over in the wind.

All pots should have drainage holes. Watering pots without drainage holes is possible, but it requires extreme skill. The danger of root rot is high unless

Keep pots off decking with feet or spacers

water can drain out of the bottom as necessary.

Each type of pot has advantages and disadvantages. Ceramic (unglazed) and clay are attractive and breathe freely. They are also heavy enough to hold plants in place in gusts of wind. However, they are hard to move around, require frequent watering, and are quite expensive.

Plastic pots are less attractive and do not breathe freely. However, they are inexpensive, easy to clean, and durable, they require less watering, and they are easier to move around.

Wood is attractive, stays in place, and retains moisture well. However, it is expensive and difficult to clean and move around, and will rot out.

Never use metal pots. They get too hot, and the soil dries out much too quickly.

If you have a pot on a deck or patio, keep drain holes exposed. You can buy little feet or spacers to keep the pots off the wood or cement. If pots are resting on the ground, provide a mulch, such as cocoa bean hulls, under the pots for good drainage.

For winter protection of container roses there are five methods: (1) If you place a perforated plastic bag in the bottom of the pot at planting time, pull the bag out of the pot and bury the bagged rose using the Minnesota Tip Method; (2) slide the plant out of the pot into a plastic bag and then bury it; (3) bury the rose, plastic pot and all, with other roses; (4) protect the plant in a garage; (5) or transfer the plant indoors, which we suggest for Miniature roses only, and even that's not highly recommended. Winter protection is covered in detail in Chapter 5.

Special Planting Decisions

To get your plants off to the best start possible and to make caring for them easier, take an extra minute for the following considerations. How the plant is oriented and supported can make a difference right from the outset.

If your plant will require tipping for winter protection in the fall, bear this in mind when planting. Look at the shape of your plant, and decide which way you're going to tip it over.

It's generally best to tip cane over the bud union, which could snap if you tip the plant in the wrong direction. This is especially important for Tree roses planted directly in the garden. Extremely experienced growers hold the base of a rose as they tip it and can ignore this advice, but most of us are better off tipping over the bud union than away from it.

If there is more cane on one side of the plant than another, plant the barest side facing south. This will encourage growth on the bare side. Bare root plants are called *one-sided* when cane is not evenly placed around the bud union.

If any part of the bud union is planted above the soil, as recommended for some types of roses, face it toward the sun. Bright light encourages new shoots (basal breaks) from the base of the plant.

If a plant is large, support it with a stake at planting time or when you raise it up in spring. This prevents wind from jiggling the roots and damaging the plant by disturbing the natural flow of water and nutrients from the roots to the rest of the plant.

Transplanting

Transplanting should be done very early in spring before buds begin to form. If possible, do this on a cloudy day. Definitely do it in the morning. Most growers report a much higher survival rates for plants transplanted early in the day.

Water the plant if it is freestanding. If it is being raised from the ground after winter protection, the soil will still be moist, so watering isn't necessary.

Have a planting hole ready before you begin to dig up a rose. Also have all the appropriate materials at hand. You want the plant out and into the ground as fast as possible to stop roots from drying out.

If a rose has been buried, raise it early in the season. The earlier you transplant the rose, the better. But follow the guidelines under "Raising Plants in

Planting Potted Roses in the Garden

AFTER YOU'VE PROPERLY PREPARED the bed or planting hole, follow the steps as pictured here for planting a potted rose. The rose in this illustration was planted the year before, then buried under soil, and has not yet leafed out. While the ones you buy will generally have leaves, the planting steps for both are the same.

Place a tarp next to your rose bed. Dig a hole larger than your pot in the prepared bed, setting the soil on the tarp.

Set the pot in the hole to check the depth. Add or take away any soil to get the bud union, if there is one, at the correct level. The exact placement of the bud union is covered in the individual sections in Part I. Remove the pot, and loosen the soil at the base of the hole.

Mix several tablespoons (about 42 g) of superphosphate into the soil at the base of the hole. If you're an organic gardener, use bonemeal instead.

Cut the bottom off the plastic pot with a utility knife. It's easiest to do this if you can rest the pot on a table or bench.

With the utility knife, cut the side of the pot from the bottom up to within 1 inch (2.5 cm) of the top. Then pick up the pot, holding it together with your hands so that it won't break open. You can hold the already-cut bottom of the pot in place underneath the soil to stop the soil from spilling out.

Let the bottom of the pot drop off just as you begin to slide the pot into the hole. Keep your hands firmly around the pot until it touches the bottom of the hole.

Once the pot is in the hole, firm Pro-Mix or a similar product around portions of the pot to anchor it in place.

Planting Potted Roses in the Garden (continued)

Sprinkle Milorganite over the Pro-Mix.

Place loam or potting soil around the plant and firm it in place.

Now that the plant is anchored in place, cut the top inch (2.5 cm) of the pot with your utility knife.

Pull up on the sides of the pot. It will slide up easily without disturbing the root ball of the rose.

Press down on the soil with your fists or fingers. Work around the plant until you've forced the soil to be as compacted as possible, but don't use your feet.

You'll often have to add more loam or potting soil. Do this as necessary and firm again with your hands. Then dump a scoop of soil around the crown of the plant. This will get washed into place after the following step.

Soak the plant thoroughly. Do this as many times as necessary until water pools on the surface. If the plant has no leaves, treat it as you would a newly planted bare root plant by misting it daily. Keep the soil around the base consistently moist until the plant is growing vigorously.

Spring" in Chapter 5 so that you don't bring it up too early.

Cut mature roses back to a few shortened canes at this time, as if preparing a new bare root rose for planting. By choosing a few healthy canes, you increase the odds of a plant's survival. Emotionally, the removal of lots of cane from a healthy rose can be difficult—just do it.

Dig up the rose with a spade, starting as far out from the plant as you can manage. Get as much of the root system as possible.

Snip off elongated roots, but leave a healthy root system attached to the cane.

Keep the roots moist by soaking them in a pail of water until you can plant the rose.

Since your planting hole is already prepared, plant the rose as quickly as possible. Saturate the soil after planting. Spray the exposed canes with an anti-transpirant such as Wilt-Pruf® to stop them from drying out.

Keeping the soil and canes moist until the plant is growing vigorously is critical to survival.

Even if you do all of this, transplanting roses is not recommended. We do not know why transplanted roses seem to suffer so badly, but we speculate that damage to the root system of established plants permanently shocks them. We do have one tip: once the transplanted rose is showing signs of recovery and new growth, dig a trench off to the side of the plant and fill it with potting soil containing slow release fertilizer. Roots tend to grow into this area and increase the odds that the plant will recover somewhat from the shock of being moved. Again, we do not know why this seems to work, just that it does. However, any transplanted rose is unlikely to match its former vigor and beauty. Seriously consider starting over with a new bare root plant (very hard to accept).

CHAPTER 5

CARING FOR ROSES

Roses are a versatile group of plants. Some seem to thrive on what amounts to benign neglect, but most need quite a bit of attention to bloom profusely. This chapter gives detailed information and tips on the proper care of roses. We think you will discover some new information here even if you have been growing roses for years.

Water

Woody plants need lots of water. Water is a nutrient containing both hydrogen and oxygen, two elements needed by roses for good growth. It causes more bloom, larger flowers, longer bloom time, better flower color, lusher leaf growth, better stem color, fuller growth, faster growth, greater fragrance, and better disease resistance.

There is no formula for proper watering. You know when to water by feeling the soil. Don't look at the soil. Dig into it with your hand or with a trowel. Water the soil whenever the top 2 inches (5 cm) begin to dry out. Don't be fooled by rain; some rainfall appears heavy but may be quite light. A light sprinkling does little for a rose. Each time you water, saturate the soil.

The basic rule is water, water, water. Remember that roses are in the same family as raspberries. Any-one who has grown raspberries knows how lush and abundant the fruit is with consistent watering. Roses respond the same way.

Roses grown in pots need careful attention. Never let the soil dry out. If soil pulls away from the side of a pot, this indicates erratic watering which stresses plants badly. If this happens, push the soil back against the side of the pot and water immediately. When watering potted roses, fill the pot to the brim. Let the soil absorb the water. Then water again. Keep doing this until water runs out the drain holes in the bottom of the pot. Most growers use plastic pots to cut down on the need for watering. Clay pots (far more attractive) are fine as long as you are willing to water your plants frequently—often twice a day in hot or windy weather.

Use overhead watering with a sprinkler or direct watering at the base of the plant from a hose. If you enjoy watering, hold the hose in place to saturate the soil around the base of the plant. If hand watering is a chore, set the hose down at the base of the plant. Let it run until the ground is thoroughly saturated. Some growers place the hose on a board to prevent the formation of a hole where the water runs out.

The use of drip irrigation or other slow-watering methods may cause salts to build up in the soil. Salts are found in both city water and inorganic fertilizers. Do not confuse these salts with sodium chloride, which is found in sea water and mixed with sand in winter to melt ice on streets. These fertilizer salts are entirely different compounds. If they build up in soil, they can harm roots. Some rose growers use drip irrigation. However, if heavy rains do not occur, they often saturate the soil with a hose or sprinkler every few weeks to carry toxic salts away. In this way they combine the benefits of drip irrigation (less time spent watering and less water used) with those of deep watering (possibly leading to less salt buildup in soil).

If you are using water from a well, turn on enough hoses to keep the pump running continually. If you don't, you run the risk of burning it out as it goes on and off repeatedly (just stand by the well and listen).

Many cities get their water from lakes, rivers, or wells. If water is treated, it may contain some harmful chemicals. However, you have to use what you've got. Even if there were some negative affects of treated water, they are minimal compared to deep and regular watering.

If soil is properly prepared, it is almost impossible to overwater roses. If the soil drainage is poor, the veins of leaves often turn yellow. With good drainage, this is not a problem.

Watering with sprinklers is most highly recommended during the spring. It keeps canes from drying out and helps the buds to develop and form new leaves. Both sun and wind will dry cane out during this period. You may want to mist the plants twice daily until buds forming leaves or branches begin to emerge.

Once leaves appear, the best time to water is early in the day so that foliage dries out before the end of the day. In extremely hot weather, consider a second watering in the evening. Many books tell you not to water at night for fear of increasing the chance of disease. If the soil is dry, it's better to water. Avoiding stress caused from lack of water is more important than avoiding wet foliage in the evening.

Overhead watering is also recommended in especially dry periods to discourage spider mites. There is a new theory that watering roses in this manner for

4 hours every 10 days gets rid of black spot spores. Overhead watering during high heat also cools plants and reduces stress from water loss.

Overhead watering can be tricky. Dig into the soil after watering to see whether or not the soil has been thoroughly saturated. Light sprinklings are not helpful. You may have to water for several hours.

In humid weather, use hand watering at the base of the plant to keep soil moist. Avoid spraying the foliage. Wet foliage at this time may encourage disease.

Always saturate the soil with water before and after the application of any chemical fertilizer. This protects the plant from possible damage, often referred to as *burn*. Deep watering dissolves nutrients and carries them to the roots.

Water also dissolves carbon dioxide, forming a mildly acidic solution. This reacts with minerals in the soil to form nutrients easily absorbed by the plant.

A continuous supply of nutrients makes plants vigorous. Healthy plants can resist disease and insect infestations more easily than weak plants. They also bloom more profusely, more often, and with better color (as mentioned earlier).

Always water before spraying plants with fungicides, insecticides, or miticides, to prevent leaf burn. This is very important.

While many Old Garden Roses, some Shrub, and some Species Roses tolerate dry conditions fairly well, it is common in cold climates to have dry spells or even drought in late summer and early fall. Late watering (not recommended in the past) is now encouraged so that plants do not go into winter in a distressed state. Water all roses well until the first freeze.

When you bury roses to protect them from winter cold and winds, the soil should be moist. Once roses are covered with leaves, sprinkle them as necessary to keep the protective layer of leaves and underlying soil moist until the ground freezes. This process is covered in detail later in the chapter under "Winter Protection."

Mulch (Summer)

Mulch is any material put on the surface of the soil around a plant to keep the soil moist and cool during hot weather. Mulch inhibits weed growth and makes it easier to pull the few weeds that do sprout. Weeds compete with roses for moisture and nutrients. Mulch also reduces soil compaction as you walk on it. It prevents soil from splattering against the plant's leaves, a cause of some disease. Mulch can actually kill some diseases. It also feeds soil microorganisms and worms which benefit the plant enormously by keeping soil loose (lots of oxygen) and fertile. Use organic mulches only. Avoid rocks or pebbles, since they get too hot during the summer and stunt plant growth. It is difficult to work soil that is covered in stone, although a few rose growers disregard this advice with remarkable success. We don't think it's worth the work or cost. Furthermore, roses like a cool, moist root run provided by organic mulches.

Place mulch around the entire plant, but do not touch the canes with it. Apply mulch after mid-May. Let the soil warm up thoroughly before putting mulch around roses. The soil should reach a temperature of about 60°F (15.6°C). If you want to be precise you can use a soil thermometer, but most growers just feel the soil to see whether it's warm, cool, or cold. Although not proven, too early application of mulch *may* be slightly responsible for an increase in powdery mildew. Replace mulch as it disappears during the summer, as soil microorganisms and worms feed on it. Remove and compost all mulch in the fall to prevent disease.

Mulch for Roses

ALTHOUGH YOU CAN USE any of a number of mulches, most rose growers have the best success with shredded leaves.*

Shredding leaves, especially large piles, is easiest with a commercial shredder. Leaves used as winter protection often get quite wet and compacted. Avoid placing too many of them in the shredder at one time.

Occasionally, a glob of leaves may get stuck in the shredder. Never put your hands into the machine. Instead, use a stick or long tree branch to push the material into the blades.

Place the shredded leaves around the roses in a nice, thick blanket. Do this when the plants have already begun to leaf out. Keep adding mulch throughout the season as necessary.

*You don't have to have an expensive mulcher to mulch leaves. Spread leaves out in a shallow layer on a hard surface (driveway) and mow over them with a rotary lawn mower until they are pulverized. Leaves are easiest to break down when they are dry.

Common Organic Mulches

SOME MATERIALS USED FOR mulch cause a nitrogen deficiency in the soil. The reason is that soil microorganisms use up nitrogen in the process of breaking these materials down. Materials high in carbon cause the most problems. Because bark and wood chips fall into this category, whenever you use these for mulch, be sure to sprinkle additional nitrogen around the roses.

Following is a list of the most popular organic mulches:

Bark (Shredded) Quite expensive. Nice color and texture, easy to apply. Looks sensational with larger plants, including Species (wild) and Shrub roses. Whenever you use bark as a mulch, add more nitrogen to the soil than you would otherwise use—no less than 1 pound (about ½ kg) per 10 cubic feet (3 cubic meters) of bark. The reason is that soil microorganisms will use up more nitrogen as they try to break down the bark.

Cocoa Bean Hulls Quite expensive. Nice color and texture, easy to apply. Has noticeable odor of chocolate. Will mildew if applied deeply, more than 2 to 3 inches (5 to 7.5 cm) at a time. Looks especially nice under a group of potted roses, such as Miniatures.

Grass Clippings Costs nothing, readily available in colder climates. Gets hot and will mildew if applied too thickly when fresh (can also cause slight odor). Apply to depth of 2 to 3 inches (5 to 7.5 cm) at a time. Some people prefer to let it dry out before using it. Compost it if grass was recently treated with an herbicide. Rose growers are about equally divided on its use—some love it; others hate it.

Leaves Should be shredded with shredder or rotary mower. Slightly acidic oak leaves are best, but all leaves are great. Apply to a depth of 3 to 4 inches (7.5 to 10 cm) at a time. This is the best and most readily available mulch for growers in cold-climate areas. Worms love it, and it keeps soil nice and fluffy (aerated). When leaves have decomposed, they're referred to as leaf mold. Decomposed leaves are similar to compost: wonderful to have, but there's rarely enough.

Pine Needles Inexpensive. Slightly acidic (good), drain freely, don't compact. Apply to depth of 4 to 5 inches (10 to 12.5 cm). An excellent alternative to leaves and very good mixed with leaves.

Wood Chips Expensive if bought in stores. Often available for free in large quantities from local utility or tree-trimming companies. Attractive if properly chipped. May temporarily take nitrogen from soil; apply nitrogen more liberally than normal, just as you would for shredded bark. Wood chips are best with extensive plantings of Shrub roses. Keep depth at 3 inches (7.5 cm).

Fertilizing

To create magnificent bloom, fertilize exactly as directed in the individual listings in Part I. Use either chemical, organic, or a combination, as suggested, for best results. The methods outlined have been developed over a period of 60 years and will provide your plants with all the food they need. In some instances, they may seem quite intensive, but bear in mind that in cold climates the growing season is extremely short. Proper fertilizing increases the amount of cane and bloom substantially. Some rose books give detailed descriptions of the symptoms of nutritional deficiencies. If you follow the feeding system outlined in this guide, you will rarely have to deal with these problems.

Fertilizer Basics

Plants, like people, need a balanced diet to grow well. Fertilizer is plant food, which consists of a number of chemicals found in the air and soil. Roses need lots of some (macroelements) and very little of others (micro- or trace elements). Loam with a high amount of organic material contains most of the elements needed by a rose.

However, some of the elements are used up fairly quickly and must be replaced regularly for healthy plant growth. If you follow the directions throughout this guide, your roses will always have enough essential elements for good health. The critical point is that if you follow the directions on proper fertilization you will never have roses with nutritional deficiencies, which are often mistaken for disease or insect problems.

You must add three major elements to the soil on a regular basis. These are nitrogen (N), phosphorus (P), and potassium (K). Phosphorus moves extremely slowly through soil. Ideally, place it just below the plant's root system at time of planting either as bonemeal (organic) or superphosphate (inorganic).

When you buy fertilizer, the label will tell you how much of these major elements are contained in the package. The amounts are always in the same order: nitrogen (N), phosphorus (P), and potassium (K). If a package reads 10-10-10, this indicates that the fertilizer is 10% nitrogen, 10% phosphorus, and 10% potassium. The rest of the material in the package is filler, an inert substance of little value.

Fertilizers are either inorganic (synthetic) or organic (naturally occurring from living creatures). Most roses grow best if a combination of these is used. A few respond best to organic fertilizers only. This information is given throughout the book. Following are some tips on the two types of fertilizers.

Inorganic Fertilizers

There are several advantages to inorganic fertilizers. They provide essential food to plants quickly. They work early in the season before the ground gets warm enough for organic fertilizers to be effective. They are inexpensive (one bag of 10-10-10 inorganic fertilizer contains as much of the three essential elements as 1 ton [900 kg] of organic material). And, some inorganic fertilizers contain trace elements, which are listed on the label.

Use any granular 10-10-10 or 20-20-20 all-purpose garden fertilizer without herbicides (weed killers). This type of fertilizer is sold in bags. The label will indicate whether the fertilizer is quick or slow release. Slow-release fertilizers consist of coated granules. The outer coating breaks down to release the fertilizer at varying intervals. Scatter granules around the base of the plant as indicated throughout the book. A rough guide is ⅓ cup (about 75 g) per square foot (78 square cm) of soil around mature plants.

This is truly an approximation. Each variety of rose responds differently to varying amounts of fertilizer. Giving an exact amount is like trying to tell you how much each person should eat. You'll know whether your feeding is correct by letting the rose tell you. If it's doing well, you'll get a lush plant with lots of foliage, new growth from the base of the plant, and plenty of flowers.

Use any water-soluble fertilizer. All brands with major nutrients listed as 20-20-20 or close are fine. You dissolve these fertilizers in water before applying them to the ground around the plant. Some growers also spray leaves (especially the undersides) with water-soluble fertilizers, but this is optional. Follow the directions on the package.

If you have a large bed, consider using a Siphon

Essential Elements for Growing Healthy Roses

Boron (B) Needed in minute quantities. Important to cell division, flower formation, and pollination. Ample amount exists in most soil. Augmented yearly by adding fish emulsion. (.005% of plant tissue)

Calcium (Ca) Needed in moderate amounts. Important to cell structure and good root growth. Gets plants off to good early growth. Ample amount exists in most soil. Found in gypsum (calcium sulphate), which is sometimes added to soil to get rid of salt deposits. Augmented by adding bonemeal or superphosphate to the planting hole. (0.6% of plant tissue)

Carbon (C) Needed in large amounts. Ample supply in air. (44% of plant tissue)

Chlorine (Cl) Needed in minute quantities. Important in transfer of water and minerals into cells and in photosynthesis. Ample supply exists in soil or city water. (.015% of plant tissue)

Copper (Cu) Needed in minute quantities. Important in stem development and color. Essential in enzyme formation, root growth, and respiration.* Ample amount exists in most soil. Augmented by addition of Milorganite to the planting hole. (.001% of plant tissue)

Hydrogen (H) Needed in large amounts. Ample supply exists in water. (6% of plant tissue)

Iron (Fe) Needed in minute quantities. Important in chlorophyll formation and for proper plant respiration. Ample amount exists in most soil. Augmented by adding bonemeal to the planting hole. (.02% of plant tissue)

Magnesium (Mg) Needed in small amounts. Important in chlorophyll formation and respiration. Essential for healthy foliage and disease resistance. Found in soil, but added with Epsom salts (magnesium sulfate) yearly. Also present in fish emulsion. (0.3% of plant tissue)

Manganese (Mn) Needed in minute quantities. Important in chlorophyll formation and the production of food through photosynthesis. An enzyme regulator. Ample amount exists in most soil. Augmented by addition of Milorganite to the planting hole. (.05% of plant tissue)

Molybdenum (Mo) Needed in minute quantities. Helps roses use nitrogen for vigorous growth. Essential for enzyme formation, root growth, and respiration. Ample amount exists in most soil. Augmented by addition of Milorganite to planting hole. (.0001% of plant tissue)

Nitrogen (N) Needed in large amounts. Critical to healthy cane growth, lush foliage, and beautiful bloom. Important in cell growth and plant respiration. Essential food for soil microorganisms. Must be added to soil on a regular basis. (2% of plant tissue)

Oxygen (O) Needed in large amounts. Ample supply exists in both air and water. (45% of plant tissue)

Phosphorus (P) Needed in large amounts. Essential to rapid root growth. Important in stimulating quick growth which improves winter hardiness. Important to proper formation of stems and to good color and solidity of petals. Must be added to soil at planting time as bonemeal or superphosphate. (0.5% of plant tissue)

Potassium (K) Needed in large amounts. Important to root growth, formation of blossoms, and bloom color. Critical in forming sugar and starches. Must be added to soil on a regular basis. (1% of plant tissue)

Sulfur (S) Needed in small amounts. Keeps soil at the right pH (slightly acidic). Important in formation of plant proteins needed for good health and root growth. Ample amount is supplied by rain. (0.4% of plant tissue)

Zinc (Zn) Needed in minute quantities. Important in stem and flower bud formation. Essential to enzyme formation, root growth, and respiration. Ample amount exists in most soil. Augmented by adding Milorganite to the planting hole. (.01% of plant tissue)

*Respiration refers to the ability of cells to produce energy using chemicals. This is often confused with transpiration, the evaporation of water from plants. Note that roses release oxygen into the air during the day as a byproduct of photosynthesis. Plants do need some oxygen but release far more than they use.

Mixer to make this job easier. You dissolve 2 pounds (about 1 kg) of 20-20-20 in a 5-gallon (19-liter) pail. For fish emulsion, the amount is 1 quart (about 1 liter). The Siphon Mixer will automatically mix the solution in the pail with water running through a hose. You'll get roughly 60 gallons (228 liters) of fertilizer mixture per 5-gallon (19-liter) pail.

Here are more tips on using inorganic fertilizers:

Always follow directions exactly. Frequent, small feedings are better than large doses of fertilizer all at once. When in doubt, reduce the amount given at any one time and feed more often.

When using inorganic fertilizers, always saturate the soil with water before and after use. This will prevent any damage to the plant's delicate root system. Think of water as you would an IV tube that carries all nutrients into the plant's roots. Consider water a nutrient as well, since it contains hydrogen and oxygen.

If you want to grow roses from seeds, never use inorganic fertilizer in the growing medium. It will kill seeds or seedlings.

Avoid the use of inorganic fertilizers in planting holes. These fertilizers may come in contact with tender feeder roots and retard the growth of the rosebush. Use inorganic fertilizers on the surface of the soil only. The exception is superphosphate. This inorganic fertilizer will not damage roots if used according to the instructions on the package. It must be placed in the planting hole to be effective.

Some growers might also argue that you could put slow-release granules (the kind coated with resin) 2 to 3 inches (5 to 7.5 cm) below the lowest point in a planting hole without harm, but we prefer using only organic fertilizers in the planting hole itself to eliminate any chance whatsoever of damage to roots.

You'll find Epsom salts (magnesium sulfate) in most drugstores. It neutralizes the soil and makes nutrients more readily available to the plant. It also induces new cane growth (basal breaks) from the base of plants. Use 1/3 cup (about 75 g) for a large plant, less for a smaller one, twice a year: once in late May and again in early July.

If you are growing roses in pots, add Sprint 330 (a source of iron) to the soil as indicated on the label.

Miniatures are often grown in pots and respond beautifully to such treatment. Potted plants indicate a need for iron if leaves yellow but still have green veins.

The French have found that yellow roses have better bloom and more intense coloration if fed iron nitrate in spring (a light handful) and iron sulfate in fall (again, just a light handful). Since iron sulfate does not contain nitrogen, it doesn't stimulate growth likely to die back in the winter.

When using hoses for foliar feeding, you're required by most state building codes to have a backflow preventer. This attachment stops fertilizers and chemicals from getting into your home water system. Use it. See Chapter 9 for more information on products helpful in feeding, such as a Siphonex, which has a backflow preventer in it.

In cold climates, avoid any fertilizing after mid-August (early August for Shrub and Species roses). Late fertilization induces new cane formation. Cane formed late in the season usually dies during hard winters. Thus, later fertilization not only damages the plants, but also is a waste of time, money, and energy.

Some growers continue to feed plants with 0-10-10 until late in the season. Since this fertilizer contains no nitrogen, they believe that it will not stimulate growth susceptible to dieback. However, recent testing suggests that roses may react to this feeding by producing cane later in the spring than normal. Our advice: don't fertilize late in the season.

Many lawn fertilizers contain herbicides (weed killers). Avoid getting any of these fertilizers into your flower beds or around the base of rosebushes. Any container or sprayer used for killing weeds should be used for that purpose only. Mark the container or sprayer with permanent marker so that it is never used for any other purpose. It's an excellent idea to have several sprayers for different purposes (see p. 230).

Organic Fertilizers

All rose beds should contain lots of organic matter (roughly one-third of the soil in total). The importance of this was stressed in the section on good soil. Most organic materials contain relatively low amounts of the three essential elements, but they are excellent as food for soil microorganisms and worms. These creatures help plants take in nutrients and enrich the soil at the same time.

Organic fertilizers are released slowly into the soil. They do not damage seeds or seedlings when applied in small amounts (read the label). They can be added to planting holes in moderation without root damage (read the label). And, they are actually preferred by certain roses.

However, they are expensive and are not effective early in the season when the ground is still too cold for soil microorganisms to be active.

The best organic fertilizers are alfalfa meal (contains triacontonal) which is sold as rabbit pellets in farm stores, blood meal (15-2-1), bonemeal (4-25-0), compost (4-1-3), cow manure (.6-.2-.3), fish emulsion (10-7-0), horse manure (.7-.3-.5), and Milorganite or treated sewage (6-3-0). A mixture of these is ideal. Again, most animal manures contain very little in the way of actual nutrients, but are excellent for feeding soil microorganisms and worms, which benefit the soil tremendously.

The fertilizers most commonly recommended in this book are alfalfa meal, bonemeal, fish emulsion, and Milorganite. The actual amount of essential elements in these products varies. Check the label for an exact analysis of the product you're buying. All of these are commonly available in nurseries.

Here are more tips on using organic fertilizers:

Alfalfa meal is readily available in the form of compressed pellets, which are used as rabbit food. You can mix the pellets into the soil or, if you prefer, dissolve them in water before pouring the mixture around the base of the plant. The pellets can also be made into what is known as *Alfalfa meal tea* (more about that later).

Bonemeal, like superphosphate, must be added to the planting hole to be effective. It contains lots of phosphorus, which moves extremely slowly through soil, so place it by the roots for good results. Just mix it into the soil at the bottom of the hole. Note that most growers in most instances find bonemeal less effective than superphosphate. We mention bonemeal throughout this book nevertheless, because many organic gardeners insist on using only natural products in their garden.

Fish emulsion contains a number of trace elements. Soak the base of each plant with it. Some growers spray the leaves as well as the soil for foliar feeding. The smell usually lasts for only a few hours as the emulsion dries out.

Spray foliage with an orchid fertilizer containing triacontanol, which is a derivative of alfalfa. A grower discovered the benefits of this substance when interplanting tomatoes with alfalfa. He got much better tomatoes. The studies of alfalfa indicated that triacontanol was the active ingredient encouraging better growth. It works with many plants (not all), including roses. Spray once when plants are growing well in spring and once again a month later.

Alfalfa meal (rabbit pellets) can be used to make a very effective replacement for orchid fertilizer, since it too contains triacontanol, as mentioned previously. Add 10 cups (about 2 kg) of rabbit pellets, 1 cup (227 g) Epsom salts, and ½ cup (about 115 g) Sprint 330 (source of iron) to a 32-gallon (122-liter) plastic garbage can filled with water. Put on the lid and let the mixture soak for 4 to 5 days. Stir it each day, but keep the lid on tightly the rest of the time since the mixture has a very unpleasant odor. Pour 1 gallon (about 4 liters) of this alfalfa meal tea around larger plants and one-third that amount around smaller ones twice a season—once after plants are growing well in spring and once one month later. You'll notice greener coloration and stronger growth within a week whether using orchid fertilizer or alfalfa meal tea. Note that there is often a residue in the bottom of the can after you've removed the liquid. Fill the can up again with water, adding the same amount of Epsom salts and Sprint 330 to start the process all over again. Keep doing this until there is no residue left in the can.

Milorganite is treated human sewage. Testing has shown that it is safe to use in any garden, although it once had a warning in the package against use in vegetable gardens. According to the manufacturer, it contains no more heavy metals than the average handful of soil. It does, however, have an odor. For this reason scratch it lightly into the surface of the soil. Also, Milorganite is attractive to dogs, which like to eat and roll in it. If eaten by dogs, it could make them sick but will not kill them. Keep it away from them.

Blood meal (15-2-1) is a good choice for organic gardeners who want to avoid using any synthetic products. It contains a number of valuable trace elements and may repel rabbits.

Never use fresh manures in the garden. Let them decompose before use. Fresh manures are very high in nitrogen and soluble fertilizer salts, which can burn plants. Burning means that the high concentration of these chemicals draws moisture out of the plants, resulting in damage and even death to plants, especially seedlings.

Urea is either concentrated urine or a synthetic fertilizer resulting in an incredibly high amount of nitrogen (42-0-0) and must be used with extreme caution. It is commonly used by professional rose growers and is best left in their hands. We mention it only because you will read about it. That does not mean you should use it, although you certainly can if you follow warnings explicitly. Considering how it is now generally made, it could just as easily be listed under inorganic fertilizers.

Weeding

Weeds compete with roses for water and nutrients. Destroy all perennial weeds, including dandelions, grass, quack grass, vines, and thistles, before you plant a bed. These weeds will often resprout from infinitesimal portions of root. So, when preparing a bed, use an herbicide such as Roundup® to kill them. This chemical is absorbed by the plant in such a way that all roots are killed. Chemicals that work in this way are called systemic (taken into the entire plant's system). Roundup® is extremely expensive, but it breaks down quickly. Follow the directions on the label exactly, and wait the specified period before planting a rose in a treated area. The best idea is to kill all weeds in late summer and fall before making a bed. Then plant roses there the following spring.

Do not use Roundup® or other systemic herbicides in areas where roses are already growing, without taking special precautions. Using an herbicide near growing roses requires extreme care. If any wind is blowing, some of the herbicide can get on rose leaves and cause damage. One way to avoid this is to soak a rag or paintbrush in the herbicide and rub or paint it on the weeds you want to kill. This takes time but prevents accidental damage. If doing this with a rag, always wear rubber gloves. Use this method on perennial weeds only. Just pull up annual weeds. You can generally tell the difference between perennial and annual weeds by their root systems. Most perennial weeds have a much deeper or extended root system than annual weeds. If a weed resprouts from a portion of its root, it's a perennial weed.

If you kill perennial weeds from the start, you probably will be able to dig or hand pull most weeds that grow later. However, if you want to use an herbicide, use one such as Preen (contains Trifluralin) or similar products designated for use in rose beds.

Deter the growth of annual weeds by lightly hoeing the upper inch (2.5 cm) or so of soil frequently early in the season. The best tool for this is a pronged hoe. Just scratch the surface. Loosening the soil keeps it from compacting, kills off young weeds, and helps oxygen get to the feeder roots. Always do light hoeing just before applying a summer mulch.

Once plants are growing, mulch stops most weed growth, since many weed seeds require light in order to germinate. The few weeds that do appear are easy to pull up by hand from the moist soil. Hand weed

Know Your Herbicides

YOU SHOULD BE FAMILIAR with the following terms used for herbicides:

Contact Kills weeds by hitting foliage.

Nonselective Kills anything it touches.

Post-emergent Kills weeds that have sprouted.

Pre-emergent Stops weed seeds from germinating.

Selective Kills only certain plants.

Systemic Drawn into the plant, killing it all the way to its roots.

around the base of plants to avoid damaging the shallow root system. Hoeing at this stage may wound roots or stems, inviting infection by disease and infestation by insects.

Finally, you should also keep living turf grass out of your garden. Often rose beds are surrounded by lawn. Grass thrives in the moist, rich soil provided by roses by a good rose grower. Its underground stems (stolons) search out such an area and spread rapidly. Each spring, use an edger to outline your rose bed. Step on it firmly so that the blade goes down as deeply as possible. Move along the edge, cutting the outline with the edger. Once you've cut the entire edge, go back to pick up the sod. Knock it hard against the ground to get loose soil off the roots. Toss the grass into a wheelbarrow. Do this along the entire edge, removing all grass and roots. Otherwise, even the tiniest piece of root will take and grow into grass.

Other ways of stopping grass from getting into the bed are to edge it with deep strips of metal, plastic, or treated wood. Grass has a way of defeating all of these strategies, but they do help.

If you see any grass getting into the bed, dig it up with all attached roots and remove it immediately. You can also kill it with Roundup® as outlined earlier, using a rag or paintbrush.

Staking

For Support

Staking, or supporting, different types of roses serves different purposes. Tree (standard) and a few larger roses need support to stop cane from breaking or roots from loosening in heavy winds. If roots get loosened, this retards growth by cutting off nutrients and water from the upper portion of the plant.

Supporting larger roses is one of the most important steps for the home grower to take. It is also one that is commonly neglected. It takes very little time if you gather the right materials during the winter.

Use electrical conduit for support. Conduit (ask for Thinwall or EMT) is made out of thin steel, which you cut with a hacksaw. Hardware stores sell it in 10-foot (3-meter) lengths, which are easy to cut to the

length desired. Conduit is lightweight and very strong, it doesn't rust, and it stays cool during hot weather (won't burn the plant). It's also easy to paint if you don't like the color.

You can also buy rigid, plastic stakes (often lime to dark green) at nurseries. These should be long enough to drive deeply into the ground and still have plenty of room for support of the rosebush at full maturity.

You can use other products as well, such as bamboo stakes. Unfortunately, these often rot out or break over time. Using metal or plastic is preferred. If you live in a wooded area, young saplings of poplar (aspen) trees work fine for a season when cut into proper lengths. These trees are easy to cut, grow straight, and will regrow after each cutting. And, of course, the stakes cost nothing. But, like bamboo, they will last only so long before turning brittle and breaking.

The easiest way to attach a stem to a support is with electrical tape. It's easy to use and relatively inexpensive, and it doesn't restrict stem growth. It is also very strong, providing just the kind of support needed.

A method that takes a little more time but is preferred by many growers is to cut out a small piece of insulation and wrap it around the stem. Use either tape or a plastic tie to clamp the support and insulation in place. This keeps the plant firmly anchored without any chance of the support's rubbing against the stem.

Casual growers can even use strips of cloth tied around the stem to the support in a figure-eight pattern. The knot should be firm. The cloth stops the stem from rubbing against the support.

Climbers need something to climb on, or they would just run along the ground. Specific information on support for Climbers is outlined in the Climbing Roses section in Part I.

To Increase Bloom

There is a special kind of staking that few rose growers know about. It is used on arching, one-time bloomers such as 'Harison's Yellow' and *Rosa hugonis*. To increase bloom on such plants, bend a long, arch-

Staking

STAKING IS ONE OF the most neglected aspects of rose growing. Here are just a few examples of staking methods:

Get your stakes in place as early in the season as possible. All Tree roses need to be staked to prevent damage to their roots and heads. Note how the stake goes up into the head of this rose.

Larger bush roses benefit from staking as well. Conduit should be placed in the center of the bush as soon as it is raised in spring.

When growing roses in pots, you may need to stake the stem of the Tree rose to a support and also anchor the pot to an additional stake to prevent it from blowing over in a windstorm.

Of the many methods of attaching roses to their support, one of the most common is simply to wrap the stem with foam and then tie it to the support with a plastic tie, electrical tape, or even duct tape. The key is to do this early in the season to stop roots from jiggling which retards growth.

ing cane over and stake it to the ground. Do this carefully to avoid crimping the base of the cane. This method causes new canes (basal breaks) to emerge from buds at the base of the plant. After the new canes are growing well (usually the following year), cut off the original cane. The purpose of this type of staking is to create a full, bushy plant from one that might normally be quite leggy.

Pegging

You can increase the amount of bloom on a number of repeat bloomers by using the technique known as pegging. It is effective on all Climbers and Hybrid Perpetuals and most long-caned roses, which tend to arch over as they mature. Simply bend canes to run parallel to the ground and attach them to a support. This causes the cane to produce numerous branches (laterals), which are covered with a mass of blossoms. It is similar to staking, but the formation of branches (laterals), not new cane from the base of the plant (basal breaks), is the goal.

Disbudding

Some roses produce a number of flower buds at the end of the cane. For larger flowers, gardeners can remove the buds around the central bud to get a larger bloom. Simply twist the buds between your thumb and forefinger and discard them. Remove buds when they are just appearing, before they have a chance to become large. All of the energy of the plant will then go into forming one magnificent bloom on the stem. This is done mainly for exhibition purposes. Disbudding on Floribundas is slightly different. In this case the central bud is removed so that the other buds surrounding it will all flower at once. Disbudding is covered in each section of Part I to avoid any confusion about when and how to do it.

Deadheading

The correct method for removing spent blossoms, known as deadheading, varies by the type of rose you're growing. But, generally speaking, removing a blossom should be considered a form of pruning. You

The removal of spent flowers on repeat-blooming roses is critical to get them to rebloom. Remove the flowers as outlined in the individual sections of Part I for best results. Here a cluster of flowers on 'Love,' a Grandiflora rose, is cut well below any individual bloom, as appropriate for this type of rose.

The plant now sends energy to the bud already sprouting from below the original cluster of flowers. The plant also creates new growth in other areas for additional repeat bloom.

cut down to a five-leaflet leaf as you remove the blossom or cluster of blossoms.

The purpose of removing blossoms is to encourage repeat bloom on roses that bloom more than once in a season. If you do not remove blossoms, the rose may form hips and think its blooming season has come to an end. No matter when you remove blossoms, always toss them into a bucket to keep the garden clean and to prevent disease. Few people bother with removing spent blossoms on roses that bloom only once early in the year.

Generally, the best time to remove spent blossoms is just as the petals are about to fall. You know you've waited too long if they spill onto the ground when you brush the cane. Blossoms on the ground invite disease and insects into the garden. If petals fall to the ground, pick them up. Keep your garden clean, especially if you're relying on organic controls. But, even if you're not, it's a good idea.

If a plant will not be buried for winter protection, stop removing spent blossoms in late summer. This allows the plant to form hips and prepare for winter dormancy. The plant will suffer less dieback (dying of cane) if you follow this simple advice.

If you do bury the plant using the Minnesota Tip Method, you can let the plant flower until the end of the season. But remove any flowers still on the plant before burying it.

Pruning

Wild roses survived for millions of years without human help. However, breeding has created roses that are much more susceptible to disease, insect infestations, and death from cold. Pruning techniques vary with each class of rose and are outlined in detail in the appropriate sections of Part I. This section deals with information applicable to all roses.

Many people are confused by and afraid of pruning. Let's clear the air. First of all, pruning serves several important purposes:

- **Health** Proper pruning keeps the center of the plant open to sun and air circulation. This helps prevent disease. Cutting off dead, diseased, or deformed cane also prevents destruction of the entire plant.
- **Vigor** Proper pruning increases the amount of healthy cane, making the plant bushy and vigorous. You get a good blend of new and old cane. Strong growth helps ward off insect infestations and disease.
- **Beauty** Proper pruning results in more beautiful shape and form. It causes the bush to produce more flowers, larger blossoms with more lustrous color, and rich coloration in the cane and foliage.
- **Longevity** Proper pruning enables rosebushes to live longer. Some bushes are extremely expensive. If properly cared for, they may live for decades. If ignored, they may die out quickly.

Correct Tools and Accessories

Pruning is as much art as science. You will learn how each plant responds over a period of years. Even the best growers occasionally make mistakes. A better way of putting this is that they learn from their experience. The tips here and throughout this book will help you avoid the most common mistakes and give you confidence that you're doing the right thing at the right time. So, the first thing you need in order to prune correctly is the courage to do it. Equipped with that, you're ready to take pruners in hand.

Buy good tools. You'll need pruning shears (the two-bladed bypass kind are best—see p. 230). Make your cuts with the center of the blades, and keep blades sharp. You'll also need lopping shears (long-handled pruners) to cut larger canes. Again, the two-bladed bypass kind are best and must be kept sharp. The best tool to keep blades sharp is a whetstone (with oil) or an appropriate file. Sharp blades make clean cuts. Clean cuts are important to prevent cane damage. Also important for pruning: leather gloves, a knife, bleach (or similar disinfectant), and something to coat cuts (Elmer's glue, nail polish, or other commercial sealants). Avoid any pruning pastes with a petroleum base. We have found that these often cause the end of canes to die back by an inch or 2 (2.5 to 5 cm). Many excellent rose growers do not seal cuts, but most do follow a rigorous spray program. Organic growers are wise to seal cuts.

What to Prune and When

Choose plants wisely according to the space available so that you are not trying to make a large plant into a small one. Pruning is time consuming. Overpruning is harmful to a plant. By choosing the right plant for the right place from the start, you save time and money.

Follow these pruning guidelines for best results:

On new plants, remove any broken or dry cane, cutting above a bud (as outlined later). Otherwise, during the first year, avoid any pruning at all on all roses grown in cold climates. This allows the plant to develop a full root system and as much cane as possible. Cutting stems for cut flowers is a form of pruning and not recommended at all during the first year. In fact, it's best not to let a plant flower in the first year. Remove the buds immediately as they appear to create a healthier bush. Remove no leaves at all, unless they are diseased. We realize that you might not like the idea of having no bloom the first year; a compromise is to let the plant bloom, remove the spent blossoms, but never remove any foliage.

In cold climates, prune as little as possible. Because the season is short compared with the South or West, light pruning is usually best. Light pruning means pruning the plant as high as possible, not cutting it way back as recommended in many guides geared to warmer climates. A rosebush needs cane and leaves to form flowers. It takes from five to seven leaves to generate enough food to form a single bloom. Food is stored in cane, so why remove it or foliage unless necessary? Also, the higher you prune, the sooner your roses will bloom.

If plants are budded onto rootstock, remove suckers that come from the rootstock. If suckers are not removed, they will eventually overwhelm and choke out the hybrid plant, the rose you really want to grow. Suckers usually have a different-colored stem, frequently lighter green. They may be more or less thorny than the upper canes. They will not yield flowers similar to those on the budded or grafted, upper portion of the plant, so they are not desirable. It is sometimes difficult to distinguish exactly where the suckers are coming from. You may have to dig around the base of the plant, carefully removing soil with your fingers to find the location of the bud union or graft. Cut as close to the stem as possible. Occasionally, roots are attached to the sucker. Remove these as well even if you have to cut into the crown. If you leave a stub, several new suckers will emerge. Paint the cut with a sealant available at nurseries. A second way to remove suckers is to push them down to break them off with force. This causes a wound which occasionally becomes infected, but in most cases this is a simple, effective way to get rid of suckers. Coating the wound prevents most infections. Remove all suckers as soon as you notice them. On Tree roses, remove suckers immediately as they sprout at the base of the stem. Also remove any branches sprouting along the standard (stem).

Each spring, remove all diseased or dead cane by cutting back to live cane. Do this by snipping off small pieces of cane at a time. Dead cane is brittle, dry, and brown. Keep snipping back until you reach live wood, usually pale or yellowish green in the center (pith) of the cane. Make your final cut just above an outward-facing bud. Do this just as buds begin to swell. If you do it earlier than this, the new growth may be damaged by a late frost; if you wait too long, you weaken the plant and increase the risk of disease.

Remove canes that crisscross so that the center of the plant has plenty of light. Canes that cross may also rub together, causing an open wound, which can become infected and damage or kill the plant. Do this also in early spring just as buds begin to swell, or even earlier if possible.

Remove old canes when they stop producing bloom in order to encourage the growth of new cane from the base of the plant. The younger canes often have more prolific and longer-lasting bloom. However, if older canes are producing numerous branches (laterals) with lots of bloom, do not cut them off. Cut out old cane in early spring just as buds begin to swell. Cut them all the way down to the base of the plant (the bud union).

When cutting off canes, especially large ones, at the base of the plant, make a flat even cut right next to the crown. If you leave cane sticking up (referred to as a *stub*), you may reduce the formation of new cane (*basal breaks*). Furthermore, stubs are unattractive, often rot down to the crown, and commonly get infested with insects and infected by disease.

Remove twiggy top growth or many canes clustered together (cane head). Cut to an outward-facing bud below the cane head. Do this just as buds swell in spring, or even earlier if possible. Do the same with forked cane tops.

If there are double or triple growth buds in one spot on the cane, remove all but one by rubbing out the other bud or buds with your thumb. Do this just as they begin to swell.

If you don't notice these buds right away, you will get two or more shoots from the same spot. Snip off all but one of these.

Sometimes, canes produce no flowers. These are referred to as *blind shoots,* possibly caused by cool temperature or frost at time of bud formation. If the rest of the plant is vigorous and flowering profusely, cut the blind shoot back so that the cane has only a few growth buds. If the plant is somewhat frail, leave the blind shoot alone. Even though it is not producing flowers, it is helping the plant produce food for healthy growth. Certain varieties are prone to forming blind shoots. These include some very fine plants, such as 'Chrysler Imperial.' If you have a plant forming blind shoots frequently, do not prune it early in spring. Delay pruning for several weeks. This is an exception to the rule, and we know only that it works—but not why or how.

Taking cut flowers from a rosebush is a form of pruning. Be careful. Never take cut flowers from young or frail plants. Simply remove spent blossoms. Leave all foliage on the plant. On vigorous plants, leave two leaves on each branch when cutting off roses. Cut back to a point ¼ inch (6 mm) above a five-leaflet leaf. The rule about five-leaflet leaves is flexible. Really healthy plants may have seven leaflets or more. You want to cut back to a leaf with a larger, rather than smaller number of leaflets.

Hedges require special attention. Keep the base of the hedge wider than the top. Keep the hedge relatively informal. If you sheer it as you would other hedges, you destroy many buds. Remove older cane each year to encourage bushy new growth which often blooms more prolifically than aging growth. If you want a formal look, use a different plant.

Yellow roses resent pruning. Prune them as little as possible. They have a common parentage that makes them ultrasensitive to pruning. Remove dead or diseased wood only.

Never prune late in the season on plants that will not be buried using the Minnesota Tip Method of winter protection. Pruning encourages new growth, which will die off in severe weather. Furthermore, roses need the food stored in cane to survive cold winters. Late pruning or deadheading also delays the formation of hips which leads the plant into dormancy. So, stop pruning by late July or mid-August. This advice applies mainly to Species (wild) and Shrub roses that are left standing through severe winters.

If you spray plants with a dormant spray before tipping, you can cut off the tops of cane before winter protection without serious consequences. A number of growers cut cane back to a length between the knee and waist. This makes the plants shorter and

Pruning

Roses need pruning throughout their entire life cycle. Detailed information on pruning is also included in the individual sections in Part I. Following are just a few examples.

Pruning begins in the preparation of bare root stock for planting. When a whole cane is cut off, it should be cut back flush with the crown. On this bare root plant it wasn't. The same flush cut is important on mature plants that lose a whole cane to dieback or winterkill.

When plants are raised in early spring, they often need extensive pruning. This plant has dead cane, forks, crossing cane, spindly cane, and so on. It needs extensive pruning to grow properly.

This rose has been pruned to a few healthy canes. The process appears ruthless, but it results in healthier plants with much more bloom.

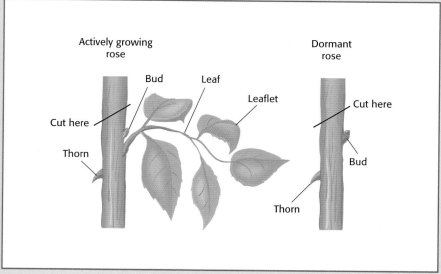

Whenever you prune, make clean cuts at an angle above a leaf with no fewer than five leaflets.

The correct cut on cane begins ¼ inch (6 mm) above an outward-facing bud or leaf and angles down to a spot just opposite the bud or leaf. When making this kind of cut, keep the thin, sharpened portion of the pruners against the cane you intend to save. This results in a cleaner cut.

therefore easier to bury. Jerry Olson does not cut his plants back, although he does snip off blossoms. He goes through the extra work of burying the entire (taller) cane. The advantage of this is extra and earlier bloom the following season. Both methods are commonly used by serious rose growers in cold climates.

Carry a bucket to dispose of cut-off portions of cane. Snip longer portions into tiny sections. Never leave cut cane on the ground because it may get infected by disease or infested with insects causing problems in your standing roses.

Winter Protection

Information on hardiness is given throughout this book. This will help you decide whether or not to buy a specific rose. Plants that can survive without any winter protection require the least maintenance. Those that require tipping, a special method for protecting them during cold winters, take a lot of time and energy to maintain. Many rose varieties will not survive at all in cold climates. Growing these is fine if you're willing to replace them each year at considerable cost. This guide includes very few varieties that have a tendency to die out even when properly protected from the cold.

In understanding the term *hardiness* it helps to know a few other terms as they're used throughout this book. *Fully hardy* or *very hardy* means that a plant will survive in your area when exposed to the elements, including cold, sun, and wind. Some fully hardy plants suffer limited dieback. This means that part of the cane may die, but the rest of it lives. Die-

Making a Proper Cut

IT IS IMPORTANT TO make proper cuts to help a wounded plant heal quickly, prevent disease, and produce growth in the right direction.

Place the sharpest part of the blade next to the portion of cane you want to save. This makes a smooth, even cut. Always use bypass pruners or lopping shears when making a cut. When cutting off entire canes, cut right at the base of the plant with a flat, horizontal cut as close as possible to the bud union, if there is one, or base of the plant.

When cutting off portions of upper cane, make cuts ¼ inch (6 mm) above an outward-facing bud. Buds look like the point of a pencil on the cane. Don't touch or rub the bud, since this can damage it. If you cut too close to a bud, it often dries out and dies. If you cut too far above the bud, you leave a piece of budless cane (hat rack). This also dies off and may become diseased. If you cut between two buds or leaves, the cane will die back to a bud below. The dead tissue above the bud is then prone to disease that could cause the death of the entire cane.

Make every cut at a 45-degree angle. Begin the cut above the bud, and angle it down to the opposite side. This helps moisture run off the end of the cane and prevents disease.

After each cut, dip blades in a solution of 9 parts water to 1 part bleach. The solution kills bacterial disease. Other disinfectants (Lysol or rubbing alcohol) may be used, but bleach is effective and inexpensive. This step is particularly important if you are using organic growing methods exclusively.

Cover every cut with pruning compound (a number of commercial types are available), orange shellac, nail polish, grafting wax, or Elmer's glue to stop insects from burrowing into the exposed cane. This likewise is especially important if you are using organic pest and disease control. Gardeners using inorganic sprays rarely bother with this step.

When cutting cane for cut flowers, follow the rules as if you were pruning the plant. Cut ¼ inch (6 mm) above an outward-facing five-leaflet leaf (or seven-leaflet leaf on some plants), not a three-leaflet leaf. The latter produce spindly shoots with small or no flowers (blind shoots); the former produce vigorous shoots with lots of flowers. (See illustration of proper cut on the previous page.)

back can affect just the ends of the cane (mild die-back) or the entire cane (severe dieback). You remove the dead portions while the rest of the plant grows and blooms. *Crown hardy* means that all of the cane probably will die and have to be removed each year, but the plant will send out new growth and form a bush in one season—similar to a perennial. *Hardy with protection* means that a plant must be buried using the Minnesota Tip Method. *Not hardy* means that a plant will not survive in cold climates, even if properly protected; almost all of these have been excluded from the book.

Minnesota Tip Method

There are a number of ways to protect roses during severe winters. Most of them are variations of the Minnesota Tip Method, the single most effective way to protect tender roses from severe temperatures.

If you have lots of roses, you can begin the process as nights begin to cool at the end of the season. A fall frost is a good signal to begin tipping. Definitely, tip roses after a hard frost and before temperatures dip to 20°F (-6.7°C) or below.

Always tip before the ground freezes.

Here are the steps:

- Rake away all mulch, and compost it. Mulch may contain disease spores or the eggs of insects. In many instances, almost all mulch will have disappeared by fall. In such cases, just make sure to pick up all fallen leaves and debris around the plant.
- Unless it's rained recently, saturate the ground around your roses with water after raking away and composting all mulch. Do this 1 or 2 days before tipping.
- Spray the entire plant with a dormant spray, containing mineral oil or a lime sulfur solution. Dormant sprays are available in most landscape centers. Spraying a second time 2 days later is very effective.
- Although optional, it's a good idea to remove all flowers and buds. Plants will continue to grow underground. Why sap the plant's strength by supplying food to buds that will never be seen or used?

- You may or may not want to cut the plants back to make them easier to work with. Jerry Olson, for one, does not cut his roses back at this time of year because cut ends do not form calluses (covers over a wound) easily in cold weather. Calluses help the plant by preventing infections and invasions by insects.

  On the other hand, obviously, shorter roses are easier to bury. So, some growers do cut back their roses. Cut roses back between your waist and knees if you want to make the job of winter protection easier. If you do this, using a dormant spray is particularly important, since the cut ends do not have a chance to heal before they are buried. Also cover the cut ends with an appropriate substance such as Elmer's glue or clear nail polish.
- Tie the branches of each bush together with a polyester twine (for smaller plants) or nylon rope (for larger plants). Polyester or nylon won't rot and break as you pull up the plants in spring. Polyester rope will often unfurl, however, so burn the ends with a lighter to melt the fibers together. Tie the twine or rope as tightly as possible without snapping any cane. If a piece of cane breaks, snip it off with pruning shears. Also try to avoid scraping off outer bark as you pull the twine against the cane, since these wounds can get infected during the winter. If you pull the cane together slowly but forcefully, it rarely scrapes.

  The easiest way to tie plants is to begin wrapping the twine around the base of the canes. Pull the plant together in a bear hug, then encircle the cane farther up, pull it together in another bear hug, and continue doing this to the top of the cane. You can also think of this as lacing a boot—only you're fully encircling the plant. The length of rope or twine used should accommodate the height of the plant. You'll have to do this exercise a few times to better judge what lengths of rope or twine to cut.

  Some growers find this technique easy with nylon rope but difficult with thin, nylon twine. If you prefer, cut twine into smaller pieces. Pull the twine tight around the lower area and tie it in a knot. Then keep doing the same thing as you work your way up the plant. Instead of one long piece of twine, you'll use six or more smaller pieces.

The tighter you can get the branches, the better. The plant will take up less space and be much easier to bury. If canes stick out from the ground, they often die off. Tight canes stay underground.

Leave extra twine from the top portion to stick out of the soil if you want to know where a plant is buried.

- Dig a trench 6 to 8 inches (15 to 20 cm) deep (or deeper if necessary), starting away from the plant and working toward it. The trench should be as long as the plant is high, and wide and deep enough to accommodate the entire rosebush. Throw the loose soil onto a tarp or into a wheelbarrow so that it doesn't get compacted on your lawn or garden soil.

Dig right up to one side of the plant. You can use a shovel on this one side. Occasionally, it helps if you dig out some of the soil underneath the stem with your hands. This way, you can clean away much of the soil around that side of the base of the plant. You'll begin to expose some roots.

- Loosen the soil around the base of the plant on the *opposite* side of the trench with a spading fork (not a spade, but rather a four-tined garden fork). Do this as deep as necessary to loosen roots. Avoid breaking the roots, but loosen the plant thoroughly. *Do not use a shovel on that side, since it will cut roots.* You want to loosen, not sever, the roots!

- Using your spading fork, dig deeply under the rosebush and "tip" it into the trench. Use the fork like a lever. To stop the plant from springing back up, use a second fork as an anchor over the bush. Or, you can make prongs to hold rosebushes down temporarily.

Another way is to kneel next to the bush and push it down into the trench with your gloved hand. You can feel it giving way. If it resists, loosen the soil some more with the spading fork. Eventually, the plant will lean down. Push it down into the trench and hold it with one hand while you scoop dirt onto it with the other hand. Get enough soil over the plant so that no cane is exposed. The weight of the soil should stop the plant from spring-

ing back up. If you've tied the cane tightly in the first place, the rose should fit snugly into the trench.

Tip the plant so that the bud union goes into the trench first. (Look for the flat spot where the original grower cut off the cane of the understock.) This puts less pressure on the vulnerable bud union which could snap (a problem mainly with Tree roses). *It also forces you to lay the plant in the same direction each year, highly recommended by all experienced rose growers.*

When you tip a rose, be prepared for some strange noises. You might hear cracking or roots giving way in a kind of tearing sound. As long as the plant is fairly loose, these sounds are normal.

- Plants can be placed over each other with soil in between without problems. You can also place potted roses in the garden area in between or alongside tipped roses. These take up space and reduce the need for additional soil. You can also bury potted roses in flower beds and vegetable gardens if you prefer. The goal is to bury plants wherever there is loose soil.

- Cover the entire plant with soil. If more soil is needed, bring it in from another garden area. The uppermost portion of the bush should be covered with from 4 to 6 inches (10 to 15 cm) of soil. As long as all cane is covered, you'll have few problems.

- Yellow and some pastel-colored roses are quite delicate and benefit from a little extra care. When you tip them, place them on leaves and not directly on the soil. Then cover the stems with leaves. After that, cover the plant with soil as outlined in the preceding step.

- Keep the ground moist until it freezes.

- Then cover the buried plant with at least 4 to 6 inches (10 to 15 cm) of whole leaves, deeper if enough leaves are available. Oak are best because they are acidic and don't mat as easily as other types, but other leaves are fine. Collect leaves from the neighborhood if you don't have enough of your own. Most people are glad to get rid of them. Pine needles are also good for winter protection. Avoid grass clippings, since they mat down and may heat up when you want the area to be cooling off. Some people use straw, but we don't recommend it for three reasons: it's quite expensive, it often attracts rodents, and it may contain weed seeds. Once you've dealt with thistles brought in by dirty straw, you'll begin to use leaves.

If you have to cover the buried roses with leaves before the ground freezes, place rodent bait in the cover. Bait may come in its own container, or you may have to place it in jars set sideways in the leaves. If you've waited for the ground to freeze before applying leaves, you can often skip this step. However, using rodent bait is highly recommended if you've ever had a problem with field mice or moles in your area.

- Spray the leaves with water. Moist leaves stay in place and freeze better when the temperature drops. These leaves act as an insulation barrier to prevent alternate freezing and thawing. **Warning:** Some gardeners cover roses only with leaves, believing that this will be enough insulation. Layering with leaves is not adequate protection. The covering of soil is critical. If you use only leaves, you may lose many valuable plants.
- Leaves will blow away in fall gusts of winds. Cover or enclose them with netting or wire to stop them from disappearing and exposing your bed in a heavy wind. Keep the corners of the wire in place with patio blocks or heavy stones.
- You may prefer to place bagged leaves, rather than loose leaves, over buried roses. This simplifies everything, works well, but is aesthetically unappealing. Or, cover the tipped roses with construction insulating blanket available from construction supply houses. The blanket must be weighed down and anchored in place. It saves on the work of dealing with leaves, but is expensive and bulky to store. It is sold in 6-foot (2-meter) by 25-foot (8-meter) rolls.

Alternative Winter Protection Methods

A number of publications promote different techniques for protecting roses during the winter. Styrofoam cones are one example. They are not reliably effective, however, and we don't suggest their use unless you're on the southern edge of the cold climate range.

Hilling up soil around the base of the plant, another alternative, is effective protection for some

Winter Protection – The Minnesota Tip Method

WINTER PROTECTION IS ONE of the most important considerations for cold-climate rose growers. The following method was developed over many years by some of the world's finest rose growers. It is the most reliable method yet developed to protect tender roses.

A severe frost is a warning sign that winter is close behind.

Though roses may have bloom and even buds on them, now's the time to act. Remove all blossoms and buds. Clean up all debris around the plant, and place it in a compost pile.

Spray your roses with a dormant spray several days before burying or tipping them. Do this twice, a couple of days apart, if possible. Spray all parts of the rose and the soil around the base. You want to kill off spores and eggs.

Starting several feet (about a meter) away from the base of the rose, dig a trench wide and deep enough to hold the entire rose once it is buried. If you have nice, loose soil, this is not a difficult job. Dig the trench so that the rose ends up buried in the direction you choose.

Winter Protection (continued)

When you get close to the base of the rose with the spade, you'll notice roots. At this point, stop using the spade and start removing the soil with your hands. Loosen the soil as much as possible, and set it off to the side of the trench. Dig in between the roots to expose the shank or main stem as best you can.

Using either twine or rope, tie up the rose. Tie it as tightly as possible without snapping the canes. It is surprising how tight you can get roses using this method. The object is to take up as little space as possible and also to stop the rose from popping up through the soil, where it could be damaged by winter cold snaps.

Now dig into the soil on the opposite side of the rose with a spading fork, not a spade, to loosen the roots. Avoid severing the roots. Keep digging around the plant until you begin to feel that it's giving way. You may have to dig and rock the plant in several places to loosen it up.

Now's a good time to add a little organic matter to the trench. It stops the soil from compacting when you raise the plant in spring.

If the plant is loose enough, you can bend it down into the trench. This is known as *tipping* the rose. The first time you do this seems scary because you hear strange noises. You don't want to break any cane, but don't expect the plant to bend over without any resistance whatsoever. On the other hand, if it just won't bend over, loosen the soil some more with the spading fork.

Each grower has his or her own way of covering roses, but one of the easiest is to kneel down next to the plant, hold it in place with a gloved hand, and push soil over the plant with the other. Keep covering the plant and firming the soil until the buried rose stays in place. The more soil you get over the plant, the better. You want at least 4 to 6 inches (10 to 15 cm) of soil covering the rose.

Buried roses can be hard to find in spring. If you leave a piece of twine or rope sticking out of the ground from the tip of the cane and tie it to a marker, you can follow the twine or rope right back to the plant easily the following year. If you have lots of roses, we suggest keeping a notebook of where each has been tipped or buried.

If you have roses grown in pots, they can be placed in a trench and covered in the same way as roses planted directly in the garden. Tree roses respond well to a bed of leaves underneath.

Also cover Tree roses with whole leaves before burying them. Many rose growers do this with yellow-flowering roses as well, since they seem to die out more easily than other roses. *The covering of leaves is in no way a substitute for 4 to 6 inches (10 to 15 cm) or more of soil to be placed over the tipped roses.*

After soaking the soil over a single tipped rose or entire bed, cover it with whole leaves to a depth of 6 inches (15 cm) or more. If you have a large bed, roll chicken wire over the leaves. Hold the corners of the wire in place with stones or cement blocks. This stops leaves from blowing off the bed.

If you don't cover the leaves with wire, stop them from blowing away by sprinkling them regularly. When temperatures dip below freezing, they become stiff with frost or ice crystals. This holds them in place until it snows. Sprinkling leaves until the ground is frozen is a good idea whether using wire or not because it keeps the soil below moist until it has a chance to freeze solid during winter.

More and more growers bury their roses, saturate the soil, and then place plastic bags filled with leaves on top of the soil. This way there is no need to rake up leaves in spring. You can dump the leaves directly from the bags into a shredder mulcher to create mulch for the garden. Until it snows, the bags are admittedly unattractive.

roses. We suggest this at times throughout the book for specific varieties or groups.

Loosening the bases of roses, bending them over, and then covering them with cardboard and a little soil is also fine for specific roses. Again, we tell you when this will work.

However, we do not agree that planting the bud union 6 inches (15 cm) deep is a good way to protect most tender roses, especially those that grow best when half the bud union is exposed, as suggested in many sections in Part I. This method, presently being touted as "new," was thoroughly tested more than 50 years ago. The idea is to avoid the work involved with winter protection by planting the lower portion of canes above the bud union deep enough so that the roses won't die off in winter. Each fall, you cut the canes back to ground level, rake up all debris, and let the plant resprout the following spring like a perennial. You will often get growth and some bloom, but, in our opinion, the resulting plants are far inferior to those protected by the Minnesota Tip Method. Almost all experienced growers are now using the Minnesota Tip Method or slight variations of it to protect their tender roses in cold climates.

We do agree that the principle of deeper planting is effective with specific types or varieties of budded and own-root roses. The deeper planting may cause budded roses to produce roots from the lower portion of their canes, and this normally increases the plant's hardiness. In effect, you create own-root roses from budded ones. When you bury own-root roses more deeply than normally recommended, you may also get healthier plants with better hardiness. We tell you exactly when to do this throughout the book.

Some of the plants in this book are crown hardy. This means all of the cane dies off, but the plant sends up new growth that blooms well. A little winter protection other than snow is highly recommended for these roses. In late fall cover the base of the plant with potting soil. Emptying a bag or two of potting soil over the crown is easy and enriches the soil. We recommend potting soil because it is looser than most garden soils and much easier to wash down and away from the plant in spring. Even better is compost, if you have enough of it. Moisten the soil or compost. Once frozen, cover it with a thick layer of whole leaves. This protection will then be covered with snow and acts like a blanket for the crown below.

Rich Hass has developed two methods of protecting roses, both standing and in pots, that vary somewhat from the traditional ones already described. So that we don't have to explain them throughout the book, we'll simply refer to the first as the Hass Teepee Method and the second as the Hass Mound Method. They cut down on the time and energy needed to protect tender roses—nice for anyone who loves roses but leads a hectic life.

The Hass Teepee Method

A number of plants in this book are best protected using the Minnesota Tip Method, which has been covered in detail. However, tipping roses involves more work than the following procedure, which can be used for any tender rose: In late fall, after the first frost, tie up the plant you want to protect with nylon rope or synthetic twine as explained earlier in the Minnesota Tip Method section. Now, place three bags filled with whole leaves around the plant at an upward angle to create a teepee-type shelter. If there are any spaces between the bags, just stuff them loosely with whole leaves. Tie them together if necessary to keep them firmly in place. These bags act as insulation against winter cold and desiccating winds. This method has been tested now over a period of years and has worked well with little loss of cane. Note that black plastic bags are preferred to the compostable green or paper bags, because they do not break down and conform to the plant's shape better.

The teepee method depicted here is simple. Tie the canes of the rose together and then in late fall place 3 bags filled with leaves against it at an upward angle (in the photo the bags have been pulled down and away from the protected rose to show you how to prepare it). In a more accurate depiction you would not be able to see the plant and would just be looking at three bags that seemed to be self-supporting.

The Hass Mound Method

This method was devised as a way of protecting potted roses without having to bury them alongside tipped roses or without having to bury them by themselves

Surround an open space with black bags filled with whole leaves. Then lay your potted roses down between the bags. The pots can be very close together.

Once you lay the pots down, cover them with a very thick layer of whole leaves. The more, the better. Then cover the entire area with bags filled with leaves leaving no area exposed. This acts like a cocoon for your potted roses. Throughout the cold-climate region bagged leaves are often lined up for pickup in the fall. This offers you an almost endless supply of leaves other than the ones available in your own yard.

in another area of the yard, such as a vegetable garden. Place bags of whole leaves on their side around three edges of open ground in late fall after the first frost, but before the ground freezes. The enclosed space should be just large enough to accommodate all of your potted roses laid down on the open ground. If you want to place the pots over an area where roses have already been tipped, that's fine, but any open area works well. The bags act as a wall for this enclosure. Now place your potted roses (canes tied up tightly) on the ground horizontally. Work out from the enclosed side to the open one, which you now enclose with bags as you have already done with the other three sides. You can get lots of pots into a relatively small area. Cover the entire enclosure with 12 to 18 inches (30 to 45 cm) of whole loose leaves. Finally, lay bags filled with whole leaves on top of the blanket of loose leaves below. This method saves you from having to dig trenches in which to bury the roses. Again, this has been tested for years now with remarkable results. Locating the plants without damaging them in spring is very easy. Until the first snow, the mound is unattractive. But, it saves you a lot of time and energy. It is also encourages you to grow more roses in pots than you might have in the past.

Raising Plants in Spring

As the weather begins to warm in spring, feel the leaves. If they are beginning to thaw, remove them in stages. This will take a number of days. Reserve the leaves to use as a mulch.

The heat of the sun will now warm the soil. As the weather continues to warm up, the soil will thaw completely. Begin removing soil when ice crystals are no longer present. You may have to do this with your gloved hands to avoid injuring the cane of the buried roses below.

You want to raise the plants up only after any danger of severe cold is past. This varies by area and by year. Giving anyone an exact date is impossible. You will get a sense of this as the years pass. Sometimes, you'll make a mistake—everyone does, even the best rose growers. A severe cold snap under 20°F (−6.7°C) in this early stage will kill some of the cane and, occasionally, may kill a plant, but raising plants has to be done before they begin to rot in the soil or before they begin to shoot out lengthy, new growth. The latter will break off and stunt growth when the plant is taken out of the trench.

Pull the plant from the ground by grabbing onto the twine or rope. Tug up on it until it's nearly upright, but be careful. Sometimes, the bottom of the plant will still be frozen in place. If you jerk the plant up, you may snap the bottom of the cane. So, pull up slowly. If you feel resistance, check the base of the rose. If it's frozen, run a hose on it until it thaws.

It is especially important with Tree roses to be patient after running water over the frozen lower area. You might wait a full day for the water to soak into the ground and for the base of the plant to thaw out. The reason for lots of patience is that the bud union on Tree roses breaks more easily than on bush roses. So, raise a Tree rose only after the ground is completely thawed and loose.

Use the spading fork for leverage in getting the plant into an upright position. Press down on the soil with your foot to anchor the plant in place. At this stage, don't be too concerned about the plant's exact position. You just want to get it up and out of the ground. You can move it a little at a later time to position it correctly. This is also a good time to add extra organic matter to the soil. It prevents compaction.

Cut off the strings holding the branches together. Tap the branches lightly to get soil to drop off.

Wash all soil off the plants with a gentle spray from a hose. This cleans the plants and saturates the ground around them at the same time. Cleaning the plants helps prevent disease. Saturating the soil stops roots from drying out after being exposed to open air.

If it does suddenly turn cold (under 20°F [−6.7°C]), run a sprinkler over your plants until the weather warms up. That might mean running the sprinkler all night long in an emergency. Extreme cold snaps are relatively rare. Snowfall is generally not a problem; roses can be covered with snow after being raised and still come out just fine.

Water the rosebushes each day with a sprinkler. This "misting" keeps the cane moist, which encourages budding. Also, the plants need lots of water in early spring to form a healthy feeder root system once again.

If frequent misting is not possible, spray the roses with an antitranspirant, such as Wilt-Pruf®. This helps prevent canes from drying out until the roots have begun to establish themselves again. Once roots are growing well, the need for misting decreases.

Once bushes are budding strongly and uniformly, prune them and get them into the exact position you want. All pruned cane must be removed from the bed. Since you'll be compacting the soil, use a pronged hoe to break up the soil surface after walking on it.

Spraying with Wilt-Pruf® again is highly recommended. Only, this time you're not only trying to stop the plant from losing moisture but also smothering insect eggs and fungal spores.

Shred the leaves that you used as a winter mulch. Then place these pulverized leaves around the base of the plant, but do so only after temperatures are warm (the soil should reach 60°F [15.6°C]).

Raising Roses in Spring

The process of bringing up the buried plants in spring is known as *raising* roses. The hardest thing for us to tell you is when to do it. There is no secret we can give you, since no two seasons are alike, but the following steps have worked for years with the loss of very few roses:

After the leaves thaw in spring, remove them. Using a pitchfork is the easiest way. Just toss them onto a tarp.

Drag the tarp to an area where you can stack the leaves temporarily. You'll be using them shortly for summer mulch.

If the soil has thawed out, get into the garden and begin to locate your roses. If you have only a few, the string method outlined earlier works well. If you have many potted roses mixed into the bed, the process is a little more difficult. The object is to find the roses without doing any damage to them. Work your way from the outside in to avoid damage to buried canes.

A spading fork helps in locating roses without cutting canes and also in prying them up from the soil. When doing this, always wear glasses and thorn-resistant goat-skin gloves if possible.

Once the bases of the plants are no longer frozen, you can pull them up by the twine or rope to a nearly vertical position.

Normally, you don't want to compact soil by stepping on it. However, when raising roses, it helps to get them standing upright if you'll step down on the soil at their base. Avoid stomping all around the plant; just press down hard enough to keep them upright.

Raising Roses in Spring (continued)

Remove all potted bush and Tree roses from the bed. Place them off to the side of the bed for the time being.

Cut off all twine at this stage. If the plants are held by rope, untie it. You want the plants to spring back to their natural shape over a period of days.

As you would expect, raised roses are covered with soil. Wash them off with a hose until the cane is completely clean.

Now that you've got all of your roses raised, cleaned, watered, and ready to grow—it could snow. A late snow is nothing to worry about. However, if it gets into the low 20s (−5°C) without snow, turn on your sprinkler and keep the roses watered until the temperature gets into the high 20s or low 30s (−2 to 0°C).

All of these roses were buried, raised, and covered with snow, but now they are growing vigorously and will bloom beautifully throughout the season.

CHAPTER 6

SOLVING ROSE-GROWING PROBLEMS

As we noted, roses have survived for millions of years unaided by gardeners. The purpose of good gardening techniques therefore is to aid nature, not fight against it. The results can be spectacular. Roses have natural enemies, such as diseases and insects, which you can prevent or control. Other enemies include deer and rodents which eat the cane to survive. However, one of the worst enemies of roses is a gardener using improper growing techniques or using chemicals inappropriately. This chapter will help you solve rose-growing problems in simple, effective ways.

Diseases and Insects

Diseases and insects are common in the rose garden. Organic methods are helpful, but they cannot prevent or control all problems. Nevertheless, some growers never use sprays in the garden, not even those derived from organic substances. The reason is that sprays are indiscriminate, often killing good as well as bad living organisms. Organic gardeners let nature take its course, helping it as best they can but allowing some destruction to take place.

Other gardeners disagree and use chemicals. Chemicals are potentially dangerous and must be handled according to the directions on the containers. Chemicals used include fungicides (kills fungi and bacteria), insecticides (kills insects), and miticides or acaricides (kills mites). The spraying schedules and steps recommended in this book will keep your plants healthy while minimizing damage to the environment and you.

A third group of gardeners compromises. They try to use as few inorganic chemicals as possible, but do use them when necessary. This compromise has the fancy title of Integrated Pest Management (IPM). The basic idea behind IPM is to learn to identify diseases and insects, to use the least toxic substance to control them, and to let nature take its course whenever possible. The idea is terrific in theory but requires enormous skill to pull off in practice. It has one major drawback: by the time you spot the signs of disease or insects in any stage of development, it may already be too late and require an even greater amount of chemicals.

Whether you use chemicals or not is a personal decision. Many very fine growers with substantial numbers of roses do not. However, most growers do. You will find both organic and inorganic remedies for specific problems outlined in Appendix A—"Insect and Disease Control."

Preventing diseases before they spread and controlling insects before their population balloons makes a lot of sense. This section includes both organic growing tips and an outline of a possible spraying program for inorganic gardeners.

Organic Prevention

Organic gardeners can increase their likelihood for success with the following practices:

Buy certified stock. This is a plant certified to be free of mosaic, a viral disease for which there is no cure.

Buy types of roses noted for their resistance to disease. Hybrid Teas and Floribundas are generally vulnerable to disease and insect attack, whereas some of the Shrubs and Species (wild) roses rarely suffer more than minimal damage, if any at all.

Buy disease-resistant varieties within categories of less resistant roses. Join a local rose society; members often know which varieties stand up best to local problems. Some roses almost never get powdery mildew or black spot in specific regions, for instance, while others are extremely susceptible.

Return packaged or mail-order plants that have swollen areas (galls) on their stems or roots.

Buy potted plants with no signs of powdery mildew, black spot, or rust—check all newly purchased plants carefully.

Segregate newly purchased potted plants for several days. Keep them cool and moist during this time. Check for any disease. Then plant them with other roses.

Healthy roses resist disease. Weak or poorly cared for plants are most susceptible. So, keep your plants well watered and well fed. Plant them only in appropriate light and soil.

Fertilize exactly as directed. Do not overfertilize. Too much nitrogen can cause excessive growth prone to insect attack.

Give plants enough space for good air circulation. This keeps foliage dry at night and prevents disease. Some roses take up a large amount of space. Plant according to the rose's potential size, giving lots of room for future growth.

Water by hand, or let water run from a hose at the base of a plant to saturate soil thoroughly. Use a sprinkler (overhead watering) only as advised in the guide.

If watering with a sprinkler, water early in the day so that foliage has a chance to dry out by evening.

This rule has several exceptions, however, which are noted throughout the book.

When you raise plants in spring or bury them in late fall, consider spraying them with an antitranspirant such as Wilt-Pruf®. This may kill spores of diseases such as powdery mildew. However, recent testing suggests that using the product for an extended time during the active growing season may stunt growth.

Use extremely sharp shears when cutting canes. The cut portion of cane will be smooth, not rough. Cuts should be at a 45-degree angle so that water slides off. Read the pruning instructions in the appropriate section of Part I for each type of rose to avoid mistakes which can kill cane and cause disease and insect infestation.

Dip pruning shears after each cut into a bucket of water (9 parts water to 1 part chlorine bleach). Or wipe blades with straight rubbing alcohol or Lysol. If you use rubbing alcohol, just screw a spray nozzle right onto the bottle (cut the siphon tube as necessary). This preventive cleaning step stops the spread of disease from an infected blade.

Seal cuts with Elmer's glue or commercial products created for this purpose. Many growers who use chemical sprays do not do this, but if you are going to garden strictly by organic methods, taking this step will be helpful.

Prune as outlined earlier in this book so that light gets to the center of the plant. Try to keep plants open and airy for better disease resistance and increased bloom.

Get rid of all weeds. Many of them serve as host plants for insects that feed on nearby roses and also carry deadly diseases.

Use mulch. It encourages healthy root growth and feeds beneficial soil microorganisms which help plants take in valuable nutrients. Mulch also stops many disease organisms from splashing onto foliage during heavy rains.

Mulch is excellent food for worms. They keep the soil loose and airy by digging through it. They also fertilize the soil with their droppings (castings) and with their bodies when they die (their flesh is high in nitrogen, an essential plant food).

Keep rose beds clean of debris. Pick up fallen leaves and flowers. Remove any dead leaves from the bush. Remove spent blossoms. Cut off damaged cane. In the fall, rake up and compost all mulch and fallen leaves, leaving the ground bare. The mulch often contains spores and immature insects or eggs.

Plant *Alliums* (any member of the onion family) with roses. Bulgarian growers insist that this prevents black spot and mildew while enhancing the fragrance of many roses. However, to many people, this combination is not aesthetically pleasing and is best done in a cutting garden only. Some people insist that marigolds (*Tagetes*) or parsley (*Petroselinum crispum*) offer protection against aphids. The entire concept of companion planting is controversial, with some growers believing that any plants growing with roses compete with them for water and nutrients, and thus become weeds.

Inspect plants frequently and carefully, checking the undersides of leaves closely. Pick off insects by hand. Shake infected plants over a white plastic garbage bag. Toss insects into salty or soapy water or vinegar.

Use products containing *Bacillus thuringiensis* (Bt) to kill caterpillars and inchworms. Use powders containing *Bacillus popilliae* or Milky Spore Disease over a period of years to kill grubs. Use other biological controls that are safe to humans but toxic to many insects in the larva and adult stages of development.

Invite birds into the garden by providing lots of nesting sites. Have birdhouses nearby and a birdbath right in the garden. They are voracious feeders and one of your best allies in the garden.

Frogs, salamanders, snakes, spiders, many predator insects, and toads are also good friends to the organic gardener. If you place the birdbath right on the ground with the rim even with soil level, you will give frogs and toads a watering hole. Replace the water every other day to kill mosquito larvae (may spread serious diseases, such as encephalitis).

Use organic traps for insects as long as they do not contain any sexual scent (pheromones). Stirrup M for spider mites is an exception (see p. 230). Sexual scents often cause more problems than they solve. Organic traps are advertised in organic-oriented publications. Some are excellent, but others are worthless. In most cases you can easily create traps of equal value at home using simple methods as outlined in Appendix A.

Some organic growers advocate a simple all-purpose spray for roses which they believe does little harm and lots of good. Every 3 days they spray their roses with a solution of 1 tablespoon (15 ml) or more of liquid detergent to 1 gallon (about 4 liters) of water. Adding horticultural oil to the mixture often makes it stick better and more effective. Worth a try, but not nearly as effective as commercial fungicides.

Provide diversity in the yard—lots of different types of plants, including trees, shrubs, grasses, and perennial flowers which offer areas for predatory (beneficial) insects to survive. These insects feed on their destructive cousins and often are great allies to organic gardeners. They don't have to be in the rose garden, just nearby, to be effective.

Inorganic Prevention

You can greatly reduce the use of chemicals if you follow the suggestions in the preceding section. The most important tip of all is that healthy roses have the best chance of resisting both disease and insect infestation.

The second most important point is that diseases and insect infestations are best prevented, rather than controlled. Once a disease has started in your garden, you may have a serious problem. The same is true for insect or spider mite infestations.

A number of sprays are available for the same purpose. Alternating the use of different chemicals seems to be the most effective approach in controlling disease and insects.

Frequent, mild applications of chemicals prevents most problems.

The number of chemicals you could use in the rose garden is diminishing. In Appendix A you will find a variety of possible chemicals outlined. Some of these may not be available to the home grower in the near future or may be replaced by more effective and less toxic chemicals as the years go by. Still, you should know the names of products currently being used.

Finally, organic compounds promoted by organic growers can often be as toxic as inorganic (synthetic) sprays. Many of the tips in this section apply to the use of organic sprays as well. Remember: organic does not necessarily mean safe.

Tips on Spraying for Rose Protection

Spraying is the most effective and easiest way to apply chemicals to roses. It is the only sure way to cover all cane and foliage, especially the underside of leaves.

Preventive spraying is highly advised for roses that are susceptible to disease and insect problems. Mild doses of chemicals sprayed on plants on a regular basis are preferred over heavier doses needed to control a problem.

If you prefer to spray only after a problem begins, use the appropriate chemical. To do this correctly you must be able to identify diseases, insects, and spider mites. You may need a magnifying lens. Controlling a problem should begin as soon as possible. The longer you wait, the more serious the consequences.

Begin with the least-toxic substances. Work your way up to more powerful chemicals only as needed. You will need a variety of chemicals to care for most roses.

Do not mix chemicals unless labels indicate that it is okay and that chemicals being blended are compatible. Consider buying a chemical that serves more than one purpose. For instance, some products will kill many fungi, insects, and spider mites with one application.

Read labels carefully and completely. Using too much of a chemical can cause leaf burn.

Labels may also indicate that the chemical will harm specific plants. Certain roses do not tolerate specific chemicals well. Keep notes on ones that react badly to any particular product. For example, Double Delight® is notorious for suffering foliage burn after fungicide use. The foliage of *rugosa* Shrub roses also is often damaged by the use of fungicides.

Saturate the soil with water before using any chemical sprays. Avoid spraying water on the foliage.

Dry soil stresses plants, making them vulnerable to damage by chemical sprays.

Mix only the amount of spray needed by adding the chemical to water. Use up the spray completely. Never pour it on the ground.

When using a fungicide, growers have found that adding 1 tablespoon (15 ml) of white vinegar per gallon (about 4 liters) of solution makes it more effective. Apparently, slightly acidic fungicidal solutions are more potent.

Avoid spraying open blossoms if possible. Use a coarse spray on foliage and stems when a plant is in flower. This reduces chemical drift, but will not protect blossoms completely. When a plant is not in flower, use a fine mist.

Spray early in the morning on a calm day to prevent unwanted drift of the spray. If sprays can kill honeybees, spray in the evening after bee activity has stopped. Evenings are often the calmest part of the day. If there is a slight breeze, stay upwind.

Spray during cool and cloudy weather, not during hot and sunny periods, since many chemicals will burn foliage at higher temperatures. During prolonged heat waves, spray in the evening, when sun is not shining on foliage.

Maintain a regular schedule of spraying, beginning within 2 weeks after raising plants in spring. Frequency depends upon the weather and amount of rain but is normally no less than once every 7 to 10 days. Do this until the end of the rose-growing season.

Add 3 drops of liquid detergent to every 10 gallons (38 liters) of a complete rose spray. This helps the spray stick to leaves and stems. Some sprays, however, contain substances (spreader stickers) that make them adhere to leaves; adding detergent to these is not necessary. Do not add detergent if the label advises against it.

Try to apply fungicides several hours before a rain. This gives the chemical enough time to dry. It will then be extremely effective in preventing disease during wet or humid periods.

Spray after a rain to kill insects. They must touch or eat different insecticides to be controlled. If it rains again shortly after spraying, reapply the spray. Or, use a systemic insecticide, one absorbed by the plant and which will then destroy insects that eat foliage or flowers.

Use dormant sprays when covering plants with soil in fall (preferred) or when taking them out of the ground in spring (okay, but second best). Dormant sprays kill bacteria, minuscule insect eggs, and the spores of many fungal diseases. However, they kill the eggs of both damaging and beneficial insects.

TIPS ON SPRAYING FOR YOUR PROTECTION

Read labels carefully before using any chemical. The labels tell you how toxic a substance is, what to do in case of a problem, and often whom to call in an emergency. Follow all directions exactly.

Spray when it's calm to avoid drift. You don't want the spray to go anywhere except on the plants—not on you, not on your neighbors.

Cover all skin with protective clothing, especially rubber gloves, from the moment you begin working with a chemical. The rubber gloves that go all the way up to your elbow are best (though expensive). Cotton or leather gloves are not good enough. Wear goggles that completely protect your eyes from all sides. Use a respirator to protect you from inhaling chemicals. The point: don't let chemicals touch your skin, and don't breathe them in.

Use equipment that is working properly. Use water for a trial run. Bad nozzles or leaky hoses will get your clothing wet. Fix old or buy new equipment in winter before you really need it.

Mix chemicals outdoors in a well-ventilated area. Be careful not to spill or splash liquids.

If you accidentally spill a chemical on your skin, wash it off immediately. Use a detergent instead of soap. Read the label for further instructions. When in doubt, get to a doctor. Take the container with you.

If you spill any chemical on your clothing or it gets wet while spraying, stop and remove it immediately. Wash yourself off wherever wet clothing came in contact with your skin. Then put on dry clothes.

Immediately after spraying, wash your gloves off in soapy water (use detergent) before removing them. Have a bucket ready ahead of time. That way, you just plunge your gloves into the water and then remove them after they're already clean.

Next, remove all clothing. Wash your entire body well, especially your hands and face.

Wash your clothing right away. Do not mix it with other clothing. Use plenty of detergent.

Wash your goggles in soapy water. Clean respirators according to directions on the label after each use.

Never eat or smoke while spraying or immediately after until you've removed your clothing and washed well.

Never spray plants if the flowers or hips will be used for food, jams, or jellies.

Avoid getting chemicals in sandboxes, in pools, or on play equipment by spraying only when there is no wind. If equipment does get contaminated, wash it off.

Protect pets and wildlife by moving food or water dishes away from the area, preferably indoors. Cover birdbaths.

Allow children and pets into a sprayed area only after all foliage is dry. If the chemical is especially toxic, no one should touch the foliage until the appropriate waiting period is over as indicated on the label.

Keep all chemicals in their original containers. They should be stored in a dry, dark place (preferably cool). The best place to keep all chemicals is in a locked cabinet. All chemicals should be out of the reach of children.

Buy only the amount of chemicals needed for one season. The shelf life varies widely by chemical, but most last a year, some two years, and a limited number longer. It is sometimes hard to judge how much spray you'll use in a season. If you have any chemicals left over, move them indoors by mid-fall. If they freeze, many types become worthless. Keep them locked up.

You can get rid of chemical containers without charge at hazardous waste sites now available in most communities. Your local government will tell you where these are, as will the Environmental Protection Agency (EPA).

Sample Preventive Spraying Program

You'll develop your own spray program from personal experience. Here is one example of one that is highly effective which you can use as a model. Since insects and disease become resistant to specific chemicals, you may have to modify the program slightly as the years go by. Hand misters are fine for a few roses, but you'll want a compression sprayer for many. When spraying, cover the tops and bottoms of leaves just to the point that they begin to drip. Systemic chemicals are best applied just as dormant buds are emerging early in the season and then again about one month later.

In the fall, spray roses with a dormant spray or lime sulfur solution, especially if they are going to be buried.

Repeat this procedure just as plants are raised in spring. Once foliage begins to emerge, spray the roses with any product containing Triforine. If plants have not been buried, spray them with this fungicide just as buds begin to form.

One week later, scatter any product containing Imidacloprid around the base of the plants and over the lawn to kill grubs and larvae. Spray the plants again with any product containing Triforine or with an all-purpose fungicidal rose spray available from local nurseries.

One week later, spray plants again with a product containing Triforine or a fungicide labelled for use on roses.

One week later, spray the plants again. Varying the product used at this time is especially effective. Now you'll begin killing insects and spider mites. Use a product labelled for this use and available from local nurseries or garden centers.

One week later spray the roses once again with any product containing the fungicide Triforine. Also, kill insects using any product containing Acephate. This systemic insecticide will be absorbed by the plant and will kill chewing insects.

One week spray the plants again. Sprinkle Imidacloprid around the base of the roses and on the lawn as you did before.

Continue varying the sprays throughout the season.

If thrips become a problem, check the label of products available as rose sprays at local garden centers. If the product you're already using kills thrips, they should not become a problem.

Spider mites may appear in hot, dry years. During these times, spray your roses with a product specifically labelled to control mite infestations. The sooner you begin to control them, the better. Again, many all-purpose rose sprays will control spider mites.

The main point is to use this type of spraying program on a regular basis. Vary it slightly according to weather conditions and what you see happening in the rose garden. This kind of regular, preventive spraying is extremely effective.

The chemicals mentioned are just some of many available. Using different ones is not a problem as long as you use a combination of fungicides, insecticides, and miticides.

The availability of specific chemicals varies by area. There is a strong trend to limit the use of chemicals in the home garden. It is extremely difficult to give chemical recommendations without hesitation. Many rose growers use preventive spraying each year. Others use no chemicals at all—this is a personal decision.

Special Problems and Idiosyncrasies

Following are tips on how to deal with specific problems or idiosyncrasies.

Not Enough New Cane (Basal Breaks)

Your goal in growing roses is to get as much productive cane as possible from every plant you grow. The more healthy cane you have, the better the chance of getting prolific bloom. Canes growing from the base of the plant are known as basal breaks or shoots.

Some roses seem reluctant to produce new cane from the bud union. When you run into a plant like this, there are several possible solutions. They may seem unusual, even wacky, but they do work:

Cut down on the use of inorganic fertilizer, especially later in the season. The theory is that it may cause prolonged dormancy the following spring.

Gently rub the bud union with a toothbrush. There seems to be no explanation for this, but it does encourage new canes to emerge from the bud union.

Another technique is to wrap sphagnum moss around the bud union. Keep it moist until new canes begin to shoot up.

A variation of the moss technique is to cover the reluctant plant in early spring with wet leaves. Cover the entire plant with a cardboard box, plastic pot, or large pail after soaking the ground with water. Anchor

The reddish shoot in the foreground has no flower bud. It is known as blind wood. Directly behind it is a shoot with a bud forming as you would expect. You will inevitably have plants with some blind wood. Follow the suggestions listed here when you do.

the box or pot in place with a stone or brick to keep it from blowing away. Keep the soil around the base of the plant moist at all times. New canes (basal breaks) often appear within a few weeks.

And, don't forget to feed the plant Epsom salts (magnesium sulfate) as outlined in the section on fertilizers.

Pegging (p. 187) also will help new canes to break on Climbers and roses with arching stems.

Blind Wood

Sometimes, a cane doesn't produce flowers. This cane is called blind wood. Some growers believe this condition is a result of cloudy or cool weather early in the season; others think it may be a chemical imbalance. Before giving up on the cane, cut it back to a five-leaflet leaf. If that doesn't work and you've got plenty of canes on the rose, cut it off at the base. If you have only a few canes on the rose, don't get rid of the blind wood. It is still producing food for the plant. Leave it alone until more canes form later in the season or the following year.

One-Sided Plants

If you buy plants through catalogs, eventually you'll get a mail-order plant with canes on only one side of the stem. It's much nicer when canes are more evenly spaced, but you're stuck. The remedy: when you plant the rose in the garden, face the sparse side of the plant toward the south. This gives you the best chance of encouraging new growth in that area.

Split-Centered Blooms

On some roses you'll get deformed blooms, ones that look as if they have two centers. Some growers believe this occurs when certain varieties of roses are budded

to a specific rootstock. That is, the combination between the rootstock and the bud is to blame; if the grower had used a different rootstock, you wouldn't have the problem. Whatever the cause, there's nothing you can do about this, except to dig up and replace the rose. If you keep track of where you purchase each rose, you can ask for a refund or a replacement on different rootstock.

Rose Reversion

When you buy a climbing rose—such as 'Climbing Peace,' which is a sport of the bush rose 'Peace'—avoid heavy pruning. Hard pruning can cause Climbers that are sports of bush roses to revert to their bush form. The word *sport* refers to a natural mutation: in this case, a bush became a Climber—just by chance. Note, however, that we have not included 'Climbing Peace' in this book as a recommended plant for cold climates. We just want you to understand why some Climbers revert to bush form.

Rootstock Taking Over

It is very important to remove all suckers that appear from the rootstock on budded roses. This is covered under "Pruning" in Chapter 5. If your rose begins to have blooms different from what you expected, check the base of the cane to see whether it is sprouting below or above the bud union. If the cane is coming from below the union, you're growing the rootstock, not the budded portion you want to grow.

Plants of the Same Variety Growing Differently

Rose growers are often surprised when plants of the same variety don't grow in the same manner. One plant may do well, another poorly, even when planted in the same area and given identical care. You may be blaming yourself for a problem caused by the original grower. The best growers use selective budding, which means they choose only the best buds on a portion of cane to propagate a new plant. If you were to cut off the tip of a cane, the bud closest to the tip would form leggy growth, the bud in the middle would be a "select bud," while the one closest to the base would grow vigorously but produce few flowers. Select buds are the ones exceptional growers use for their stock. This explains why roses of the exact same name can be so different. If budded, roses can also vary depending upon the type and quality of their rootstock.

Roses with Varying Length of Shanks

You will notice that the space between the roots and bud union on roses can vary greatly. The reason is that some growers raise rootstock from seed, while others (most) grow their rootstock from cuttings. The stock grown from seed will produce rootstock with shorter shanks. The length of the shank is not important. This is just the answer to a common question.

Buds Not Breaking in Spring

When roses bud out in spring, they are said to "break." If all of your roses seem to break late compared with other roses in nearby gardens, cut down on the use of fertilizer late in the season. There may be a connection between too much inorganic fertilizer and late breaking.

If only one or two or your roses are not breaking while the rest are budding out nicely, cover them with moist leaves. Then cover the entire bush and mound of leaves with a bucket or 5-gallon (19-liter) pail to create a "sweatbox." Check the rose each day, making sure that the leaves stay moist. This treatment often induces buds to begin forming. Once buds break, take off the pail and the leaves, but keep the soil moist at all times until the plant is growing well. This also may encourage new canes to emerge from the base of the plant as explained in Chapter 4.

All of the roses around this plant are growing well, while this one plant is not producing shoots or breaking properly. Cover it with a "sweatbox" to get it to break (see p. 169).

Marauders

Roses are vulnerable to many animals. Following are a few tips.

Deer

Deer love eating rose canes.

A 7- to 10-foot (200- to 300-cm) fence with an electric wire on top will generally keep deer out of an area, but they can jump these. Encircle the entire tall fence with a single strand of wire several feet (about a meter) away. Deer shy away from this double fencing. Unfortunately, this is the only reliable method for keeping deer out of a garden, but it is expensive and unsightly. Less conspicuous but still expensive is 7½-foot (225-cm) polypropylene deer fencing available from Deer-resistant Landscape Nursery, 3200 Sunstone Court, Clare, MI 48617 (800) 595-3650.

An alternative method is to tack several feet (about a meter) of welded wire fencing horizontally onto wooden stakes 4 to 6 inches (10 to 15 cm) above the ground. Deer can easily jump this wiring but are reluctant to do so.

Roses left standing during the winter are especially vulnerable to deer. Surround individual plants with wire fencing. This can be a real chore.

Some growers claim that Milorganite used religiously prevents deer damage during the summer. Spread Milorganite (dried and processed sewage) on the soil surface after every rain. Yes, it does smell.

Other growers alternate the use of human hair, flakes of soap, and reflectors as deer deterrents. Owners of apple orchards insist that deer get used to scents and unusual sights quickly—so, keep changing the repellent.

Commercial repellents are available for both summer and winter use. These products are available in most garden centers. They can be effective, but are expensive and need to be reapplied regularly.

We have found that if deer are hungry enough, they will eat just about anything, no matter what home or commercial repellents are used. Dogs and high fencing remain the best overall deterrents.

Field Mice

Mice eat the roots and outer protective layer of canes on rose plants. This often stunts growth and may kill plants altogether. Place poisoned baits around the base of cane in summer. Make sure that pets and wildlife cannot get into them. Traps work well, especially in late fall. Bait them with peanut butter or American cheese. Place poisoned baits or mothballs next to canes buried using the Minnesota Tip Method for winter protection.

Moles

Moles can eat the roots of roses during the summer and both the roots and outer layer of canes buried in the winter. The damage can result in the death of individual canes or the entire plant. The most effective control for moles are traps specially designed for them. These are available in some nurseries. You need to locate mole tunnels and place the traps early in the season for best control.

There are also many poisonous baits available at local garden centers. Again, location of tunnels and proper placement of baits is critical for control during the growing season. Read instructions carefully to avoid harming pets. Frankly, traps are effective, fast, and pose less danger.

Some gardeners swear by offbeat ideas to get rid of moles. They include pushing bits of sponge soaked in ammonia, mothballs, laxatives, and fruity gums into tunnels. Some mole deterrents contain castor oil and are believed to be effective. Home remedies include mixing it with water and pouring it into tunnels at least twice early in the season. All of these deterrents are controversial. They only repel moles. Traps kill them.

Although moles are a nuisance, they do kill many destructive soil insects. Some gardeners regard them as more beneficial than harmful. That would depend, of course, on how much damage they are doing to your roses, other valuable plants, and lawn.

Rabbits

Mature rabbits like to nibble on cane during the winter when food is scarce. Surround standing roses with chicken wire. Take expected snowfall, which raises elevation, into consideration when putting up wire. It may have to be much higher than you think.

Baby rabbits come into the garden in early spring. Stop rabbits from getting into the garden by surrounding the area with a fence. The bottom of the fence should have very narrow openings or holes so that even the tiniest rabbit cannot get through. However, rabbits are really much more of a problem in winter than in summer for rose growers. They much prefer other spring and summer plants to thorny cane.

Cats are extremely effective at controlling rabbits, but they also reduce the bird population proportionately. Birds are a great ally in the rose garden.

CHAPTER 7

PROPAGATING ROSES

reating new rose plants from old ones is one of the most exciting aspects of rose growing for some people. In certain instances it can be extremely difficult, almost impossible, even for commercial growers to use a specific method for a specific rose, but in other instances you can just stick a piece of stem in water, and it will form roots. So, propagation of roses varies from mystifying to magical. The huge benefit of propagation is that you can get many roses from a single plant, *although legally you cannot propagate plants that are under patent without permission and appropriate payment to the patent holder.*

Plant labels and mail-order sources listed in this guide should provide you with up-to-date patent information. If you discover that the rose you want to propagate is still under patent (patents last 20 years), you must get permission from the patent holder and pay the required fee for duplicating the rose.

A number of roses produce little plantlets off to the side of the mother plant. These runners or suckers are easy to dig up with a spade. Do this in early spring.

Division

Unlike perennials (herbaceous plants), most roses (woody plants) cannot be divided to produce new plants. Some mature roses can be divided, however, but this is relatively rare. However, a number of the Shrub and Species (wild) roses produce runners, or suckers, off to the side of the mother plant. The original plant sends out long underground stems, or stolons, and from these emerge the suckers, the new plants.

Dig up these plantlets in early spring as soon as the ground can be worked. Plant the sucker as you would a bare root rose, giving it lots of water to help ensure its survival. The easiest way to cut off the root is with a pointed shovel. Aim it straight down, and step on the shovel. The blade will cut through the root. Then dig around the runner, keeping as much of the soil around the root as possible. Sometimes, the soil falls off. This is okay as long as you plant the runner immediately and give it lots of water.

A good sucker has several stems and a healthy root system, like the one illustrated here. Plant it immediately as you would a bare root plant.

Air Layering

Anyone who has propagated woody indoor plants will be familiar with air layering, another way of creating new rose plants from a mother plant. Air layering is an ancient art for propagating woody plants. It is often, but not always, successful.

Following are the steps:

- Choose a healthy cane, and remove leaves along an 8-inch (20-cm) portion halfway up the cane. With a razor, make an upward, ¾-inch (18-mm) cut at an angle to a point approximately ¼ inch (6 mm) below the spot where a leaf was attached to the cane.
- Cut halfway through the cane. Hold the wound open with a ⅛-inch (3-mm) piece of toothpick. Dust the wound with rooting hormone to encourage the formation of roots at the cut. These hormones are sold mostly as powders in local nurseries.
- Moisten sphagnum moss, long and stringy, with warm water. Place a handful of the moss around the cut portion so that the moss is 2 to 3 inches (5 to 7.5 cm) above and below the wound. When doing this, wear rubber gloves, since the moss can cause a rare fungal disease.
- Wrap the moss with clear plastic. Secure the bottom of the wrapping with tape. Secure the top of the plastic with a piece of twine. Open the twine every few days and mist the moss if it begins to dry out. Keep the moss moist at all times.
- Do all of this carefully. If you make too deep a cut, the cane will break. If you bend the cane, it will also break. Many growers support the cane with a piece of bamboo to stop it from breaking in heavy wind or by careless handling of inquisitive visitors.
- Over a period of weeks or months, roots will begin to form where the cut was made into the cane. When the roots are 1 inch (2.5 cm) or longer, remove the plastic. Leave the moss in place, since roots are growing in and through it.
- Sever the cane below the rooted area. Then plant the rooted portion as you would a bare root plant.

Although air layering is possible with some roses, it's not a common technique for this group of plants. Ground (or soil) layering which follows is more common.

Ground Layering

One of the easiest ways to propagate roses with long, arching canes is ground layering, which is the same as soil layering. Climbers, certain Shrub roses, and Species (wild) roses lend themselves to this technique. This is a variation of air layering and is most effective with plants that do not require winter protection, since it may take a full year for roots to form.

- In early summer bend a long cane to the ground. The cane should be mature but flexible. When bending the cane, be gentle; avoid crimping the base. Anchor the end of the cane at its tip. Tie a portion 12 inches (30 cm) long to a stake. Tie it firmly, not tightly, using fabric or soft tying material. Where the cane would touch the ground, either loosen the soil or place a pot filled with soil into the ground.
- Strip all leaves off of an 8- to 12-inch (20- to 30-cm) section of the cane. Make a slanting cut as explained in the air layering method, or wound the cane with a long, shallow, 3-inch (7.5-cm) cut. Place a piece of toothpick in the wound to keep it open. Dust with rooting hormone. Push the wound into the loose soil and cover with 3 inches (7.5 cm) of soil after pinning the cane into place with a piece of wire. Or, just place a rock on the soil covering the cane to keep it anchored in place.
- Keep the soil moist at all times. Over a period of weeks or months (depending on the variety), roots will form at the cut. In some cases, root formation may take a full season.
- Once roots form, sever the cane from the parent plant. Then plant the tip as you would a bare root plant or just let it mature in the pot placed in the ground.

Pegging

As discussed in Chapter 5, pegging refers to the practice of bending the tips of long, arching canes down so that the tip is lower than the origin of the cane. The tips are usually held in place with some sort of tie on a support. On repeat-blooming varieties, pegging causes the cane to produce numerous laterals which bloom profusely. Pegging can also be used to propagate new plants. Some arching canes naturally touch the ground. The tips send out roots and create new plants. If you have seen a patch of wild blackberries growing, then you have a natural model for pegging.

You can help nature by forcing and fastening the tips of arching canes to the ground. These will then sprout and create new plants. Sever the tips off and plant as bare root plants once a solid root system is established. Keep the plants well watered until they're growing vigorously.

Pegging is extremely effective with 'Frau Karl Druschki' and ground covers such as 'The Fairy.'

Cuttings

There are two types of cuttings used to propagate roses. Cuttings taken from mature wood are called hardwood cuttings because the wood is hard, while cuttings taken from new wood are called softwood cuttings because the wood is more pliable. Hardwood cuttings are taken during dormancy (after leaves fall off), while softwood cuttings are taken during a plant's active growth (when buds or leaves are growing).

Professional growers often take hardwood cuttings and are successful getting these to root. The home grower often finds this method extremely difficult, however. Softwood cuttings are much easier to work with. Still, roses vary in their likelihood to form roots from cuttings, and it is common to lose a number of cuttings even under good conditions. Commercial growers report losses of up to 30% even under ideal conditions for some varieties. Moreover, some roses simply will not root using this method.

The home grower has the most success with Climbers, hardy Floribundas, Miniatures, Species (wild) roses, and some Shrub roses.

Being able to take cuttings has a number of advantages. Plants are quite expensive, and making a number of plants from a single parent saves a great deal of money. Cuttings form plants that grow on their own roots. These can be more valuable than plants grown on rootstock. They are hardier and will not form suckers (canes from the rootstock). In the case of Miniatures, the small size of the plant is easier to control if plants are grown on their own roots. Most, not all, Miniatures are. All should be. Expect mature flowering plants in 2 to 5 years depending on the variety being grown.

Hardwood Cuttings

Remember that this technique is difficult even for professionals, so don't get discouraged if it doesn't work on a particular variety. Try it on a number of types of roses. Consider joining a local rose society to have someone show you exactly how to do this, when, and with what roses.

Following are the steps:

- In the late fall, choose a healthy cane from the current year's growth. Push against the thorns. If they snap off easily, the cane is at the right maturity. The upper portion of the cane should be no smaller in diameter than a pencil. Cut the entire cane off cleanly at its base.
- Now cut the whole cane into several cuttings or more, each with three to four growth buds (little bumps found just above the spot where leaves join the cane). Use a sharp knife (not pruning shears) to cut the cane. For each cutting, make a horizontal cut ¼ inch (6 mm) above a growth bud on the uppermost portion of the cane. Make a diagonal cut ¼ inch (6 mm) below the third or fourth growth bud farther down the cane. The slanting cuts are simply a way to identify the bottom of the cuttings. If dried leaves are still on the cane, simply snip them off with pruning shears.
- With some roses you increase your chance of success by taking cuttings from the lowest portion of

the plant containing a heel. The heel is at the base of the stem, usually encircled with narrow rings. Lower portions of the cane are most likely to develop roots, although upper portions of the cane will develop roots under ideal conditions. The reason for this is believed to be that the lower portions contain more food reserves (carbohydrates) and less nitrogen than new growth on top of the cane. A cutting tends to root better when this is the case.

- By making the cut just above a growth bud, you reduce the chance of rot on the upper portion of cane. Cane rots down to a growth bud because roses send all of their energy to buds, not to cane. Making the upper cut in the right place reduces the chance of disease and loss of the cutting.
- Cuttings taken from the upper portions of cane should be wounded to increase their chance of rooting. Cut out all but the top two growth buds with a sharp knife. Or, make a slanting cut so that the bottom fourth of the cutting is exposed. Or, cut up through the bottom third of the cane, slitting it open.
- Spray cuttings with dormant spray or dip them in a solution containing insecticidal soap. Bury them outdoors under 8 inches (20 cm) or more of moist peat. If you take a number of cuttings, you can bundle and tie them together before covering them with peat. Keep the peat moist until the first hard freeze. Do not let the peat dry out.
- In early spring, remove the cuttings from the peat. Wash them off with water. The wounded ends will have formed calluses (protective coverings over the wound). You may want to dip the diagonal cut ends in either liquid or powder rooting hormone.
- Fill plastic pots with loose potting soil made of equal amounts of loam, peat, and perlite. Match the size of the pot to the length of the cutting. Moisten the soil slightly. Plant one cutting per pot. Poke a hole into the soil as wide as the diameter of the cane of the cutting. Insert the diagonal end into the soil with just the top node exposed. Firm the soil around the cutting with your fingers. Then water the soil lightly to secure the cutting in place and get rid of air pockets.

- Place the pots in indirect light in a location sheltered from wind and direct sunlight. If you have a cold frame placed correctly or painted to filter sunlight, it will provide good protection for the potted cuttings.
- If you have only a few cuttings, you may prefer to start cuttings indoors enclosed in a tent, as described in "Softwood Cuttings" later in this chapter. Do not place potted cuttings that are covered with plastic in direct sun.
- Keep the exposed portion of cane misted and the soil moist at all times. If canes could be exposed to drying winds, spray them with an antitranspirant, such as Wilt-Pruf®, to stop them from drying out too quickly. Misting the canes twice or more daily is highly recommended. Keep the humidity high around the cuttings during the early stages of rooting.
- Roots may form anywhere along the buried section of individual canes. This may take a few weeks or longer. It is critical to keep the soil moist at all times to prevent the loss of cuttings.
- Once cuttings bud out and begin to grow well, you can fertilize them lightly every week with an extremely dilute water-soluble fertilizer. You can also begin to expose them to ever-increasing amounts of sunlight over a period of two to three weeks, eventually growing them in full sun.
- Naturally, at the end of the season you'll want to give these young plants optimal winter protection. Just bury the roses—pots and all—alongside other roses being protected using the Minnesota Tip Method.

Softwood Cuttings

Rooting softwood cuttings can be either extremely easy or next to impossible. Some Miniatures will form roots in water just like Geraniums (*Pelargoniums*). Place cuttings in an opaque jar in indirect light, change the water frequently, and plant when roots are several inches long. It's as simple as that.

But, it's not that simple for many other types of roses. Cuttings often fail for Hybrid Teas. They are more successful with Alba, Bourbon, Gallica, hardy Floribunda, Miniature, Shrub, and Species (wild) roses. Expect the unexpected; some varieties simply root more easily than others. Frankly, some won't root at all.

Here are the steps to follow with the more finicky roses.

Have individual plastic pots filled with proper rooting medium already prepared before taking cuttings. Good rooting medium is a mixture of perlite and peat with slow-release fertilizer mixed in. Some growers use 2 parts perlite to 2 parts peat; others prefer 3 or 4 parts perlite to 1 part peat. Sterile sand is also very good but may be difficult to find. The rooting medium should be moist, but not soggy or wet. The best slow-release fertilizer is made up of small coated granules which break down over time with moisture. Sprinkle a small amount of these granules in the bottom of each pot, matching the amount to the size of the pot. Match the pot size to the potential length of the cutting. Pots must be sterile. If they have been used before, sterilize them by washing them in a solution of 1 part bleach to 9 parts water.

Take cuttings from wood that is firm, not too young or flimsy. The latter will just dry out and flop over. The width of the cane depends on the type of rose being propagated. Vary the cuttings to determine which take best for you.

Take cuttings at different times to increase the odds of success. Take some cuttings just as leaf buds form, take a second batch of cuttings just as flower buds begin to form, and a final batch just after flowers fade. Take cuttings from both the main cane and side branches to increase your chance of success.

Each cutting should have three to four growth buds, little bumps found just above the spot where leaves join the cane. Use a sharp knife (not pruning shears) to cut the cane. Make a horizontal cut one-quarter inch above a leaf on the upper portion of the cane; make a diagonal cut one-quarter inch below a leaf.

The length of cuttings will vary in size, depending upon the type of rose. To get three or four growth buds per cutting, you may need only a few inches or more for Miniatures, up to 10 inches (25 cm) or more for some Species (wild) or Shrub roses.

If making several cuttings from one cane, always make the upper cuts (the ones farther up on the cane) horizontal, the lower cuts (the ones farther down on the cane) diagonal. You will always plant the diagonal or lower portion down.

Remove all thorns (press them down with your thumb) and lower leaves (clip off with shears). Leave the top leaf on. If cuttings are taken after flower buds or flowers have formed, pinch or snip these off as well. Immerse cuttings for a few minutes in a solution containing insecticidal soap. Then place them in a tray of water. Never let them dry out; work quickly.

Tip: For years, growers soaked branches of willows in water to create willow water, which induced root growth from cuttings. An analysis of this water indicated the presence of salicylic acid. To create a similar solution, dissolve two aspirin per gallon of water. Then soak the base of cuttings overnight in this solution. This procedure is helpful, but not essential to get cuttings to take root.

Equally helpful is to dip the diagonal end of each cutting in either liquid or powder rooting hormone, which is available at many garden centers and nurseries. One of the most common is the powder Rootone®. If using it, tap off any excess powder.

Poke a hole in the rooting material with a pencil. Bury the cutting up to the base of the leaf (or bud if taken in spring). Firm the medium around the cutting before watering lightly to settle the medium in place.

Now you can grow your potted rose either indoors or out:

• **Indoors:** Place three cedar or rot-resistant sticks, pieces of wire, or straws around the edge of the pot. Cover the entire pot with a clear plastic bag. Then tie a piece of twine around the bag to keep it in place. The plastic keeps humidity high to help the cuttings take root. Place the pot in indirect light or under artificial light. Remove the bag occasionally to check that the medium is moist. Water as necessary, never allowing the growing medium to dry out. Mist the cane frequently; consistently moist soil and high humidity are secrets to success with softwood cuttings.

- **Outdoors:** Place the pots in a cold frame. The potted plants must be protected from direct sun. Keep the humidity high by misting the plants daily and keeping the growing medium moist at all times. Vent the cold frame as necessary to keep temperatures between 70° and 77°F (21.1° and 25°C).

If the weather is cool, heating pads indoors and heating cables outdoors can be helpful in maintaining the correct temperature.

When the cuttings begin to show new growth from the exposed cane, they are usually taking root. At this point, remove the plastic bag or lift the cover of the cold frame for several hours each day to harden off the young plants. Over a period of several weeks, lengthen the time that the cuttings are exposed. Then move the potted cuttings into full sun gradually over a period of two or more weeks. Expose them, to just a little more sun each day. Don't let the growing medium in the pots dry out; keep it moist at all times.

Once the plant is growing well in full sun, you can plant it in the garden. However, most home growers keep it growing in the pot for a full year. We suggest repotting at this time. Disturb the root ball as little as possible as you replace some of the growing medium with potting soil.

Young plants need optimal winter protection. After the first frost in late fall, bury the plants, pot and all, in the ground. Cover the soil with a thick layer of whole leaves.

Root Cuttings

It is possible to grow roses from portions of roots, as it is for a number of perennials. However, the lower root and upper woody tissues differ significantly. The result is plants with different rose coloration. We suggest using other methods of propagation.

Budding

Some roses are difficult to start from cuttings. A different technique, budding, has been developed for them. Budding, or shield grafting, is the process of taking a bud from one plant and putting it on the rootstock of another, completely different rose. It is a time-consuming technique for the home grower. Considering the cost of rootstock, the length of time necessary to get flowers, and the success ratio, it makes sense to buy professionally budded roses. However, because understanding the process and even trying it can be enjoyable for the avid rose grower, we include it in this book.

Here are the steps to budding your own roses, if you would like to give it a try:

- The first thing you're going to need is rootstock (understock) which serves as the host for the bud. There are three families of commonly used understock (*Caninae*, *Indicae*, and *Synstylae*) for bush roses. In the *Caninae* family there are at least 15 named varieties possible for use, but the one we recommend for the home grower in cold climates is *Rosa canina inermis*. In the *Indicae* family, most of the roses are best suited to the South or greenhouse culture. In the *Synstylae* family, you'll find a number of named varieties. The one we recommend, although it's prone to mildew and not particularly hardy, is 'Dr. Huey.'

    You can grow rootstock from seed or buy plants, the latter usually about 6 to 8 inches (15 to 20 cm) tall with roots 5 to 6 inches (12.5 to 15 cm) long.

    Tree roses can consist of rootstock either budded just at the top of a tall cane or budded twice, once at the base and then again at the top. The most common practice in the United States is not to bud only at the top of the cane. The most common rootstocks for Tree roses in the United States are 'Dr. Huey' and 'RW,' producing long single canes as the main stem. Other rootstocks, hopefully hardier, are being tested. *Rosa multiflora* 'De la

'Grifferaie' was once used between the rootstock and top portion of the cane, but this is now uncommon. Budded on the top of the long cane is the named variety that appears in catalogs. Numerous other roses are used in creating Tree roses in other countries or warmer climates, but these are the ones we recommend for use in cold climates.

- Plant the rootstock as you would any other bare root plant. Get it growing. Since you'll be making an incision in the cane just above the roots, keep that area as clean as possible, including rubbing off any tiny roots growing there. Space plants at least 12 inches (30 cm) apart.
- Consider buying a budding knife. It should have a plastic or bone quill for peeling back bark. A really sharp knife, scalpel, or utility knife with a razor (used for cutting Sheetrock) also are fine and less expensive.
- In midsummer, just as flowers begin to bud or just as they die off, cut off cane from the desired rose. Choose canes that produced the finest flowers. Cut off new cane from the center of a shoot about the thickness of a pencil. Pieces of cane should be at least 12 inches (30 cm) long and contain several buds, or eyes. These are called *bud sticks* (scions). Take these bud sticks when the weather is favorable (neither during nor directly after rain nor on dry, windy days).

  You are doing what is known as *selective budding*. If you take buds from the end of the cane, they will produce plants that are weak and spindly. If you take buds from the base of the cane, they will often be vigorous but with little bloom. Buds taken from the middle of the cane produce good growth and excellent bloom.
- Get rid of thorns by pressing them sideways with your thumb. If they snap off easily without taking bark off the cane, the cutting is at the right stage.
- Clip off all leafstalks back to ½ inch (12 mm) with pruning shears.
- Dip the canes in insecticidal soap solution for a few minutes. Rinse. Then submerge the cane (bud sticks) in water for 3 hours. Never let the cane dry out.

- Using the budding knife or other tool, cut ½ inch (12 mm) above an eye (node or bud), scooping into the cane with as shallow a cut as possible. Move downward and under the eye, coming out ½ inch (12 mm) below it. Some growers find it much easier to start from the bottom and work up. It's a matter of opinion, but you want to take the bud with as little material under it as possible. This flap of wood is called a *shield*. The bud is directly in the middle of the shield and is normally separated from the woody tissue underneath.
- Hold the shield gently by the ½-inch (12-mm) leafstalk. Pry off the bark from the bottom of the shield while gently pulling the bud from the underlying wood. If the bark won't come off, the cane is too old. Cut another piece of cane from a younger portion of the plant. If the bark does come off, you now have a bud to attach to the rootstock. **Note:** Some growers insist that with a really shallow cut you don't need to take off this sliver of wood at all.
- On a cool, cloudy day, prepare the rootstock for budding. Remove the soil around the base of the rootstock to expose the area just above the roots. Wipe off the stem with a moist cloth. Get it completely clean. Now make a T-shaped cut into the stem with your budding knife. Cut straight down ¾ to 1 inch (18 to 25 mm), cutting only through the bark, with the cut as close to the roots as possible. Then make the top of the "T" by cutting straight across ½ inch (12 mm), again only through the bark and never more than one-third of the circumference of the cane. This forms your T. The area should be moist, not dry, indicating that the plant is growing well and not stressed by lack of water (water plants well ahead of time).
- Peel the bark back gently along the vertical cut, easing the two flaps open.
- Hold the shield by the ½-inch (12-mm) leafstalk. Slide the shield down into the T-cut, until the bud is lined up with the horizontal cut.
- Fold the bark over the shield. Cut off any portion of the shield above the bud, without cutting into

the bud itself. Wrap the bud into place with budding tape (elastic ties). Or, use thick rubber bands, ¼ inch (6 mm) wide and 5 inches (12.5 cm) long. Wrap the tape several times below and above the bud. This keeps air out of the incision and, hopefully, will prevent the bud from drying out. Slide the band under the last loop to secure it in place on the opposite side of the cane from the bud. The bud must be completely exposed but held firmly in place with the band. Replace soil around the base of the plant almost to the level of the bud. If the bud shrivels and turns black, remove it immediately and try budding in a new place on the rootstock. You'll know the bud has taken when it begins to swell and turn green. When this happens, remove the tape.

- Depending on when the bud is inserted into the rootstock, it may begin to grow soon or the following year. Naturally, you must winter protect these young plants. When the rose forms a new shoot, cut off the entire portion of the original rootstock above it. Make the cut with pruning shears approximately ½ inch (12 mm) above the shoot. If the rootstock suckers, remove the unwanted cane immediately. If the new cane forms any flower buds, remove these immediately to direct all energy into the formation of a new rose plant. This new rose is called a *maiden*. In some varieties of roses, these maidens produce the loveliest blooms they'll ever produce and are prized by exhibitors. However, most growers remove buds so that all of the plant's energy will go into forming a new, healthy plant. In fact, they often cut the new growth, when it reaches 5 to 6 inches (12.5 to 15 cm) in height, back to 2 inches (5 cm) so that several shoots form from the base of the plant. Once a maiden has been pruned, it becomes a *cut back*.
- Following a season of full growth, winter protect the plant as well as possible. After all this work, you don't want to lose it now.
- The following year, replant the budded rose so that the bud union is half under the soil if you're growing a Floribunda, Grandiflora, or Hybrid Tea. This will help prevent suckering from the rootstock. If you're growing a Species or Shrub rose, place the bud union 4 to 6 inches (10 to 15 cm) under the soil to encourage root growth from the buried cane.

Explaining the process of budding is difficult. We suggest you ask a local person to show you how to do it. Seeing it done makes it a lot easier.

Seed and Hybridization

Species (wild) roses produce seed which in turn produces plants similar to the parent. To propagate these plants, collect seeds from hips as they change color or mature at the end of the season. Miniatures are also popular plants to start from seed.

Hybridization is more complex. Crossing one rose with another to create an entirely new variety of rose is exciting for the expert grower. **Note:** Hybridization works only on fertile plants, ones that form hips each year. Most roses form pollen (sperm), but a number will not form the plant equivalent of eggs.

For those that do form hips, here are the steps of crossing roses to create an entirely new plant:

- Remove all petals from a flower bud of one of the two varieties selected for crossing. Do this on a bud just about to open.
- Remove the anthers (yellow pollen sacs) from the tips of the stamens (male organs) to prevent self-fertilization. The easiest way to do this is with a pair of tweezers. You do not want any pollen to drop into the female organ (pistil). Pluck out the little supporting stems (filaments) as well in case a little pollen got on them.
- Cover the denuded flower with aluminum foil.
- Wait 24 hours. Open the foil. Check the pistil to see whether it is slightly sticky. Once it is, it is ready for pollination. This may take another 24 hours or so.
- From the second variety, tap pollen into a glass container. This can be done before or after denuding the first flower, since pollen can be stored in a glass jar in a cool place. Dip a brush, preferably camel

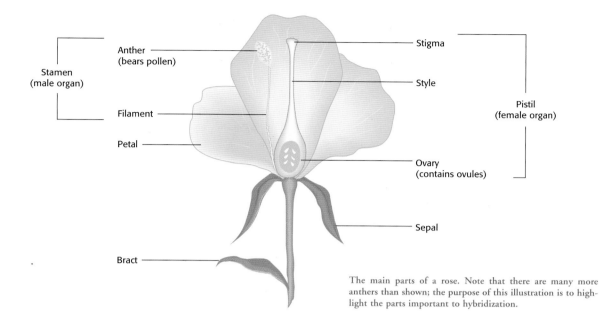

The main parts of a rose. Note that there are many more anthers than shown; the purpose of this illustration is to highlight the parts important to hybridization.

hair, into the pollen and paint it onto the stigma (female portion) of the emasculated bud of the first flower, after removing the foil, of course. Do this by lightly touching the female organ two or three times. The minuscule grains of pollen will fertilize the sticky female organ (pistil). If you are going to cross a number of roses, clean the brush each time. Swish it in alcohol and let it dry before attempting a second cross.

A second method of fertilization is to remove all petals from the second flower, and rub it on the first. Of course, the two have to be blooming at the same time and be at just the right stage of maturity for success, which is why the first method is the most commonly used, since pollen can be stored for a limited time in the glass jar.

Now cover the fertilized bud with aluminum foil. If you're worried about its coming off, tie it in place with a piece of string. This protection stops stray pollen from fertilizing the bud. Normally, you can remove the foil after a week or so. But don't do it until a hip begins to form (it will swell). On some plants this could take several weeks.

If your efforts are successful, buds will ripen into hips (fruits) containing hybrid seeds. Hips are fully ripe when they turn color and get brittle toward the end of the season. The seeds inside are dark tan or brown, not white or green. Remove and wash the seeds. If any seeds float, discard them. Keep only seeds that drop to the bottom of the bowl.

Some growers pick hips at varying stages and claim to get good results from hips just beginning to change color in the fall. Try picking hips at various stages of maturity to see which gives the best results for a specific cross.

Following are the steps for planting seeds:

• If you pick hips early, plant the seeds right away in a sterile growing medium. Some growers are lucky with quick germination. But, of course, with this procedure you are going to have to care for the seedlings indoors throughout the fall, winter, and early spring. If you're really careful, giving the young plants enough light, moisture, and mild fertilizer, you could get a small bloom in about 8 weeks.

An alternative is to mix the seeds with moist peat moss in a locking bag. Place the bag in the crisper of your refrigerator for 3 to 4 months, if you are not storing fruit there. Some fruit gives off ethylene gas, which can destroy seeds. Keep the peat moist at all times. Then plant the seeds in late winter or early spring indoors. **Warning:** If seeds begin to sprout, you must plant them right away. So, check the peat moss regularly.

- To plant seeds, press them into moist, sterile growing medium (peat, perlite, vermiculite, or a combination of the three). Cover them with a ¾-inch (18-mm) layer of vermiculite.

  The temperature should be about 70°F (21.1°C). Some growers germinate seedlings at much lower temperatures. Depending on the variety, they may be more or less successful. You might try two sets of seeds at different temperatures to weigh the odds in your favor.

  Light in a north window or weak light from grow lights is fine. Keep the growing medium moist at all times. Seeds may sprout in a few months or may take as long as 2 years. As mentioned, this process is for the ardent grower.

- When seeds sprout, place them 6 inches (15 cm) away from grow lights for 18 hours a day. Again, keep the growing medium moist at all times.

  Seedlings have dark green leaves. If you look at them with a magnifying glass, you will see leaves with saw-toothed edges and reddish stems with tiny bristles.

- Watch for *damping off,* the condition in which the seedlings topple over, their bottoms crimped and mushy. If damping off occurs, use a fungicide to kill the disease-causing organisms, and cut back on watering. Captan was once the most commonly used fungicide, then was taken off the market, and is now available in a reformulated version.

- When plants have a third set of leaves, carefully dig them out of the growing medium and pot them up in individual pots at least 4 inches (10 cm) deep filled with sterilized potting soil. Keep the pots under lights. Fertilize lightly with every watering. Keep daytime temperatures around 70°F (21.1°C) and nighttime temperatures no lower than 50°F (10°C).

- After all danger of frost, move the pots outdoors into increasingly brighter light over a period of 14 days (hardening off). Plant as desired outdoors in a protected area. Winter protect the plants during the first few years. Once mature, hardier varieties will survive harsh winters. Tender varieties, however, must continue to be protected for their entire life span. Expect mature flowering plants in 3 to 5 years.

Note: It's often easiest for the home grower to plant seeds directly in a cold frame outdoors. Soil in the frame should be moist and contain lots of peat. Keep it moist until the soil freezes. Then continue to keep it moist in the spring. The critical point is to keep the soil evenly moist for as many months as it takes for the seeds to germinate and, once they germinate, not to let the soil dry out at all. If the soil dries out when seeds begin to germinate, they will die.

CHAPTER 8

SPECIAL USES
FOR ROSES

Roses are among the most versatile flowers, with many potential uses. Each brings with it a special set of questions or problems. By reading the following information, you will have great success using roses in as many ways as you choose.

Cut Flowers

Many varieties of roses produce excellent cut flowers. You can arrange roses in almost any way you like. One rose floating in a shallow bowl can be lovely. A few or dozens of roses in deep containers are breathtaking and often exude an exquisite fragrance. Containers can be china, glass, metal, pottery, or wood. You can mix them with other flowers, but many growers prefer roses all by themselves. This section gives you some tips on using roses for ornamental purposes.

If you want long-stemmed cut flowers, the best are Hybrid Teas listed on page 42. Most roses with few petals are lovely, but do not last long. Many flowers with an extremely high petal count, including a number of Old Garden Roses, may or may not last well. Plant descriptions throughout the book often include information on the desirability of the rose as a cut flower.

Some varieties open quickly once in water but have a long vase life; others stay closed but really have a short vase life. Many stunning and longer-lasting varieties have quite an open look—one that is now in vogue.

Carry a bucket of tepid water into the garden with you when you take cuttings. Many guides tell you the bucket should be metal. There is nothing wrong with metal buckets, but it's a myth that they increase the vase life of cut roses.

Take cuttings in early morning or in late evening. The moisture and sugar content in the stems are said to be highest at these times.

If you want roses for fragrance, cut them in late morning just as it begins to warm up, preferably on a calm day. High humidity often increases fragrance.

Cut roses with fewer petals (4 to 16) when the outer (guard) petals just start to unfurl. Cut roses with more petals when the rose has unfurled slightly more. Avoid cutting fully open flowers because they will not last.

Cut just above a five- or seven-leaflet leaf, using the traditional diagonal cut described under "Pruning" in Chapter 5. Use pruning shears for a clean cut. Wear leather gloves. Support the stem by holding it just under the blossom. Leave as many leaves on the cane as possible for new bloom, preferably no fewer than five mature leaves, never fewer than three. Take cuttings only from mature plants.

Flower Food

Good flower food (frequently called a preservative) will increase the vase life of cut flowers. Good ones contain sugar (giving the flower energy), acidifiers (keeping the solution at the correct pH), respiration inhibitors (reducing food intake), and stem-plugging inhibitors (keeping fluids flowing smoothly). You can buy flower food at floral shops. Follow the label directions exactly. Too little of the food does nothing, while too much can harm the cutting.

Tell the florist what kind of water you have—hard water from wells, city water, purified water, or softened water. This helps to determine the right product for your situation. All of these waters vary dramatically in pH (acidity). You need to get the pH of your water down to 3.0 to 4.5 for longest vase life. This is very important and commonly ignored, by florists and customers as well. But, it is absolutely critical if you want to get the longest life from your cut roses.

If a commercial food is unavailable, try one of these mixtures: 1 quart (liter) 7-Up soda (not diet) with 1 drop of liquid bleach (to keep it sterile); 2 tablespoons (30 ml) lemon juice, 1 tablespoon (14 g) sugar, and ½ teaspoon (2.5 ml) bleach per 1 quart (1 liter) water; 2 ounces (60 ml) Listerine per 1 gallon (4 liters) water; or, 4 tablespoons (64 g) sugar and 1 drop bleach per quart (liter) of water. Citric acid, available in pharmacies, is also a product used to preserve roses—vary the amount to see how well it works for you. These will all do in a pinch but are not as good as commercial flower foods.

Before placing the cutting in the bucket, strip off any leaves that would end up under water in the final arrangement. Do this with a quick, downward stroke of your hand. If you're wearing leather gloves, this is easy and also dulls the thorns. You never want to leave any leaf on that would end up submerged in the final arrangement, since they infect the water with bacteria and shorten vase life.

Place the cut rose immediately into the bucket of water. The longer you delay, the greater the chance of air getting into the stem, which can harm the rose. If possible, keep cut roses shaded until you take them indoors.

Once indoors, wash off the stems. Place roses up to their necks in warm water (105°F [40.6°C]) containing flower food for about 20 minutes at the minimum and up to 2 hours if possible. Keep them in a cool, dark place during this time. Using the correct water temperature is important, since this reduces the chance of air getting into the cut stems.

Some people put cut roses in refrigerators, but as mentioned earlier, some fruits (especially apples) and vegetables produce ethylene gas which can damage blooms. A few varieties change color when chilled (especially the hybrid teas 'Mirandy' and 'Tiffany').

Just before arranging, recut the stem under tepid water (100°F [37.8°C]). Use sharp shears and cut off 1 to 2 inches (2.5 to 5 cm) of the lower portion of stem. This helps the stem absorb water once placed in the arrangement. Whether the cut is straight or at a diagonal makes no difference.

Do not scrape the stem or smash it as you would certain other woody plants, although this has been recommended in older guides. Do not use stem strippers to remove leaves or thorns.

Have your container filled with water and ready for the arrangement. When you take a stem from the water, get it into the arrangement immediately to stop air bubbles from getting into the bottom of the stem.

If air gets into the stem, it blocks water from going up the stem to the blossom. This will cause the blossom to fade or droop quickly. Often the blossom bends over as if hanging limply. This is because the air bubble in the stem acts as a clot and prevents moisture from reaching the bloom. Florists commonly refer to this as "bent neck" or "rose neck droop." If this happens, remove the stem and recut it immediately under water before replacing it in the arrangement.

If that fails, take the stem out of the arrangement. Cut ½ inch (12 mm) off the base and place the stem in water just hot enough to put your hand into (no hotter). Leave it there until the water is cool. Then shift the stem into a container filled to the brim with cool water, which should come right up to the bloom. You can do this with several at a time. It should alleviate the problem. Rearrange after the blossoms take on new life.

City water is preferred over hard water. Avoid softened water if possible. But, don't worry about this too much. You can overcome most problems caused by poor water by adding a commercial flower food to it.

All containers used to hold cut roses at any stage must be kept sterile. The simplest method is to clean them regularly with a solution of 1 part bleach to 9 parts water. Scrub the containers until they are immaculate. Do this every time you use the container. Bacteria and fungi are often responsible for short vase life.

When displaying cut roses, keep an eye on the water level. Add water and food as necessary, generally changing the water every day. Each time, cut the stem back by an additional ½ inch (12 mm) while the stem is underwater.

Keeping floral arrangements as cool as possible at all times does make them last longer. Putting arrangements in the refrigerator, again, is fine as long as no ethylene is present and as long as roses are not types that change color under artificially cool conditions. While this sounds good in theory, in practice it is quite difficult, since there is limited space in most refrigerators. Often, the best you can do is to move the arrangement out of direct sun during the day and to keep it in a cool, dark place at night.

Dry wrapping is a process of cutting and preserving roses before you really need them, as for a party in a week or two. Cut the flowers, strip off all leaves that will end up submerged in the final arrangement, and keep the stems in water until you get them indoors. Once inside, remove one stem at a time and wrap it in three layers of plastic wrap. Refrigerate the stems in the crisper of your refrigerator. The ideal temperature is 36°F (2.2°C). You may have to cut the stems shorter than you might like to get them into the crisper. When you want to use the flowers, remove the wrap, cut off 1 inch (2.5 cm) of each stem while it is submerged under water, then immerse the stems up to their necks in warm water (105°F or 40.6°C). Now let the water cool down slowly in a shaded area.

Flowers will often perk up in about 12 hours. Dry wrapping works for the home grower about 75% of the time. Experiment with different varieties to see which ones survive this grueling treatment best.

In general, if you get more than 4 to 5 days of life from a cut rose, you're doing very well. Many Shrub roses last only a day or two but are lovely in arrangements because of their abundance of blossoms and memorable scent. Hybrid Teas may last up to 6 or 7 days under ideal conditions. Each type of rose varies in its lasting capacity. The vase life also varies with the temperature and humidity. Low temperatures and high humidity increase vase life considerably.

Some roses are grown primarily for their foliage. *Rosa glauca* is outstanding for this purpose (see p. 129). Condition the cane as you would for cut flowers.

Many roses produce hips with distinctive shape and coloration. Size varies by variety. When placed in arrangements or on wreaths, these become fascinating focal points. The hips of *Rosa moyesii* and *Rosa virginiana* are among the most commonly recommended (see pp. 130 and 133).

Potpourris (Sachets)

Rose petals can be easily dried for potpourris or sachets. Collect petals when they are about one-third to one-half open after all dew has dried in the morning. Choose fragrant varieties if possible (one of the best is *Rosa centifolia*). Some people prefer to wait until the petals are about to drop, so that they can enjoy the beauty of the rose in the garden and the scent indoors.

Place the petals loosely on paper-lined trays in a dark (preserves petal color), dry place with lots of ventilation. Just a plain cardboard box works fine. So does newspaper spread out on a folding table. Or, you can use screen (preferably not metal). Keep petals dry to prevent mildew. Place them in a single layer on the bottom of the tray or box. If you get them too thick, they tend to stay moist and will mildew. Turn and fluff them frequently to keep them drying evenly. If any appear to have mildew, remove them immediately. The drying process generally takes about 7 to 10 days. It can take longer in humid weather. Using a dehumidifier in the drying room is highly recommended.

Once petals are dry, place an inch (2.5 cm) of petals in an earthenware jar and sprinkle a tablespoon (14 g) of salt on top. Add layer after layer in this fashion. Keep the jar sealed. Let the petals and salt settle for one week before opening the jar. Stir the mixture. More petals and salt may be added later, even the following year. Keep the earthenware jar sealed at all times until you want to use the potpourri in small, open dishes.

Some people prefer to put petals in covered plastic containers. This also works well if the petals are completely dry. You can add salt as outlined or buy scented oils (commonly referred to as *essential oils*) and spices to mix with the petals for your own special creation. Commonly added ingredients include allspice, cinnamon, dried lemon or orange rinds, scented geraniums, mint, and other highly scented herbs. Adding essential oil is highly recommended, since it helps the potpourri retain scent over a long period of time. Fixatives, such as orris root, are also commonly used to preserve the petals.

Dried Flowers

To dry whole flowers, use silica gel, or, if you want to save money, sandy or clay based cat litters such as Scoop Away®. Use a container that can be sealed tightly.

Cut flowers off stems when they are in the finest stage of bloom. Cut each flower with 1 inch (2.5 cm) of stem.

Insert a small piece of florist wire about 4 inches (10 cm) long into the base of the rose and bend it up to the side. This way the rose dries with the wire already inserted. Inserting the wire after the flower has dried can be difficult. The short wire can be attached to a longer one later on. Wrap the base with floral tape.

Place the roses in the gel according to instructions on the package (usually on top of 1 to 2 inches [2.5 to 5 cm] of material). It's easiest to shape the flower if it's facing up. Fill in around the petals to shape the blossom just the way you want it to dry. Cover the flower with an additional ½ inch (12 cm) of desiccant.

If you really want a rose to retain its natural color, the drying process must be quick. Some growers go to great lengths to speed up the process. They place the roses surrounded by gel in a warm oven (after cooking dinner, for example). The warmth of the oven draws moisture out of the petals rapidly.

When removing the rose from the desiccant, be gentle. Gently shake out the drying material. Brush any remaining grains off with a soft brush.

Roses will also cure on their stems if hung upside down in a dry place. The drier, the better. Use a dehumidifier if possible. Remove all the leaves. Hang the roses so that there is good air circulation between each rose. They should never touch. Again, the faster a rose dries, the better the color. Some people even hang roses in the back of their car from a string attached between the hanger holders. The heat in the car dries them quickly. Even though they are in bright light, they still retain their color.

Do not expect dried roses to be identical in color to the original blossoms. Reds darken a lot. Pinks are often quite good. Whites may turn yellowish unless they're dried extremely quickly.

Roses are dry when they become extremely firm, almost inflexible. The base of the flower should be almost brittle. Spray the roses with a commercial sealant at this point, such as Kraft's Dri-Seal or Clear Life. Names of stores carrying Clear Life are available from Design Master Color Tool, Inc., P.O. Box 601, Boulder, CO 80306, (800) 525-2644 or (303) 443-5214.

The spray not only preserves the flower but also kills off organisms (bacteria, spores, insect eggs) that sometimes damage flowers at a later time. Once the flowers have been placed in an arrangement, spray them again. This will prolong the life of the flowers and the form of the arrangement.

Finally, some growers have been experimenting with glycerine as a preservative. The advantage is that leaves can be left on the stem, which should be placed

3 inches (7.5 cm) deep in a solution containing 1 part glycerine to 3 parts boiling water. Let the water cool off and the stems remain in the mixture until the entire stem has absorbed the solution. Leaves will change color and may begin to drip at their tips. The flowers will also absorb the glycerine. Once the entire stem shows signs of absorption, remove it from the solution. Pat it dry and hang it upside down to cure. It will dry out slowly. It does not need to be sprayed with a sealant before or after arrangement.

Roses as Food

This book is about growing roses, not eating them. However, you can use roses in foods in many ways: to flavor dishes; as parts of salads; as decorative elements in oils, jams, and jellies; frozen in ice cubes for summer drinks; on desserts, such as cakes and custards; or to make teas rich in vitamin C. Following are a few tips.

All roses are edible.

Never eat any portion of a rose sprayed with a pesticide. The fact that a spray is labeled organic or recommended as an organic control does not make it safe to eat.

To eat roses or rose petals, pick them just as they begin to unfurl. Rinse them gently under water and drain.

Although all rose petals are edible, each has its own distinctive taste, ranging from metallic to sweet. Taste different roses to see which ones you prefer.

The lower portion of petals on many varieties is often quite bitter. Cut the lower area off and then taste the rose petal or rose.

Many of the Old Garden, Species (wild), and Shrub roses produce bountiful crops of hips for jams and jellies. Pick the hips when they are mature and slightly soft. Wash them well. Some people even rub them smooth with a cloth. Cut off prickly or hard ends. Boil the hips in water for about 20 minutes. Then press them through cheesecloth to get pure juice. Boil the juice with sugar for about 10 minutes or as long as necessary to get the desired consistency. Then seal in a jar with wax or lids. The ratios of hips, water, and sugar are equivalent to most jams and jellies. Experiment to get the taste and texture most preferred. In general, 1 pound (about ½ kg) of rose hips is needed to make enough juice to mix with 4 pounds (1.8 kg) of sugar. Recipes vary, some calling for a little lemon juice. Rose petals may also be used to make jam.

Use pasteurized eggs to minimize any chance of infection in the following recipe: paint petals with beaten egg white. Sprinkle these with sugar and allow to dry. They're delicious this way on cakes and custards. This is one of the most fanciful ways to add color and unique form to desserts.

Add rose petals to vinegar, and filter for a unique taste. Boiled and strained hips added to honey is also delicious.

If you're really interested in roses as food, you can turn to a number of books with dozens of recipes for petals, whole roses, and hips. Check bookstores and libraries as well as local rose societies.

Rose Water

For centuries women bathed in water perfumed with the essence of roses. Rose water is made by floating petals on top of water or through a distillation process described in most rose cookbooks. The petals exude an oily substance. This substance can be separated from the water to make a buttery substance known as *attar of roses*. The oil is used as a basis for many of the world's finest perfumes. It takes 30,000 roses to make 1 ounce (30 ml) of attar. Attar is produced primarily in Bulgaria, France, Morocco, and parts of what was once the Soviet Union. Rose water is also an ingredient in such edible delicacies as baklava. You can use it as well to give fragrance to potpourris and rosaries.

While the following may not be the traditional way to make rose water, it is the most practical for the home grower:

Pick highly perfumed petals at their peak moment (usually late morning or early evening). Collect a total of 2 to 3 pounds (about 1 kg). This may take a long period of time. Lay the petals out on a screen (preferably not metal) in shade in an area with as low humidity as possible. Let the petals dry. Remove any that turn moldy. Once petals are dry, push them into an airtight container with 1 teaspoon (5 g) salt mixed into 1 quart (1 liter) water. Stir the mixture each day for 5 or 6 days with a plastic or wooden spoon. Then pour the liquid through a fine filter so that only rose water comes through.

For minuscule amounts of attar, you can float rose petals in water in full sun all day. Look for drops of oil glistening on the surface at the end of the day. You can absorb these by touching them lightly with cotton balls or swabs. The cottony material can be used in potpourris or rubbed on rosaries to give them scent.

Rosaries (Rose Beads), Jewelry, Art Objects

Collect fresh, fragrant petals. Collect 5 pints (about 2½ liters) altogether—packed tightly in 1-pint (½-liter) containers.

Pulverize the petals in a food processor. Do this several times until the petals are completely shredded. Add ¼ ounce (8 ml) of tincture of iodine (buy it at a drugstore). You now have a paste to shape as desired.

Or, crush petals in a mortar. Let them dry, and then add rose water. Crush them again, and let dry. Add rose water again. Keep crushing until you get a thick, even paste. If you do this in a cast-iron pan, the paste turns black (highly prized coloration). Now you have a paste ready to shape.

Wearing rubber gloves, shape the soft mixture made in either way into beads the size of marbles (they will shrink to one-half their original size). Place them on a cookie sheet or wax paper. Of course, you can create other shapes and objects, but the rest of this section deals with making beads for rosaries.

One day later, just as they begin to harden, pierce every bead with a pin or needle. Leave the pins in place for a full day, then remove them.

If you want smooth beads, place them in a soft fabric bag. Rub them together every so often, not all at once. To get them completely smooth may take a total of 5 to 6 hours of intermittent rubbing.

Lay the beads in olive oil for 36 hours.

Remove them from the oil, and dry them in the sun. Put them back in the bag and rub them again until they're completely shiny and smooth. Adding scented oils at this stage is highly recommended for extra aroma.

String the beads when they are completely hard.

CHAPTER 9

TOOLS AND SUPPLIES

Have the right tools and supplies at hand before starting any gardening project. This will save you lots of time. Also, if you're well prepared you won't be tempted to skip important steps in getting roses off to a good start. Good tools are often expensive, so you want to get full value. If you take care of them, they'll last for years. Keep blades sharp and clean. Oil metal regularly to keep it from rusting. Also clean out any sprayers or apparatus that could get clogged after each use.

The following alphabetical listings briefly explain many of the basic tools and supplies you're likely to need, along with purchasing information for items that can be hard to find.

Atomist Atomist is the brand name of the finest applicator of rose spray. It applies a fine mist on all leaf, stem, and soil surfaces. It's very expensive, however, and geared to growers with many roses. It is available online. However, you can find less expensive applicators in local nurseries and garden centers. These are geared for use by the average home rose grower and work fine.

Backflow preventer Whenever you use a hose to spray roses with fertilizers or chemicals, attach a backflow preventer to your outside faucet to prevent any chemicals from getting into your water system. It is required by code in many areas. **Caution:** *If you ever use a hose to apply fertilizer or chemicals to outdoor plants, never drink from it again.*

Bonemeal Added to the planting hole, bonemeal provides the rose plant with lots of phosphorus, one of the three essential plant foods. Bonemeal is now steamed and not as good a plant food as it once was. It is also slow to release into the soil. Many rose growers believe that superphosphate is more effective, but organic gardeners will not use it.

Bucket A large plastic bucket is helpful in many ways. It's good for hauling small amounts of soil. It's also excellent for picking up debris, such as weeds and spent blossoms. Some household soaps come in large, white buckets that can be adapted for garden use. These are lightweight, durable, and easy to carry. You can also buy solid plastic buckets in many lumber stores or garden centers.

Chicken wire If you have a large bed of roses that need winter protection, you may need chicken wire to keep the leaves in place. This is discussed in Chapter 5 under "Winter Protection" (Minnesota Tip Method). If plants are left standing in the winter, you may have to surround them with chicken wire to prevent rabbits and deer from nibbling on them.

Conduit Electrical conduit, used for supports, is sold in most hardware stores. It is light and very strong. Cut it into desired lengths with a hacksaw. Paint it any color you want if it seems obtrusive.

Dandelion digger The little- or longer-forked diggers make it easy to dig up long-rooted weeds.

Edger Useful tool to cut grass along the sides of a rose bed. Keep grass out at all times. It spreads rapidly in the moist, cool conditions of properly prepared beds.

Electrical tape One way of attaching the standards (stems) of Tree roses or large bush roses to supports.

Epsom salts The common name for magnesium sulfate. Most pharmacies sell it. It neutralizes the soil, allows plants to take in more nutrients, and induces lots of new cane to sprout from the base of the plant.

Fertilizer Inorganic: 10-10-10, water soluble 20-20-20, and superphosphate. Organic: blood meal, bonemeal, compost, cow manure, fish emulsion, rotted horse manure, and Milorganite.

Fish emulsion Although it does smell somewhat, fish emulsion is one of the best organic fertilizers, used either at the base of plants or for foliar feeding (read labels carefully). It contains valuable trace elements. Fish emulsion was once somewhat hard to find, but its use has increased and now it's stocked in many local garden centers.

Foam Thin pieces of foam placed around the stem as it's being anchored to a support protect the bark from scrapes.

Fungicide You'll need several fungicides to prevent diseases in the rose garden. Varying the type used is highly recommended and effective.

Garbage can (plastic) Get a 32-gallon (122-liter) can with a lid to make alfalfa meal tea. It's also useful in many other ways for hauling weeds or garden materials around.

Glasses (safety) Get the kind of safety glasses that are enclosed and cover your eyes completely. Wear them when spraying. When working with roses, wear glasses to protect your eyes from thorns. The latter don't have to be enclosed, just regular glasses.

Gloves (leather and cotton) Rose thorns are extremely sharp and strong. Working with thorny plants is difficult without a good pair of goatskin (not cowhide) gloves. Fabric gloves do not stop thorns from penetrating but are excellent for working with soil or tools. They are also a lot less expensive.

Gloves (rubber) If you'll be spraying with chemicals, rubber gloves are essential. The farther up they go toward your elbow, the better.

Hacksaw Used to cut conduit for support.

Herbicide Use one of the newer weed killers that decompose quickly in the soil and don't cause problems with future planting. Roundup® is one of the best to kill perennial weeds. You can use herbicides in the beds, but we think routine weeding is easy if you use a thick mulch to prevent annual weed growth.

Hoe (pronged) The best tool for scraping the top surface of soil after walking on it or just before applying a mulch. Just scratch the surface of the soil. Avoid deep hoeing.

Hose Watering is critical to good rose culture. Water roses whenever the soil starts to dry out. Place hoses at the base of plants, and let water saturate the soil. Or, attach a hose to a sprinkler to cover a wider area during hot, dry weather. Avoid overhead sprinkling during cool, humid periods when plants are most susceptible to disease.

Hose hangers The kind you can attach to your house right by the faucet are easy to use, inexpensive, and sturdy. They save a tremendous amount of time.

Insecticide You'll need several insecticides to prevent problems. Changing the type used is highly recommended. Bugs seem to adapt to the use of just one brand.

Kneeling cushion A soft pad that protects your knees from contact with the ground. Some people love them; others see them as a nuisance.

Knife A regular knife comes in handy. If you plan to propagate roses through budding, you may want to buy a special budding knife, although a razor blade or utility knife will work equally well. A scalpel is even better. These should be kept completely sterile during the budding process (see p. 216).

Lopping shears Get the kind with an angled blade (bypass, not anvil type). Note that there is a sharp side and a dull side on the curved blade. The sharp side should always be placed against the portion of cane to be saved, for a smooth, clean cut.

Magnifying glass (or **photographer's loop**) Sometimes it's hard to identify disease and insects without one of these.

Milorganite One of the finest organic fertilizers. We recommend its use in many instances throughout this book. Buy it by the bag at most garden centers. The pelleted material is easy to spread around plants, but dusty and a bit smelly. Wearing gloves and a mask is suggested when working with it.

Miticide A chemical that will kill spider mites, a common problem on roses.

Needle Stick a needle in a cork to keep it handy. Use it to clean out the hole in the **Siphon Mixer** whenever it becomes clogged.

Orchid Fertilizers Orchid growers often spray diluted solutions of fertilizer directly on the leaves of their plants. This foliar feeding can be quite effective on roses as well. A number of these water-soluble fertilizers are available at local garden centers.

Pails (5-gallon [19-liter] plastic) Pails are used to cover plants in early spring for "sweatboxes," to carry tools and soil, and to pick up garden debris. To save money and avoid last-minute searches, collect large pails throughout the year. Most restaurants and building contractors end up with many of these. Plaster-boarders (Sheetrockers) and painters likewise often have many extra pails. If using any pail that once contained paint, clean it thoroughly so that there is no paint residue whatsoever. Pails used by Sheetrockers are much easier to clean.

Peat Moss Buy peat moss in bales; the larger ones are the best value.

Planting table If you have space for a work table in a hidden area, you'll find they really are helpful.

Plastic bags Gardeners have many uses for both black and white plastic bags. Black are excellent for lining pots, carrying leaves, and so on. White ones are most useful laid out under a plant when shaking insects off the canes. The fallen bodies show up much better against the light background.

Pots Why buy when you can recycle? Black plas-tic pots are routinely discarded at many garden centers; ask if you may take some. Be sure to disinfect used pots with a solution containing 1 part bleach to 9 parts water. The size of the pot should always match the size of the plant.

Pro-Mix® This light, fluffy material is used in planting roses both in pots and in the garden. It consists of 75–85% sphagnum peat moss, along with perlite, vermiculite, calcium and dolomitic limestone, wetting agent, and nutrients. Buy it in bales, the largest available being 3.8 cubic feet (107 liters). The availability of Pro-Mix® can be a problem in some areas. If you'd like more information on this product and where to buy it, contact the following: Premier Horticulture, 127 Fifth Street, Suite 300, Quakertown, PA 18951, or Premier Horticulture, 1 Avenue Premier, Rivière du Loupe, Quebec G5R 6C1 Canada. They can also be reached by phone at (800) 667-5366. You can also replace it with other soilless mixes.

Pruning saw For larger stems, a pruning saw makes cutting much cleaner and easier. These usually cut on the pull stroke. Loppers (lopping shears) work fine on most larger canes, but extremely large canes are best cut with a saw.

Pruning shears Felco's, made in Switzerland, are best. Japanese imitations are much less expensive and quite good, but not as good as Felco. Buy the bypass (not anvil) type. These make much better cuts.

Rakes (garden and leaf) Metal, pronged garden rakes are excellent when you're leveling soil. Leaf rakes are essential for cleaning up the soil surface and gathering leaves in the fall.

Respirator It's wise to wear a respirator when spraying to prevent inhaling toxic substances. Clean it off after each use.

Rope (synthetic) Heavy synthetic rope is useful in tying up larger rosebushes before laying them under the soil for winter protection. You can pull on this rope in spring to lift the plants out of the ground. It won't rot out and lasts for many seasons. When you're cutting this rope, it may start to come apart at the end; hold the end over the flame of a lighter to stop the fraying.

Rototiller Use this machine with rotating blades to dig up and loosen soil. A rototiller is very highly recommended for someone with a larger garden. The ones with tines in the back are easier to use but considerably more expensive.

Roundup® (glyphosate) Roundup® is one of the safer herbicides on the market and effective on most annual and perennial weeds. It will not kill burdock and some thistles. Sold with various concentrations of glyphosate, it is then mixed with water and sprayed on weeds. It is also available pre-mixed. The latter is fine for limited use, but the concentrates make more sense for larger areas. Use this herbicide for bed preparation, but be extremely careful when using it around established roses. Use a paint brush to apply it to weeds in this situation. The herbicide will not be effective if left outdoors over winter. For consumer information, call (800) 225-2883.

Scissors Having a pair just for the garden makes sense. You'll use them frequently. An absolute must for cutting twine or polyester rope when tipping roses for winter protection.

Seed-starting mix You can use vermiculite, perlite, sterile sand, sterile soil, or a combination of these to start seeds. Or, you can buy sterile seed-starting mixes locally or Pro-Mix® PGX from Premier Horticulture listed earlier under Pro-Mix®.

Sharpeners Keeping tools sharp requires a variety of sharpeners. Different growers prefer different types, but most growers will have some sort of sharpener for pruners and a different type for larger tools, such as spades. Sharpeners include various types of files, grindstones, hones, and sharpening stones (corundum is very popular). Most hardware stores sell sharpeners or have a professional tool sharpener on call to sharpen tools for a small fee.

Shredder A leaf shredder is extremely helpful in making a summer mulch for a rose bed. Since leaves are commonly available in cold climates, this is one of the least expensive ways to apply mulch. If you don't have a shredder, a rotary lawn mower will also pulverize leaves quickly. Dump leaves close to a wall, then run across them with a mower; the pulverized particles will accumulate in a pile.

Siphon Mixer This handy gadget allows rose growers with lots of plants to mix large amounts of fertilizer at a time. This is then run through a hose to dilute it. It should have a backflow preventer built in for safety. Exactly how to use it is explained on the package. It saves an incredible amount of time and energy. Tip: If your rose bed is far from the spigot, attach the siphon mixer between hoses within 50 feet (15 meters) of your garden. Check in garden centers for it. If unavailable locally, go online where there are a number of brands advertised.

Spade A pointed spade is necessary for digging holes and the trench needed for winter protection using the Minnesota Tip Method. Keep it clean by washing it off after each use. Some growers mix sand and oil in a box and then dip the end of the spade in it to prevent rust. File the blade to keep it sharp.

Spading fork A spading fork (four-pronged garden fork) is less likely to cut roots when you're loosening soil around the base of a rose plant just before tipping for winter protection. Having two spading forks is especially useful; the extra one is used to hold a plant down during tipping.

Sprayer If you don't use an **Atomist** to spray roses, you'll need to buy a sprayer for larger gardens or a hand mister for a few plants. Tanks that you pressurize by pumping are easy to use and relatively economical. If you use an herbicide in a sprayer, get a second sprayer for pesticide or fertilizer application. To clean sprayers, run a solution of 1 quart (1 liter) white vinegar to 1 quart (1 liter) water through the nozzles.

Sprinklers The little metal kind that cover small areas in either circles or rectangles are inexpensive and easy to attach to and remove from hoses. They control water application so that you're covering just the area you want to. If you have a large rose garden, you'll want oscillating sprinklers to cover broad areas.

Stirrup M This product, and similar ones, sexually excites spider mites and makes them more susceptible to death from a miticide. For such a product, contact Kimbrew-Walter Roses, Route 2, P.O. Box 172, Grand Saline, TX 75140, (903) 829-2968.

Superphosphate This, or bonemeal, should be added to the planting hole to provide phosphorus to a young rose plant. Superphosphate works well in early plantings when soil is cool.

Supports Larger roses and Climbers need some sort of support. For bush and miniature Tree roses, we recommend using electrical conduit cut into the appropriate lengths with a hacksaw. Plastic supports are fine. For Climbers, fences, trellises, wires, and the like, are all good. Climbing roses do not have tendrils and must be tied to their support.

Tarp Either plastic or canvas is fine. Shovel soil onto the tarp to avoid the mess and the time necessary to clean up grass at planting or winter protection time.

Ties Climbing roses, larger bush, and Tree roses need to be tied to a support. Electrical tape is easy to use if there is a stake close to the plant. Plastic ties tightened around a foam pad are preferred by many growers. Other common ties are polyester twine, long twist'ems that come with plastic bags, pieces of nylon stocking, and strips of cloth. Tying soft material in a figure-eight knot around the stem to the support is highly recommended.

Tire pump Get a good one to keep the tire on your wheelbarrow properly inflated.

Trowel Dozens of uses.

Twine When you're using the Minnesota Tip Method for winter protection, polyester twine or rope is needed to pull the branches together. Always use polyester; any other twine will rot out. If you need lots of twine, buy it in large rolls, often sold in farm stores.

Watering can Helpful where a hose may be cumbersome or unavailable.

Wheelbarrow Get a good, solid wheelbarrow with a large, inflatable tire. This type of tire moves over rough surfaces easily. Keep a tire pump handy to fill the wheel as it looses air. Larger wheelbarrows make more sense and save a lot of time. The extra expense is worth it. Flimsy wheelbarrows are a nuisance.

Wilt-Pruf® Wilt-Pruf® is one of several brands of antitranspirant. Antitranspirants prevent canes from drying out and also help reduce disease caused by fungal infections. You'll find them in most garden centers. For information on Wilt-Pruf®, contact Wilt-Pruf® Products, Inc., P.O. Box 469, Essex, CT 06426, (860) 767-7033 (accepts collect calls from consumers).

INSECT AND
DISEASE CONTROL

This appendix gives detailed information on protecting roses from diseases and insects. Growers are divided on the use of chemicals in preventing and controlling rose problems. What you decide to do will depend on your personal opinion. A great deal of experimentation is going on to reduce the use of pesticides, and new solutions are being devised. However, most roses are susceptible to diseases and insects. A consistent and conscientious use of chemicals is still preferred by many growers. If used exactly as instructed, they pose less risk to you and the environment.

Nevertheless, there are a number of people who will not use synthetic chemicals under any circumstances. Therefore, in this section you'll find both inorganic and organic remedies for common problems. In this period of transition chemicals come and go. Remember that prevention is always better than control. The latter does not necessarily mean complete eradication of a problem, just preventing it from getting out of hand. The critical points in using any chemical control as a "cure" are these: What is the problem? Will it get worse without the use of chemicals? Is the loss tolerable? What is the best chemical to use? What is the correct dosage to control the problem? And is the organism at the right stage of development to be killed effectively? Although the information in this chapter will help you a great deal, we suggest joining a local rose society to get help from people who have been growing roses for years under nearly identical conditions to those found in your yard.

Chemicals

Following are synthetic chemicals used in preventing and controlling problems with disease and insects in the rose garden. These vary from relatively non-toxic to highly toxic. The law requires relative toxicity to be on all labels. These are the key words to look for (from mildly toxic to most toxic) Danger (Category I), Warning (Category II), Caution (Category III), and Caution (Category IV). A fungicide is a chemical that kills diseases caused by fungal infections. An insecticide kills insects. A miticide or acaricide kills spider mites. A molluscicide kills slugs and snails. Some sprays mix chemicals together for protection against two or more of these problems. However, you should never mix chemicals unless the correct procedure for this is described in detail on the label. Some of these chemicals may be off the market by the time you read this guide with new ones introduced in their place. An asterisk (*) by the chemical indicates that it is systemic and actually absorbed by all plant tissue. Many of the following products are locally systemic, meaning that a smaller portion of the plant absorbs the chemical. Always cover all skin, wear a respirator, and use long rubber gloves (the kind that go up to your elbow) when working with these chemicals. Other safety tips have been mentioned throughout this guide. Note that many of the chemicals listed below are equally effective as the ones mentioned in individual write-ups describing control of specific diseases and insects in this section.

SYNTHETIC CHEMICALS USED AS PESTICIDES

Abamectin	Miticide (kills adults)
Acephate*	Insecticide
Acetamiprid*	Insecticide
Bifenazate	Mirticide (kills adults and eggs)
Bifenthrin	Insecticide
Captan	Fungicide
Carbaryl	Insecticide
Chlorothalonil	Fungicide
Compass®	Fungicide
Copper sulfate	Fungicide
Cyfluthrin	Insecticide
Dimethoate*	Insecticide/miticide
Dinotefuran*	Insecticide
Disulfoton*	Insecticide/miticide
Esfenvalerate	Insecticide
Ferric phosphate	Molluscicide
Folpet	Fungicide
Glyphosate*	Herbicide
Hexythiazox	Miticide (kills eggs)
Imidacloprid*	Insecticide (effective on grubs)
Lambda-cyhalothrin	Insecticide
Malathion	Insecticide/miticide
Mancozeb	Fungicide
Metalaxyl*	Fungicide
Metaldehyde	Molluscicide
Myclobutanil	Fungicide
Permethrin	Insecticide/miticide
Piperonyl butoxide	Additive (not pesticide)
Propiconazole*	Fungicide
Tau-fluvalinate	Insecticide/miticide
Tebuconazole*	Fungicide
Thiabendazole*	Fungicide
Triadimefon*	Fungicide
Triforine	Fungicide

Organic alternatives

Organic alternatives exist for synthetic sprays. The term organic is used rather loosely and generally means non-synthetic. For example, sulfur is not really organic, but it is often used by "organic" gardeners. The word organic does not necessarily mean safe. Organic alternatives vary from relatively non-toxic to quite toxic. In fact, some synthetic sprays are safer than some organic ones. Read and follow labels carefully. A few of the organic alternatives, such as some brands of Rotenone, have synthetic compounds (Piperonyl butoxide) added to make them more toxic. Just as some synthetics, such as Carbaryl (Sevin®), can kill bees, so can some organics, such as Sabadilla. Use such products only when there is no bee activity, generally in early morning or late evening. Proper storage of organic materials is important to prevent deterioration. Buy fresh products and store them as indicated on the label. Finally, wear a mask and appropriate clothing, especially rubber gloves, when working with organic pesticides. Some irritate the skin or mucous membranes.

Antitranspirant (Wilt-Pruf® and others)	Anti-fungal properties
Bacillus popilliae (Milky Spore Disease powder)	Insecticide (kills grubs)
Bacillus thuringiensis (commonly known as Bt)	Insecticide (kills caterpillars)
Baking soda	Fungicide
Bordeaux Mixture (Copper sulfate, lime)	Fungicide (considered inorganic by some)
Diatomaceous earth	Insecticide/molluscicide
Dormant sprays	Fungicide/insecticide
Horticultural oils	Fungicide/insecticide
Insecticidal soaps	Insecticide
Lime sulfur	Fungicide/insecticide/ miticide
Neem products (Azadirachtin)	Insecticide (effective on larvae)
Pyrethrum (Pyrethrin)	Insecticide
Rotenone (liquid form best)	Insecticide
Sabadilla	Insecticide

Soap (detergents)	Insecticide
Spinosad	Insecticide (kills thrips)
Sulfur (in varied forms)	Fungicide/insecticide/ miticide
Traps (sticky ones)	Insecticide

Diseases

Anthracnose
(*Elsinoë rosarum* or *Sphaceloma rosarum*)

While not as common as black spot or powdery mildew, anthracnose is a serious problem. Colored circular spots, usually reddish brown purple, form on the surface of leaves (more rarely stems). Grayish white openings then appear in the center of the spot. Leaves eventually turn yellow and fall. Thorns and canes may also show similar sores. Common in cold climates, especially on Climbing roses with glossy foliage.

ORGANIC: Prevent by careful watering so that soil does not splash onto foliage. Control by cutting off and burning all infected cane.

INORGANIC: Prevent and control by routine use of a product containing Chlorothalinol. Most preventive spraying for black spot is equally effective against anthracnose.

Black spot or Blackspot
(*Diplocarpon rosae*)

This is a fungal disease commonly encountered by cold-climate rose growers, especially in humid or wet seasons. The fungus creates small, circular black spots with irregular edges on foliage (generally upper surface) and, sometimes, on stems or buds. It needs at least 7 hours of moisture to grow. Spots vary from reddish-purple to black and look like circular lesions. Leaves eventually turn yellow and fall off. Plants can be totally defoliated. Generally, lower leaves are infected first because humidity is higher there. The disease spreads rapidly and can be serious because leaf loss weakens the plant. There are a wide variety of types (races) which explains contradictions in reports of resistance by variety in different areas of the country.

ORGANIC: One of the simplest and most effective ways to deal with black spot is to buy roses known to be resistant. Plants may be resistant (vertical resistance) to specific types (races—about a dozen) of black spot or may get infected but fight it off (horizontal resistance). Check store-bought, potted plants carefully, since they may already be infected with this disease. Prevent in the garden by giving plants plenty of space for good air circulation. Water at base of plant only—wet foliage at night can be major cause of black spot (wet leaves for more than 7 hours is necessary for the disease to begin). Ironically, prolonged overhead watering at night every 10 days may remove spores and is highly recommended. Avoid splashing leaves with water bouncing off the ground. Remove lower leaves that could get splashed. Remove spotted, yellowing leaves immediately (check every day). Use a summer mulch, believed to kill spores. Keep tools clean and sterile. Each fall clean up all mulch and debris at the base of the plant (spores called *condia* overwinter there). Burn or dispose of it. David Slezak, Jeff Gillman, and others, tested many commonly recommended home remedies for black spot over a period of years. Most did not work. The few that did were copper, milk, and mouthwash (Scope® was used in their research). Although effective, copper is a heavy metal and quite unsightly when sprayed on plants. Its use is not encouraged. Mouthwash caused minor damage on new growth but afforded some protection. Not all mouthwashes were tested, but the home remedy holds promise. Surprisingly, spraying plants with 1 part whole milk to two parts distilled water may prove to be the best home remedy over time. Spray plants once a week. Let the milk stay on foliage for a day before spraying it off with water. Other remedies proved either ineffective or destructive to the rose plants themselves. Note that black spot is carried by wind on damp, overcast days and needs moisture to survive. Its types (races) are evolving and can build up resistance (as bacteria has done to antibiotics).

INORGANIC: The best control of black spot continues to be the routine preventive use of a fungicide. When spraying, be sure to cover the undersides of leaves.

Alternate types of fungicide used (the disease can become resistant to some sprays). Use products containing Chlorothalonil or an all-purpose rose spray labelled for use against black spot.

Botrytis blight (*Botrytis cinerea*)

This fungus causes gray-black sores to develop below buds, killing them (they often droop or hang over). Gray spores may cluster on infected tissue. Most common in cold, wet weather. Most commonly affects roses with higher number of petals (over 40). Common in cold climates. When petals won't open and stick together, this is called *Rose balling*.

ORGANIC: Prevent by proper pruning (cut ¼ inch/6 mm above a bud) and control by cutting off and destroying infected plant parts at the first sign of the disease.

INORGANIC: Spray with products containing Chlorothalonil or Mancozeb on routine basis if this has been a problem.

Cankers:

Brand canker (*Coniothyrium wernsdorffiae*),
Brown canker (*Cryptosporella umbrina*),
Common canker, rose graft canker,
or **stem canker** (*Leptosphaeria coniothyrium*
or *Coniothyrium fuckelii* or *rosarum*),
and many others.

Reddish-purple spots appear on cane (less often on leaves). These turn gray-white over a period of time and will eventually kill cane and entire plant. Cane turns brown and brittle, mimicking die back. There are several types of cankers (fungal diseases), but the general description fits most of them closely. Common in cold climates.

ORGANIC: Prevent canker by making clean cuts ¼ inch (6 mm) above buds and covering all cuts with Elmer's glue or a commercial product made for that purpose. Avoid damaging cane by improper hoeing or careless contact. Cut out crossing canes which can rub against each other in the wind. This results in open wounds. When raising plants in spring, spray with lime sulfur (1 part lime sulfur to 9 parts water)

before buds have formed. Control by cutting canes back 3 inches (7.5 cm) below infected area. Cut ¼ inch (6 mm) above an outward facing bud. Dip pruning shears in bleach solution after each cut to prevent spread of disease. Paint the wound with Elmer's glue.

INORGANIC: When raising plants in spring, spray with products containing Chlorothalonil or Maneb.

Cercospora leaf spot (*Cercospora rosicola* aka. *Mycosphaerella rosicola*)

This along with Black spot and Powdery mildew is one of the more common fungi to cause leaf spots. Look for purplish round spots with grayish to tan centers, primarily on leaves. The disease has a tendency to work its way up the plant (like Black spot) and is commonly confused with it. If you are spraying regularly with a fungicide to control other types of fungal infections, you will already be controlling the spread of this disease. Note that there are many other diseases that cause leaf spots, but we have covered the main ones in this section and they too are controlled by preventive spraying.

Crown gall (*Agrobacterium tumefaciens*)

A bacterial infection caused by cuts in the roots or lower stem of the bush through improper cultivation or insect damage. Growths (round, rough spots) form on the stem or roots at the base of the plant just underneath the soil. Young galls are greenish and soft while older ones darken and become tough. These stunt growth and can kill the plant. Rare in colder climates, but it does occur (most often from already infected bare root plants sold by mail order companies).

ORGANIC: If bare root plants have galls, get a refund. Don't plant them. You'll infect the soil for up to 2 years. Prevent by dipping bare root plants in bleach solution as outlined throughout the guide (not proven but may be helpful). If galls form on new plants, break as much of the growth off as possible. Cut out the remaining portion of the gall from the stem with a utility knife or scalpel (dipped in chlorine solution). Clean off the cut with a solution of ½ bleach and ½ water (just dip a rag in this and rub if across the cut several times—don't drench the area with this strong chlorine solution). This often takes care of the problem. If severely affected by many galls, pull up and destroy the plant. Do not plant roses in the same spot.

Dieback

The tips of stems die back, turning brittle and brown. This is sometimes caused by various fungal diseases. However, it is most common after severe winters, when the upper portions of cane are most vulnerable to freezing or drying in cold winds. It is also caused by improper pruning. Rose bushes need a bud to send nutrients to. If you cut too far above a bud, then the cane will die back to a healthy bud below the cut. Cure by cutting off dead cane to a healthy bud. Make a proper cut as described in the pruning section. Very common in cold climates because of severe winter freezes and hard winds.

Damping off

When you're growing seeds, the base of young seedlings sometimes turn soft and the seedling flops over. This is caused by a wide variety of diseases, collectively referred to as Damping off. Use Captan to avoid and cure this problem.

Downy mildew (*Peronospora sparsa*)

Many people believe that this fungal disease is not found in colder climates—not so! It is not as common as powdery mildew, but you may have to deal with it at some point. High humidity of 85 percent and temperatures in the low 60s encourage the growth of rose downy mildew, a very serious fungal disease. Irregular purplish-red to brown spots form on leaves. These turn yellow and fall. Leaves on top of the plant are infected first, not at the bottom as in black spot. Sometimes, the undersides of leaves have gray, fuzzy growth or down-like fungus (*mycelium*). This growth is described by some as brownish gray pustules. Some gardeners confuse this with spider mite webs.

ORGANIC: Prevent by dipping all bare root plants in the correct bleach solution before planing. Give

plants plenty of space for good air circulation. Remove infected foliage. When watering, avoid splashing soil onto foliage. Spray foliage with hot water. Spray foliage with an antitranspirant that stops spores from penetrating leaves. Spray foliage with Tide® (2 teaspoons per gallon). Avoid overhead watering so that leaves stay dry much of the time.

INORGANIC: Most common chemicals used are Bordeaux Mixture and products containing Chlorothalonil, Mancozeb, Metalaxyl, Triadimefon, and Triforine. Prevent in several ways. Method 1: Try Bordeaux Mixture. If it works, use nothing else. Method 2: Use Chlorothalonil with a product containing Metalaxyl. Method 3: From May to mid-June spray plants every 10 days with Chlorothalonil and a product containing Mancozeb. From mid-June to mid-August spray every 14 days with Chlorothalonil only. From mid-August to mid-October spray with Chlorothalonil and a product containing Mancozeb every 10 days. If downy mildew appears, apply a product containing Mancozeb every few days on infected plants. Method 4: During the spring alternate sprays every other week of the following mixtures. The first week mix 1 tablespoon Chlorothalonil and 1 teaspoon Triforine per gallon and spray. The second week mix 1 tablespoon of a product containing Mancozeb and 1 teaspoon Triforine per gallon and spray. Continue until the weather gets hot.

Mosaics (Rose mosaics)

Yellow streaks or mottling appear on leaves. Blossoms are small and distorted. Foliage turns yellow and drops. Plants often die. Viruses (two forms—typical and yellow) come from buying already infected plants. Relatively rare in cold climates, but extremely serious when it occurs.

ORGANIC: Prevent by buying plants certified to be mosaic free. If infected, destroy the plant. There is no cure. Insist on getting your money back—there is no reason why growers should be selling plants infected with this deadly virus. It is only transmitted in the budding process, so it is the commercial grower's (not your) fault.

Powdery mildew (*Sphaerotheca pannosa* var. *rosae* or *rosa*, aka. *Podosphaera pannosa*)

This is the second most common disease encountered by rose growers in cold climates. It attacks emerging foliage, buds, and stems first. It looks like a whitish powdery film on plant tissue. It emerges during periods with high humidity at night and cooler temperatures during the day. Leaves may curl up, wrinkle, and turn purple. Plants rarely die (although they could), but the disease curbs and stunts growth. Check store-bought potted plants carefully before purchase—they are often infected with this disease because they're crowded together in unsanitary conditions. Plants most susceptible to powdery mildew are Chinas (rarely grown in colder areas), Hybrid Teas with deep pink or red-colored blossoms, Polyanthas, and *Rosa wichuriana*. Note that its spores are carried by wind.

ORGANIC: Buy resistant varieties. Prevent by providing lots of space between plants. Keep plants evenly moist since drought stress may make plants vulnerable. Use a mulch around all plants to keep soil moist and cool and to prevent soil from splashing up on leaves. Mist the plants regularly in the morning. If using inorganic fertilizers, switch to organic ones. High nitrogen content in the soil *may* induce mildew. Pick off diseased plant parts as soon as you see the disease developing and burn or throw them in garbage. Spraying horticultural oil on foliage may be the most effective organic control at this time. Spraying plants with an antitranspirant every 7 to 10 days provides some protection, but can also stunt new growth. You may also have some success with a solution of 3 tablespoons baking soda to 1 gallon water (add 1 tablespoon ammonia if aphids are present). Try 2 tablespoons Lysol per 1 gallon of water as an alternative (spray in the middle of the day). These first two methods often work best if 1 tablespoon spreader sticker (ask for these at nurseries) or 2.5 tablespoons of horticultural oil are added to the mix. Or use wettable sulfur at the rate of 2 tablespoons per gallon. Never use sulfur-based products if the temperature exceeds 85°F. Clean up all debris in the fall.

INORGANIC: More formidable is regular application of any products containing Acephate, Chlorothalonil, Mancozeb, Propiconazole, or Triforine.

Rose downy mildew (see **Downy mildew**)

Rose mosaic (see **Mosaic**)

Rose rosette
(unknown virus or virus-like pathogen)

First sign is overly rapid growth with the formation of dense thorns or prickles on the canes which become unusually large and turn purple to deep red. Leaves have an abnormal distorted, crinkly appearance, may turn brittle, and often turn reddish purple like the canes. The plant forms an abnormal number of distorted canes—a 'Witches' Broom' (rosette). An infected rose eventually deteriorates and dies. The disease is carried by an eriophyid mite (*Phyllocoptes fructiphilus* Keifer or Kiefer), not a spider mite. These mites are most common in cool, moist seasons. Rose rosette has decimated wild stands of *Rosa multiflora,* but roses grown on this stock do not seem any more susceptible to the disease than those grown on other rootstocks. It is spread only by the mite, not by contaminated tools.

ORGANIC: No cure.

INORGANIC: No cure. Chemicals (most off the market now) provided only marginal protection against the mites even with weekly sprayings. Dig up the plant with the soil around as soon as the first symptoms appear. Burn the plant and contaminated soil or toss it (soil and all) into the garbage can.

Rust (*Phragmidium*—nine species)

This is a fungal disease that needs 2 to 4 hours of moisture to develop. It tends to grow on cool, damp days. Reddish orange spots (often powdery) first appear on the underside of leaves and on young stems. Very rare in cold climates. Would not occur if healthy plants were shipped here in the first place. Unfortunately, it can be fatal and spread rapidly—a good reason to isolate new plants from an existing rose bed. Spores are carried by the wind.

ORGANIC: Prevent by inspecting plants carefully before or after purchase. The disease comes in from plants grown in other parts of the country. Control by cutting off all infected plant parts (better yet, send or take the plant back). Provide good air circulation. Remove infected and fallen leaves placing them in a bag to be thrown in the trash. Clean up all debris around plants in the fall (they harbor spores). Lime sulfur may work, but use it sparingly.

INORGANIC: Use fungicides containing Chlorothalonil, Mancozeb, and Triforine. Use these early in the season if rust has ever been a problem in your garden. Spray both plants and the soil around them.

Spot anthracnose (see Anthracnose)

Verticillium wilt
(*Verticillium albo-atrum* or *dahliae*)

Plants slowly die from this fungal infection. Foliage generally drops off from the bottom up as the plant wilts. Cut off a piece of cane. The tell-tale sign of wilt is a purple to black ring in the core. Wilt is not common in roses, but will appear where there is an abundance of wild blackberries. No cure. Dig up and destroy the plant. Plant roses in a different area since the soil is now contaminated with the virus causing this disease. This wilt may be brought in from infected stock purchased locally or through mail order. However, most reputable growers are extremely well-trained in regards to this disease and do everything possible to avoid it.

Insects

Ants (*Formicidae* family)

Found mainly in sandy soils, ants can loosen the soil around roots and cause damage to the plant. This is a rare problem. However, they do protect aphids which harm plants and carry disease. Common in cold climates.

ORGANIC: Mulch around the base of the plant and keep it moist at all times. This discourages ants. If black ants are the problem, open up the hill and pour boiling water into its center. Sprinkling a detergent such as Tide® on hills also may move them on.

INORGANIC: Most stores carry lethal ant killers if mulch does not solve the problem. Don't let pets get into them.

Aphids:
**Potato aphid (*Macrosiphum euphorbiae*),
Rose aphid (*Macrosiphum rosae*),
Small green rose aphid (*Myzaphis rosarum*)**

Tiny sucking insects (called plant lice or greenfly) in a variety of colors congregate in colonies typically on new growth, usually on buds, young shoots, and the undersides of leaves. The ones attracted to roses are usually green or pinkish green and generally less than ⅛ inch (4 mm) long. Types of aphids attacking roses vary by geographic areas. They all secrete a sticky substance called "honeydew," a favorite of ants. Honeydew often becomes infected with mold (sooty gray). Leaves curl and dry up. Growth may be distorted. Aphids carry about two-thirds of viral diseases that infect plants. Very common in cold climates.

ORGANIC: If they appear in early spring, do nothing other than to remove infested plant parts if that's feasible. Spray plants down with a jet of water. Or spray them with Windex containing ammonia. Waiting allows lady bug larvae to hatch. It also protects other predators of aphids, such as gall midges, spiders, and syrphid fly larvae. Also, place a mulch around the base of the plants and keep it moist. This stops ants from protecting the aphids. Planting Alliums, any member of the onion family, may also prevent severe infestations. If colonies start to get out of hand after early spring, spray them with soapy water. If this doesn't work, use a pyrethrum/rotenone combination or horticultural oil. Insecticidal soaps and Neem products work, but kill beneficial insects as well.

INORGANIC: For severe infestations use products containing Acephate, Carbaryl, Imidacloprid, Malathion, or Tau-fluvalinate.

Borers (see Carpenter bee, Rose stem girdlers, and Rose stem sawfly)

Bristly rose slug (see **Rose slug**)

Cane borer (see **Rose stem sawfly**)

Carpenter bee (*Cerantina* species)

The black to bluish green, miniscule carpenter bee (½ inch/12 mm) bores out a hole in the top or sides of cane to lay eggs there. The eggs mature into yellowish, curved maggots. They feed on the inside of the cane and slowly move downward. Cane will wilt and die. These insects are attracted to dead cane, so keep all roses properly pruned by removing dead cane or stubs immediately.

ORGANIC: Snip down the cane until you reach healthy tissue. Make a slanting cut just above an outward facing bud. Burn or throw the dead cane into the garbage.

INORGANIC: Chemical control is not usually recommended. If you cut off dead cane and destroy it, you'll be getting rid of the problem.

Caterpillars (larvae of butterflies and moths)

The many varieties of caterpillars feed mainly on leaves and fresh shoots.

ORGANIC: Pick them off the plant. Either kill or place them somewhere else if you like butterflies. Kill them with *Bacillus thuringiensis* (Bt).

INORGANIC: Any product containing Acephate or Carbaryl is effective.

Chafer beetle (see **Rose chafer**)

Cuckoo-spit (see **Spittle bug**)

Flower thrips (see **Thrips**)

Froghopper (see **Spittle bug**)

Gall wasp (**Cynipidae family**, commonly *Diplolepsis spinosa*)

Tiny orange or black wasps burrow into the stem to deposit larvae there. This causes cane to swell creating spiny "golf balls," commonly called Mossy Rose Gall. Occasionally, wasps deposit larvae on leaves, forming a mossy ball there as well. Fairly common in cold climates, especially on Species roses and *Rosa rugosas* hybrids.

ORGANIC CONTROL: Cut below the gall and destroy the damaged cane. Cut off leaf and destroy. Generally, handled without resorting to sprays.

INORGANIC CONTROL: Chemical controls have proven ineffective. You would have to kill the adult wasps, and that is simply not practical.

Inchworms (larvae of varied moths)

Little worms which creep along and feed on leaves. Fairly common in cold climates.

ORGANIC: Either hand pick or kill with *Bacillus thuringiensis* (Bt).

INORGANIC: Spray with a product containing Acephate or Carbaryl.

Japanese beetle (*Popillia japonica*)

This is quickly becoming the number one pest for rose growers in cold climates. Shiny ¼ to ½-inch (6 to 12-mm) insects, usually coppery brown with green heads with white tufts of hair on their bellies—very beautiful, but devastating. Grubs, found under lawns, are grayish white with brown heads and curled into a C-shape. Adults cause severe damage rapidly. Normally, begin by eating flower buds and flowers first, then skeletonizing leaves. They are attracted mainly to yellow and pastel-colored roses. Once rarely found in colder climates, now becoming a menace. Large nurseries and sod farmers may inadvertently be carrying them into new locations.

ORGANIC: Once recommended, Milky spore disease has proven to be less effective in cold climates than originally hoped. In warmer areas it does protect lawns, but the beetles still fly in from surrounding areas. Remove adults by hand as soon as they appear (tap them down into a can of soapy water, salty water, or vinegar). Place cloth under plant and shake vigorously to get all bugs—they drop on the cloth. Best time to do this is early in the morning when the insects are lethargic. For just a few roses use Windex with ammonia every 5 days. *Never use traps laced with sexual hormones (pheromones)*. They cause greater, not lesser infestations by attracting beetles to the rose garden. Products containing Neem oil may be effective over a period of years if sprayed regularly according to directions on the label. But, hand picking daily remains the most effective control. Some organic growers insist that squishing the bugs and throwing them into plants reduces infestations.

INORGANIC: For best control (97%) of grubs use Imidacloprid (Merit®) in July. Kill adults with products containing Carbaryl and Imidacloprid. Time the use of all of these insecticides properly by following local advice and reading labels carefully. Avoid spraying flowers if at all possible (hand pick beetles from these) to protect bees. Unfortunately, some spray is likely to drift into flowers.

Leaf-cutting bee (*Megachile* species)

Bees that make clean, little cuts from the margins of leaves. They use these to build nests. No cure. Unfortunately, fairly common in cold climates. However, they do kill aphids and other damaging insects and may be considered beneficial in that regard. They are also good pollinators.

Leaf hoppers:
Apple leafhopper (*Empoasca maligna*), Potato leafhopper (*Empoasca fabae*), Red-banded leafhopper (*Graphocephala coccinea*), Rose leafhopper (*Edwardsiana rosae*), White apple leafhopper (*Typhlocyba pomaria*)

Little springing, winged insects. They are small, tinged pale green or white. They leave white skeletons (skins) on the underside of leaves. Clusters may be found on the undersides of leaves. Shake the leaves and the leaf hoppers spring up in a cloud. They feed on foliage. Some foliage may fall off. They are known carriers of viral diseases. Fairly common in cold climates.

ORGANIC: Raise healthy, vigorous plants. They can withstand attacks by leaf hoppers. Tolerate the mini-

mal damage done. Or, spray with insecticidal soap or products containing Pyrethrum.

INORGANIC: Kill with products containing Acephate.

Leaf rollers (Leaf tiers)
(*Platynota flavedana, Platynota stultana*) or *Choristoneura rosaceana*

Little caterpillars (usually smaller than ½ inch) look like green or yellow maggots and are the immature stage of a moth. They roll themselves up in a leaf and eat their way out. Sometimes, they will eat tiny holes in flowers. Fairly common in cold climates.

ORGANIC: Prevent by good fall clean-up. Encourage birds to feed in your garden. Control by squeezing rolled up leaves to kill them. Pick off and burn the leaves. Use *Bacillus thuringiensis* (Bt) to control in safe manner. Pyrethrum, Rotenone, and insecticidal soaps will destroy them as well.

INORGANIC: Kill larger infestations with products containing Acephate, Malathion, or Tau-fluvalinate.

Mealybug (*Pseudococcus* species)

Mealybugs look like white cotton where leaves join the stems on plants. They are soft, oval insects covered with a white powder. They suck plant juices and are aided by ants. They are really more of a problem indoors than out.

ORGANIC: Clean area off with fabric. Dab area with alcohol, rub off, and rinse. Kill ants.

INORGANIC: Kill with products containing Acephate, Malathion, or Tau-fluvalinate.

Midge (see **Rose midge**)

Mites (see **Spider mites**)

Mossy rose gall (see **Gall wasp**)

Pithborer
(see **Carpenter bee** and **Rose stem sawfly**)

Rose bug (see **Rose chafer**)

Rose chafer (*Macrodactylus subspinosus*)

Little grayish to yellowish brown beetles with spiny legs. Most are under ½-inch long. Feeds on leaves and flowers, especially white ones. Flowers often look like they've been eaten in half. Occasionally a problem in colder climates.

ORGANIC: Handpick and drown in can of soapy water. If done diligently, this works wonders. Kill in grub stage by applying Milky Spore Disease to lawn and gardens over a period of years. Kill adults with insecticidal soaps.

INORGANIC: For severe infestations by adults, spray daily with a product containing Carbaryl. If daily spraying is impossible, spray with products containing Acephate. Prevent by using Imidacloprid (Merit®) on lawns.

Rose curculio (*Rhynchites bicolor*)

A reddish beetle, about ¼-inch (6 mm) long. It has a black snout and drills holes into buds, which then don't open. Its larvae are white and feed on flowers, especially yellow and white ones. Major problem in North Dakota.

ORGANIC: Spread Milky Spore Disease over lawn and garden to kill the grubs. Hand pick adults and toss into soapy water or vinegar. Remove and destroy infected buds.

INORGANIC: Imidacloprid (Merit®) applied to lawns is effective over a period of years.

Rose midge (*Dasineura rhodophaga*)

Tiny yellow-brown red fly lays eggs. These turn into creamy white maggots which slash into buds and new shoots, that wilt and turn black. Tips of new growth look like they've been burnt with a match. Increasing rapidly in cold climates.

ORGANIC: Remove and burn damaged cane as soon as there is any sign of an infestation. Spray with insecticidal soaps or Pyrethrum based products.

INORGANIC: Kill the minute insects with with products containing Acephate. Treating the soil as well as the foliage is extremely important.

Rose scale (see Scale)

Rose slug
(*Cladius isomerus* or *Endelomyia aethiops*)

Not really a slug. Hairy, slimy creature about ½ inch (12 mm) long. Looks like a yellowish, green caterpillar. Hides under and then skelotonizes leaves in early spring. Will eat into open wounds. Rose slugs are most commonly found in greenhouse cultivation.

ORGANIC: Pick off by hand and squash. Spray with an insecticidal soap or product containing Rotenone. Spinosad, if available., may be effective. Bt, once recommended, has not proven helpful.

INORGANIC: Spray with products containing Carbaryl.

Rose stem girdlers (*Agrilus aurichalceus*)

After Japanese beetles, this is probably the second major concern for rose growers in cold climates. A ¼ inch (6–8 mm), metallic bronzy to coppery green beetle bores into cane and lays eggs. The cream colored larvae create concentric tunnels just under the bark. This stops the flow of water and nutrients through cane, causing the tops of stems to die off. You'll often see a swollen area at the base of the dead cane, which may topple over.

ORGANIC: Cut below the swelling at the base of the dead wood. Make your cut just above an outward facing bud on live cane. Burn or throw the infected dead cane in the garbage

INORGANIC: Uniform spraying with an insecticide (Malathion) from spring into early summer may kill adults before they lay eggs, but timing is so difficult that this is not recommended. Once the larvae are in the cane, chemical control is ineffective.

Rose stem sawflies (*Hartigia trimaculata*)

Sawflies (non-stinging wasps) deposit eggs in cane. These develop into creamy grub-like larvae less than ½ inch (12 mm) long. They burrow their way into cane causing it to wilt and then die. You'll often see a swollen area just below the dead cane.

ORGANIC: Cut off the dead cane a few inches below

the affected area. Make a slanting cut just above an outward facing bud on the live cane below. Seal cuts with household glue or nail polish. Burn or toss the dead cane in the garbage.

Inorganic: Chemical control is not commonly recommended since killing the wasps is difficult and reaching the larvae impractical.

Scale (*Aulacaspis rosae*)

There are many different species of rose scale. All look somewhat the same and do similar damage. These little insects with hard shells encrust stems, looking like little round or oval growths. Insects suck sap from the plants which eventually wilt and die. Fairly common in cold climates.

Organic: Prevent by spraying plants with dormant spray in early spring. Control by cutting off and burning infested cane. Or rub off the scale with your fingers or a cotton swab. Horticultural oil is sometimes effective.

Inorganic: Spray with products containing Acephate, Carbaryl, or Malathion in late spring. Kill them as early as possible. Mature colonies are difficult to eradicate. Using a systemic insecticide makes sense.

Shoot borers (see Rose stem sawfly)

Slugs (*Deroceras reticulatum* and too many others to list)

Shell-less snails that feed at night. They are slimy and shiny. They skeletonize leaves with rasping tongues covered with tens of thousands of miniscule teeth. They often leave a shiny, silvery trail behind. This trail can often cause damage as well as the slug. There are 3,000 species and the average garden hides 6,000 of these slimy creatures. Very common in colder climates, especially in wet seasons, but not very common on roses.

Organic: Prevent by digging soil deeply in the fall after removing all debris. Look for and destroy eggs in spring (clumps of up to 400 are colorless to milky white). Sprinkle ash, diatomaceous earth, egg shells,

or shredded paper around the pest's favorite plants (in a small garden). Pick mature slugs off plants by hand at night while they're feeding and dump into can of salt water or vinegar (use tweezers if you can't stand touching them). Since they hide during the day, set boards, flat rocks, or pieces of melon rind under plants at night. They'll hide under these. Turn their hiding places over and squash them the next day. Set shallow bowls of beer, grape juice, or diluted brewer's yeast in the garden so that the edge is level with the soil. Slugs slide in and die.

Inorganic: Kill with slug baits. These must be kept away from pets. They may also kill chipmunks. The active ingredient in these baits is Metaldehyde.

Spider mites:
Europen red mite (*Panonychus ulmi*),
Four-spotted spider mite (*Tetranychus canadensis*),
Southern red mite (*Oligonychus ilicis*),
Two-spotted spider mite (*Tetranychus telarius* or *urticae*)

Miniscule spiders (eight legs). Some are reddish, others vary in color and have spots on their backs. Magnify them with a lens or tap them onto piece of white paper from the undersides of leaves. Delicate webs are visible. Mites damage leaves, which yellow and sometimes die. Severe infestations can kill plants. Common in hot, dry weather. They often carry disease. Major problem in cold climates.

Organic control: Prevent by using dormant oil in both late fall and spring. Unfortunately, oil kills off helpful predator mite eggs as well. Clean up all debris and weeds in fall. Prevent and control by misting foliage in dry weather. Don't miss the undersides of leaves. Keep plants well watered for vigorous growth. Shake plants vigorously, since spider mites will not get back on the plant once off. Encourage predator mite colonies and lady bugs by not using chemical sprays. Insecticidal soaps, horticultural oils, and Neem products may kill spider mites, but also beneficial insects as well.

Inorganic control: Kill adults with Bifenthrin or Tau-fluvalinate. Remove damaged foliage before spraying. Spray underside of leaves well. Frequent application of a miticide may be necessary. Begin application in late May and continue every 2 weeks. Changing sprays may be necessary. Stirrup M (see p. 230 for source), a product which sexually excites mites, often helps kill larger colonies. Mix 1 drop per 2 gallons of spray. Spray first with miticide alone. Then 5 days later with miticide laced with Stirrup M. Use chemicals only if spider mite colonies are doing significant damage.

Spittlebug, Cuckoo spit, or Frog hopper (*Cercopidae* family)

Miniscule yellow insect hides inside frothy spittle. May cause a few leaves to wilt by sucking out juices, occasionally causes distorted shoots. Rarely causes serious damage. Fairly common in colder climates.

ORGANIC: Prevent by misting plants in hot, dry weather. Rub off spittle containing insect with cloth or cotton. Or spray off with jet of water. Kill with Rotenone.

INORGANIC: Kill with products containing Acephate (really overkill consider how little damage these insects typically do).

Spotted cucumber beetle (*Diabrotica undecimpunctata howardi*)

Small beetle with black spots on its back. Very common in vegetable gardens. A nuisance as well to rose growers. Likes to nibble on blooms. Carries disease and should be immediately destroyed. Major problem in vegetable gardens, less of a problem in rose gardens.

ORGANIC: Pick off by hand and drop into soapy water. Kill with Neem products.

INORGANIC: Spray with any common insecticide. Tau-fluvalinate is highly effective. So are multipurpose rose sprays available in local nurseries or garden centers.

Thrips or Flower thrips (*Frankliniella* species)

Tiny, winged insects with slender, brownish yellow bodies. Hard to see because they are miniscule and move rapidly. Their larvae attack buds and flowers by scraping petals with their mouths. Will cause discolored (brown streaked) petals or distorted buds and flowers as well as tip damage to young canes. They are most common in extremely hot summers. Open buds gently and look for these tiny insects, that like pastel colors but love white. Found on many indoor plants, especially African Violets. Outdoors, they love Gladiolus. The insects carry viral diseases. Very common in cold climates.

ORGANIC: Remove and burn damaged buds and tips. Weed entire surrounding area well. If available, try Spinosad (a soil bacteria) to kill these insects.

INORGANIC: If infestation is severe, kill with products containing Acephate, Malathion, or Tau-fluvalinate.

Whiteflies (*Trialeurodes vaporariorum*)

Whiteflies are winged insects that suck juices from plants and often infect plants with viral diseases. They lay eggs under leaves. Damage usually consists of leaves turning color, curling up, and dying. Whiteflies are generally more of a nuisance to vegetable growers. They especially like Tomatoes grown on patios or decks. Very common in colder climates.

ORGANIC: Cut thin strips of plastic from yellow antifreeze bottles or paint thin boards yellow. Cover these with oil or any sticky substance. Hang these on infested plants. Whiteflies will be attracted to the color and will get stuck on the oil. Clean or change these regularly to reduce the population dramatically. Kill with pyrethrum/rotenone spray.

INORGANIC: Kill with products containing Acephate, Malathion, or Tau-fluvalinate. Many other insecticides are equally effective.

Summary

Most roses are prone to diseases and vulnerable to a wide variety of insects. A few are quite resistant. It is estimated that there are over 750 possible disease and insect-related problems with roses. Yet, almost all of these can be prevented with a regular spray program as outlined in this guide. Prevention is much easier than control. If the program is followed carefully according to directions provided on the labels of each chemical, you will have few problems with diseases or insects in the rose garden. However, more is not better. Sprays can damage plants when used improperly. Read labels carefully. Also, if you follow the precautions outlined throughout the guide, you can use these chemicals safely and do little damage to the environment.

If you're an organic gardener, you can use many of the strategies outlined throughout this and other chapters to avoid the use of synthetic sprays. This is sometimes referred to as intelligent neglect. Growing roses using nothing but organic techniques is more difficult, but a number of growers do it successfully by choosing their plants wisely and growing them well. However, follow all directions when using organic pesticides as you would with inorganic ones since some of them are quite toxic. Contrary to public perception, some organic products are more dangerous than non-organic ones. The term organic is not necessarily synonymous with safe.

ROSE CULTURE
CHECKLIST

Following is a chronological cultural checklist for the entire rose-growing season. Everybody would like exact dates for each of these steps—some regional associations even publish dates—but such precision is impossible because weather patterns change every year. Also, even the most experienced cold-climate rose growers make mistakes. So, don't expect perfection. Just do your best.

☐ As soon as it begins to warm up in spring, begin removing protective covering of leaves from the rose bed. Do this in stages as ice crystals thaw.

☐ When the soil below is free of ice crystals, raise any buried roses.

☐ Add organic matter to the soil.

☐ Wash soil off all plants with a hose.

☐ Cut off the twine holding the branches together.

☐ Mist the plants, and soak the soil.

☐ Spray the plants with a dormant spray if not done the previous fall and if the plants are not budding out. Otherwise, spray with a combination fungicide (to kill spores, especially those of canker) and insecticide (to kill insect eggs). Spray the soil as well as the plants.

☐ If surprised by a late cold snap—temperatures below 26°F (−3.3°C)—run a sprinkler over the plants to keep them from being damaged.

☐ If surprised by a late snow, do nothing. The snow rarely causes damage.

☐ Keep the roses moist by sprinkling frequently. Misting all cane twice a day encourages rapid bud growth and stops canes from drying out. If you're unable to do this, then spray the canes with an antitranspirant available at garden centers.

☐ If the roses do not begin to send out new growth (i.e., they do not "break"), follow the steps outlined on p. 166.

☐ As weather warms up, begin to prune. Cut off all dead cane, and follow the guidelines throughout this book.

☐ Cover cut ends with a sealant, especially if you're an organic gardener. If you're using chemical sprays, this is less important.

☐ Keep soil evenly moist.

☐ Within a few weeks, begin the recommended feeding program.

☐ At the same time, begin spraying plants to prevent disease and insect infestations. Follow the schedule as recommended throughout the book.

☐ Apply a summer mulch around the base of the plants once the ground has warmed up thoroughly to 60°F (15.6°C).

☐ Continue watering and spraying as necessary. Never let the soil dry out.

☐ Remove spent blossoms.

- ☐ Keep the garden clean. Pick up all leaves and petals that drop to the ground.
- ☐ Stop feeding roses at the time indicated in the rose sections of Part I. Feeding schedules vary considerably by the type of rose grown. Late feeding may encourage growth which will die back winter.
- ☐ Continue watering and spraying as recommended in Part I.
- ☐ Watch for the first frost. Begin to protect roses requiring special treatment before the ground freezes or temperatures dip below freezing.
- ☐ Keep the ground moist until tipping.
- ☐ Spray plants twice with a dormant lime-sulfur solution to kill off fungal spores and insect eggs before tipping plants.
- ☐ Tie plants up so that canes are close together.
- ☐ Bury tender plants using the Minnesota Tip Method, covering them with enough soil and leaves to prevent winter damage.
- ☐ If rodents have been a problem, place poisoned bait in the bed next to the buried roses.
- ☐ Make notes on where you buried each type of rose. It's easy to forget by the following spring.
- ☐ Keep the winter protection moist until the first freeze.
- ☐ Protect standing roses from deer and rabbits by encircling them with chicken wire. You'll need less wire if you pull and tie the canes together. Don't leave any portion of cane exposed, or it will be nibbled off by spring.
- ☐ Also, protect more tender standing roses, such as Old Garden Roses and some Shrub Roses, by placing 3 black plastic bags filled with leaves around them to form teepee-like protection. Tilt the bags in and against the plants. Fill in spaces between the bags with whole leaves.
- ☐ The canes on a number of roses die back each year, but the crown survives. On these types of roses cover the crown with soil (preferably loose potting soil) and a thick layer of whole leaves.
- ☐ Sit back and relax, reading rose books and catalogs during the winter.

GLOSSARY

This glossary contains most of the more technical words that you may encounter while reading rose-growing publications. In some instances, a listed word is a complicated way of saying something simple; other words are simple (if not widely understood) terms to explain complicated things quickly. The explanations are intended to be as direct and easy-to-understand as possible.

AARS (All-American Rose Selections) An association which once sponsored field trials of roses to determine outstanding named varieties (cultivars). It no longer does (past selections are excellent).

Acaricide (miticide) A substance that kills spider mites.

Accent plant A plant that stands out, usually because of its color, foliage, form, or texture. Often planted by itself like a shrub.

Acclimation The process of rose cane beginning to go dormant in late summer and early fall. Triggers for this are less light, lower temperatures, and less moisture. Water begins to move out of cells at this time so that they will not burst when temperatures drop below freezing (process called supercooling). The genetic makeup of each rose determines how well they can acclimate. The ones that do it best have greater cold hardiness.

Acidic Descriptive of soils with a lower than neutral pH. Roses like slightly acidic soil.

Active ingredient Material that has a killing effect in an herbicide or a pesticide.

Aeration The presence of lots of oxygen in well-drained and properly composed soil. Essential for root growth and the survival of soil microorganisms.

Aerobic bacteria Bacteria that thrive in the presence of oxygen. These are best in compost piles, since they do not give off any odor.

Aiglets The "hinges" that attach petals to the flower. Also, cells in this area that exude an oily substance causing roses to have scent.

Air layering A technique of propagating plants by deliberately wounding a cane and covering it with damp sphagnum moss to create a root ball.

Alfalfa meal tea (see **Triacontonal**)

Alkaline Descriptive of any soil with a higher than neutral pH. Roses do not like alkaline soils.

Amendment (see **Soil amendment**)

American Nursery & Landscape Association This organization was the best source of patent information on roses and other plants. It no longer publishes a patent directory. Information on patents may be found online or in printed rose catalogs. *It is illegal to propagate any plant for sale if it is under patent.*

Anaerobic bacteria Bacteria that thrive without oxygen. These often cause compost piles to smell.

Anther The upper portion of the stamen (male organ) containing pollen in a flower (see **Flower**).

Antitranspirant Any product that stops cane from drying out. Also may reduce fungal infections, although it is not registered for this purpose. Wilt-Pruf® is one of the most commonly available. It is a mixture of pine oil and water.

Cloud Cover is a lesser known brand. These products must be exposed to natural light to be effective (they do not work indoors or in greenhouses).

Aoûter The French word indicating how rose wood begins to prepare for cold weather at the end of the season (August). Implies that pruning at this time will generate new wood which cannot prepare itself for the wind and cold to come.

Apical domination The tendency of rose plants to send energy to upper buds first. These will grow and expand into laterals before lower buds.

Armed A horticulturist's way of saying that a plant is very thorny.

ARS (American Rose Society) An organization that offers members numerous publications on roses as well as free admission to specific gardens. Contact: American Rose Society, P.O. Box 30000, Shreveport, LA 71130-0030.

Asexual Method of reproducing roses without seeds.

Attar of roses A pale oil distilled from rose petals and used in making perfumes.

Axil The point or angle at which a leaf joins the stem.

Bacillus thuringiensis **(Bt)** Deadly disease used in killing caterpillars on roses. Not toxic to mammals. Most commonly known as Dipel®, although sold under other names as well.

Bacteria Minute organisms usually lacking chlorophyll and that can be either beneficial or destructive to roses.

Balling Condition of a blossom that fails to open as petals cling together after rain.

Bare root A plant dug up, pruned, and shipped without any soil to the consumer.

Bark The outer woody tissue around the cane, consisting of both dead cells (on the outer portion) and living cells (on the inner). Protects the inner, living cane and should not be damaged, to prevent disease, insect infestations, and possible death.

Basal The lower portion of a rose cane.

Basal break (see **Basal shoot**)

Basal shoot Any cane (stem) that forms from a bud at the base (crown) of the plant.

Bedding plant One grouped with other plants to give a mass effect, often of the same color.

Beetle A flying insect with a hard shell.

Bent neck A term used by flower arrangers to describe the condition in which the top of the flower flops over. Generally caused by lack of water getting to the upper portion of the blossom.

Biological control Use of naturally occurring microorganisms to attack disease-causing organisms or insects.

Bleeding The loss of sap during pruning late in the season, resulting in loss of plant vigor.

Blight A disease typified by quick death of flowers, leaves, and young shoots. Rare in roses.

Blind shoot (Blind wood) A mature cane that fails to produce a flower. It may be caused by cold weather early in the season. Don't confuse blind wood with rose midge, an insect that destroys young buds. The best thing to do is cut the cane back to a five-leaflet leaf. The bud will produce a branch which often blooms well.

Bloom A flower. Also a whitish substance on leaves and stems which can be rubbed off.

Blown bloom Any flower that has opened so far that it is past its prime.

Blueing The discoloration of a mauve, deep pink, or red flower as it matures. Also refers to discoloration of petals on a few varieties of roses that react badly to artificial chilling in a refrigerator to lengthen bloom time.

Bonemeal Pulverized bones in powder form. A good slow-release source of phosphorus. Must be added to the base of the planting hole to be effective. Superphosphate (not organic) releases phosphorus quickly, is effective, and less expensive, but organic gardeners will not use it.

Borer The larvae of flying insects that dig into cane or branches, often causing dieback.

Boss A ring of stamens at their most conspicuous stage. Many are bright yellow and extremely attractive.

Bottom heat Heat applied to the bottom of a bed or container to speed up germination of seeds or encourage root growth from cuttings.

Bract A leaflike growth usually an inch or 2 (2.5 to 5 cm) below the petals on a flower stalk (actual position varies greatly by variety).

Breaking bud A flower bud just as it begins to open.

Breaks Any new canes growing from the base (crown) of the plant. The faster a plant breaks, the better. More breaks (new growth) means a better rosebush.

Bud There are two types of buds on a rosebush. A *flower bud* is a bloom not yet open. A *growth bud* is the beginning stage of a shoot. It is found in the axil of a leafstalk and looks like a small pimple. It is called a *basal bud* at the base of the plant, a *lateral bud* on the cane or shoots off the cane, and a *terminal* or *tip bud* when at the tip of the cane. Growth buds are also called "eyes." The term *bud* also refers to a method of propagation in which a growth bud is inserted into the rootstock of a completely different plant.

Budded A rose not growing on its own roots.

Budding The process of grafting a bud from one plant onto the rootstock of another type of rose. This is the process most commonly used by commercial growers to create stock.

Budhead A bud or graft enlarges over time into an expanding cane. The bud head is this enlarged growth at the base of the plant where the bud or graft was originally made with the rootstock.

Bud shield A portion of cane containing a growth bud. It is scooped out with a sharp knife and then budded to a different plant.

Bud union The spot where a bud or graft is connected to rootstock at the base of a bush.

Burning Damage to leaves caused by contact with chemicals (fertilizers or pesticides) at the wrong temperature or time. Also, scorching or discoloration of immature plants placed in direct light before hardening off is complete.

Bush A woody plant that forms cane that ages and eventually dies off. The term applied to most groups of roses.

Callus Scar tissue over any cut portion of cane.

Calyx The green cover that breaks away from a flower bud as it opens. It is made up of five sepals. 'The Green Rose' is really made of these sepals, not petals. It's an oddity in the rose world.

Cambium The living portion of cane under the bark.

Cane Main stem of a bush. Same as **Basal shoot**.

Cane head A number of branches all coming out of one spot on the cane. Should be removed to promote better growth and greater flowering. Do not confuse a cane head with the bud union on the upper part of a Tree rose. If you cut below the bud union on these, you'll destroy the plant.

Canker A sore or lesion on the cane that kills plant tissue.

Caterpillars Larvae of butterflies and moths which feed on plant foliage.

Chelating agent Chemicals such as citric acid or Sequestrene that make iron available to a plant to reduce chlorosis.

Chlorophyll The green coloring substance in plants essential to photosynthesis.

Chlorosis Yellowing of leaves from lack of iron or other minor elements. Most common in roses kept in pots over a long period of time, including indoor use.

Clay Substance made up of minute particles which hold nutrients in the soil. Needed in limited amounts for good soil.

Climbing rose A type of rose that forms long, stiff stems which under proper conditions would continue to grow longer each year. The wood does not die out every 2 years as on Pillars and Ramblers, neither of which are hardy in cold climates. Flowers are quite large and appear throughout the season. Although Climbers are grown mainly in the South, a few varieties will survive in colder climates. In colder areas, Climbing roses are often replaced by Shrub roses with a climbing tendency, since Shrub roses are much hardier. The descriptive word "Climbing" in front of a named variety indicates that a bush rose (e.g., 'Peace') produced a sport (mutation) with climbing tendencies. The sport of 'Peace' is thus called 'Climbing Peace.' 'Don Juan,' a Climber, does not have the word "climbing" in front of it because it is a Climber by nature, not a mutation from a bush rose. This confuses a lot of people, even experienced rose growers.

Clippers Same as **Pruning shears**. A tool with sharp blades to cut through cane.

Cluster A group of flowers all connected by their foot stalks, or pedicels, to the same stem.

Code name In catalogs, roses are often listed with a name and then a code name behind it in small print. For instance, "'Snow Owl' var. UHLensch." The code name tells you the first three letters of the name of the company or person who created or found the plant (here, UHL stands for the hybrider named Uhl). The last letters equal the name given the rose by the hybridizer ("ensch" equals 'White Pavement'). Distributors often change names in an effort to sell a plant that didn't sell well under its original name or was an inferior plant to begin with.

Cold frame A boxlike structure, usually with movable glass or plastic lid. It is commonly used outdoors for propagation of roses from seed or cuttings.

Compost Any organic matter that decomposes into a soft, brown, earthy-smelling substance known as *humus*. Humus is produced in nature at the rate of 1 inch (2.5 cm) per century. See p. 154 for a discussion of composting.

Creeping Growing close to the ground and often spreading out and taking root as a ground cover.

Cross A plant created by interbreeding two plants of differing parentage.

Cross-pollination The transfer of pollen (male sex cell) from one flower to the stigma (female part) of a different flower.

Crown The lower portion of a plant from which cane or basal shoots emerge. This is the point where roots and stem join, but the crown itself is really stem tissue.

Cultivar The horticultural term for a "cultivated variety," meaning a plant specially bred and not occurring naturally in the wild. Most people use the term *variety* when talking about cultivars.

Cut back To reduce the size of a plant to induce new growth. A **cutback** is a young plant, or "maiden," that has been pruned for the first time.

Cutting Any piece of cane specially cut and prepared for propagation. Hardwood cuttings are taken from dormant cane, and softwood cuttings are taken during active growth. See pp. 213–214 for detailed information.

Damping off Death of young seedlings which topple over from their base. Caused by a variety of disease organisms.

Deacclimation The process triggered by longer light and warmth in spring that begins to take canes out of the stage needed for winter protection. If canes begin to deacclimate and then suffer a severe drop in temperature, they may die back or die out completely, depending on the intensity of the late freeze.

Deadheading The removal of spent flowers from a bush. On repeat-flowering varieties, encourages continuing bloom and is extremely important.

Decorative A flower in which petals form a loose, rather than a tight bloom. Okay in the garden, not for exhibition. Occasionally, decorative flowers are exhibited and have won awards, but this is considered unusual in the rose world.

Desiccant Material used to withdraw moisture from a flower.

Dichroism An unexplained change in flower form or color that seems to occur spontaneously in nature from time to time.

Dieback The death of cane from the tip down. Some diseases and improper pruning may cause this. However, dieback is most common in cold climates after severe winters affect exposed plants.

Dimorphism The spontaneous change of a bush rose into a Climber for unknown reasons.

Dipel® (see *Bacillus thuringiensis*)

Disbudding The removal of side buds to create one extremely large central bud for exhibition purposes. Or, in Floribundas, the removal of the central bud to increase the size of buds around it. Not at all essential for the health of the plant.

Dogleg A stem that grows up and outward from below a hat rack.

Dolomite Limestone used in powder form to reduce soil acidity and add valuable calcium and magnesium.

Dormancy The period when a plant stops growing because of lower temperatures and reduced light. All leaves

drop off, and the cells in stems undergo significant change to withstand harsh winter conditions.

Dormant spray Generally, a lime-sulfur solution applied to plants just before winter or before buds appear in spring. Kills both fungal spores and insect eggs.

Double A flower with 17 to 25 petals. Very double (full) 26–40. Extremely double (very full) 41+.

Drainage The ability of water to move rapidly through the soil. Critical to good rose growth.

Drip irrigation Method of watering roses through soaker hoses or similar systems to cut down on water loss through evaporation. Not recommended in colder climates because of the potential of salt buildup in the soil. However, in skilled hands it works fine (requires occasional deep watering by regular hoses to remove salts).

Emasculation The removal of petals and anthers from a flower during the process of hybridization (see p. 218).

Enzyme Protein responsible for biochemical reactions in plants.

Epsom salts The common name for magnesium sulfate, which adds magnesium to the soil and reduces acidity. Helps roses take in nutrients by correcting the soil pH and neutralizing soluble salts created by the breakdown of inorganic fertilizers.

Established Descriptive of a plant mature enough to withstand some mistreatment.

Exhibition A flower with classic, high-centered form. Long petals form a lovely central cone.

Eye (see **Bud**) Also refers to the center of a bloom with different coloration from that of the petals.

Fasciation An unexpected flattening of a rose stem.

Fertile A rose that can form seed. Not all can.

Fertilization The moment at which pollen enters the stigma (female organ) of a flower. The net result on fertile plants is the production of hips containing seed to produce a new generation of plants. Also refers to the application of fertilizer to the foliage or ground around roses.

Fertilizer Any substance that provides nutrients to a plant. Synonymous with plant food. Generally, fertilizers are broken into two broad groups: inorganic (synthetic) and organic (natural).

Fertilizer burn Damage caused to plants by applying too much fertilizer or applying fertilizer on dry ground. Also refers to death of seedlings from application of inorganic fertilizers to the starting mixture.

Filament The stemlike, lower portion of the male organ

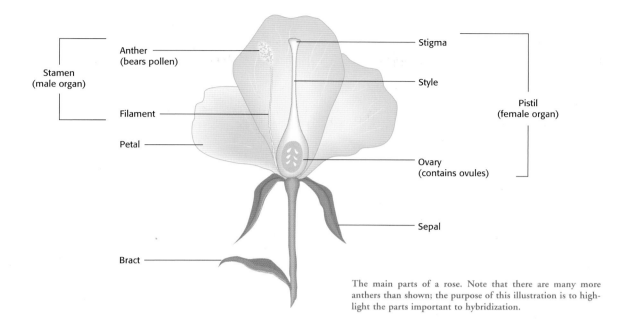

Stamen (male organ)
Anther (bears pollen)
Filament
Petal
Bract
Stigma
Style
Pistil (female organ)
Ovary (contains ovules)
Sepal

The main parts of a rose. Note that there are many more anthers than shown; the purpose of this illustration is to highlight the parts important to hybridization.

(stamen) which supports a pollen-bearing sac (anther) at its tip (see **Flower**).

Fish emulsion Ground-up fish used as a fertilizer for roses. May be applied to ground or leaves. Does have some smell, but provides many trace elements.

Floriferous A bush that produces lots of bloom. The same as "free-flowering."

Flower The reproductive organ of seed-bearing plants. Not all roses produce seeds, but this is an unnatural condition. A typical rose is illustrated above.

Flush The moment when a plant bears the most bloom. Plants may have several flushes in a given season.

Foliar feeding The application of fertilizer in liquid form to leaves and stems. These absorb the fertilizer directly.

Foot stalk (see **Pedicel** and **Peduncle**)

Forked terminal The tip of a cane with two small canes at its top extending in opposite directions. Cut one off.

Formal A style of design that emphasizes control and strict contours. Often applied to individual plants as well, as for miniature Tree roses.

Fragrance The scent or smell of a flower. It comes from oil evaporating from cells (aiglets) at the base of petals.

Free-flowering (see **Floriferous**)

Friable Refers to soil properly prepared with the right structure for good growth.

Fungicide Any substance that kills fungi. Most fungi are beneficial, but a few cause serious diseases in roses.

Fungus (Fungi) Primitive, parasitic plants which include black spot, powdery mildew, and rust, three major rose diseases.

Gall A swelling of unorganized cells caused by damage to plant tissue.

Germination Sprouting of seeds from starting mixture.

Grading Commercial rose growers classify bushes into different grades according to number and size of stems. A grade 1 is excellent, 1½ good, 2 not so good, and a cull is no more than a discard. Most mail-order houses try to send out grade 1 roses. Most prepackaged roses are no worse than 1½. But, you must look at roses carefully when buying them to select the best, whether packaged or potted. Try to buy plants with several nice stems, all about the same diameter.

Grafting Joining a stem (scion) or bud of one variety of rose to the stem of a different variety. The purpose is to get more vigorous growth. In some cases, it's the easiest and least expensive way to propagate a rose that is difficult to grow from seed or rooted cuttings.

Granule A particle of fertilizer, herbicide, or pesticide.

Ground cover Any plant used to prevent weed growth, stop erosion, or cover a large area with beautiful foliage and bloom.

Growing medium Sterile and weed-free material in which seeds or cuttings are started. Usually, sterile sand or a combination of peat, perlite, and vermiculite. Once seeds have sprouted and grown several sets of leaves, or once cuttings have taken root well, they are usually potted in sterile potting soil. Technically, the original growing medium is a starting medium or mix. In this book we use "growing medium" in a general sense to mean any mixture used to start and grow seeds, seedlings, and cuttings, whether rooted or not.

Grub The immature stage of a beetle (generally found in the ground).

Guard petals The outermost petals that ring a flower.

Gypsum Hydrated calcium sulfate often used to break up clay soils.

Habit (Growing habit) The style of growth of an individual rose, varying from compact or rigid, to spreading or arching.

Hardening off The process of moving a plant slowly into full sun to prevent damage to leaves. Also includes acclimating the plant to varying outdoor temperatures and drying winds (lack of humidity) after it has been raised indoors.

Hardpan Usually a layer of rock or heavy clay under the topsoil. Roots have a difficult time penetrating hardpan. If you can break this up while preparing your bed, you'll get better roses.

Hardwood cutting A section of cane cut from a rose during its dormancy to be used in creating a new plant.

Hardy A plant's ability to survive winters in a given region. Truly hardy plants need no winter protection. Many roses will survive only with good winter protection. Catalogs are often overly optimistic about hardiness.

Hat rack A portion of stem left above an improper cut (one made between nodes). It will often die back. Essentially, same as **snag**.

Head A cluster of stems at the top of a plant. Desirable for standards (Tree roses). Normally removed from other groups of roses to create better form and increased bloom.

Heaving (Frost) Lifting up of plants from the ground caused by alternate freezing and thawing. Often exposes roots to drying winds and severe cold. May result in death of plants.

Heel A stem cutting usually including a portion of the crown at its base.

Heeling in Burying bare root plants briefly in a trench until they can be planted in a permanent location in the garden.

Hep (see **Hip**)

Herbicide Any chemical used to kill weeds. Roundup® is one of the best, since it destroys both annual and perennial weeds.

Hip The fruit of a rose, varying from fleshy to quite dry inside. Hips contain seeds and come in many colors, shapes, and sizes. Hips contain 20 times the amount of vitamin C as an orange. They are attractive in winter landscapes and a favorite food for birds.

Hoe A bladed tool used to scratch the surface of soil to remove weeds and aerate the earth.

Honeydew A sticky, sugary substance secreted by some insects, especially aphids, and which is attractive to ants. It often becomes infected.

Hormone powder Chemicals available from nurseries applied to ends of cuttings to get them to root faster.

Host Generally a weed that feeds insects which eventually invade the rose garden.

Humus The light, fluffy, brown material produced as organic matter decomposes in nature. It is essentially the same as fully digested compost.

Hybrid A plant created by crossing two plants with different backgrounds.

Hydration The process by which a cutting takes in nutrients and water through its capillaries (comparable to miniature straws in plant tissue).

Inflorescence The flowering at the end of a stem.

Inorganic Synthetic or nonorganic. Generally, refers to sprays or fertilizers that do not come from once-living creatures.

Inorganic fertilizer Synthetic nutrients. Uses for granular (10-10-10 or 20-20-20), water soluble 20-20-20, and superphosphate are covered in detail. Slow release fertilizers are also fine.

Integrated Pest Management (IPM) A fancy name for using pesticides only when really necessary in a responsible manner.

Intermittent A rose that blooms off and on during the season, often depending on changes in the weather. Not the same as a repeat bloomer, which has more consistant bloom throughout the entire season.

Internode The area on a stem between two buds (eyes, or nodes). Never make a cut here. The tissue will only die. It does *not* have the ability to regenerate. All food goes directly to the nearest bud, not to cane.

Invasive Tending to spread rampantly. Has a negative connotation except when a plant is used as a ground cover.

Joint (see **Node**)

Knuckle (see **Bud union**)

Lanky Descriptive of a plant that appears spindly or weak.

Larva (Larvae) The first stage of life for a beetle, butterfly, or moth. May be found in or on plants or ground.

Lateral Any branch that springs off of a main cane or stem.

Layering A method of wounding and placing a cane in the ground to create a new plant (see p. 212).

Leaching The loss of nutrients from the soil from rain or watering. This is a common problem in sandy or light soils. Also refers to chemicals coming off the foundations of buildings and causing problems in nearby soil. Also refers to a method of saturating the soil with water to carry off toxic salts.

Leaf axil The point at which a leaf joins the stem or cane. Buds which produce new cane are just above these.

Leaf burn Spotting or death of leaves from high temperature or careless application of sprayed chemicals.

Leaf mold Fully or partially decomposed leaves added to the soil to make it loose and moisture retentive.

Leaflet One of the segments of a full leaf. Roses have from 3 to 11 leaflets on individual leaves.

Leafstalk (see **Petiole**)

Leggy Describes plant that looks spindly, overly tall, or weak. Also may refer to early loss of leaves at the base of a plant, giving it a bare look.

Lime Processed (burned) limestone (usually calcium carbonate) used to add calcium to soil and lower soil acidity.

Limestone Usually calcium or magnesium carbonates used to add nutrients to the soil and reduce soil acidity.

Liner A bare root plant consisting of a single stem with roots attached. Not a fully developed plant. Therefore, much less expensive than developed bare root plants that are 2 or more years old.

Loam Ideal soil, containing the right amounts of clay, sand, and silt with a plentiful supply of organic material.

Loppers (**Lopping shears**) A cutting tool larger than pruning shears with two sharp edges to cut through thick cane.

Macronutrient A chemical needed in high amounts for a plant's health.

Maiden A budded rose in its first year of growth before being cut back.

Manure Animal waste used to improve soil structure and fertility.

Micronutrient Same as **trace element**. A chemical needed in extremely small amounts for the health of a rose.

Microorganisms Microscopic living creatures found by the billions in healthy soil. Essential for the intake of nutrients by roses. Fed by organic material added to the soil each year.

Milorganite Treated human sewage which provides valuable nutrients to plants.

Minnesota Tip Method A well-defined technique of protecting roses from severe cold and wind in northern climates. Developed over a period of decades in a cooperative effort of the Minnesota Rose Society.

Miniature A rose with small leaves and flowers. Prized for delicacy and small size.

Mist To shower plants with a fine, delicate spray of water.

Miticide A chemical that kills spider mites.

Mold Visible fungal growth on the surface of plant tissue. Color often indicates potential for damage.

Mulch Any material placed on the surface of the soil to keep it moist and cool or to inhibit weed growth. Organic mulches are much preferred for roses. A winter mulch is one used to cover roses for protection from cold and wind.

Mutation Any unexpected change in the genetic makeup of a plant that alters its appearance in some way and may be transmitted to a new generation of plants.

Neck (see **Shank**) The portion of a plant above the roots but below the cane or bud union.

Neutral Neither acidic nor alkaline, as referring to soil.

Node The point on the cane where a leaf or growth bud is located.

Organic Derived from something that was once alive. All organic substances contain carbon as an essential element.

Organic fertilizer Any material derived from a living creature (animal or plant) and added to the soil to provide nutrients.

Organic matter Any decomposed material from plants or animals added to the soil to improve its structure (texture) and make nutrients more readily available to roses. Most commonly refers to compost, peat, and animal manures or mulches such as grass, leaves, or wood chips.

Ovary The lower portion of the female organ (pistil) in the flower containing the ovules, the parts that form seed when fertilized (see **Flower**).

Overwinter To survive the winter.

Ovule The portion of the ovary that becomes a seed when fertilized (see **Flower**).

Own-root Any rose growing on its own roots (generally from cuttings, occasionally from seed). This is extremely important to cold-climate gardeners. Plants grown on their own roots are hardier and more likely to maintain the appropriate size. They never produce suckers from alien rootstock (the latter must be removed) as on budded or grafted plants.

Pathogen Something that causes a disease in a rose plant.

Peat moss Partially decomposed sphagnum moss used as a soil amendment. Found in bogs in northern areas. Peat moss is quite sterile, weed free, and moisture retentive.

Pedicel The narrow stalk at the base of a flower in a cluster or truss. These smaller stems come off the larger stem (peduncle).

Peduncle The main stem supporting a single flower or a series of side branches (pedicels) which support a spray or cluster of flowers.

Pegging The process of bending the tip of a cane lower than its point of origin to increase the formation of laterals along the cane. These laterals in turn produce more bloom. The technique is used only with repeat-flowering varieties with long, arching canes or with Climbers.

Perlite Light, fluffy volcanic material used to aerate soil. Often looks like little white balls in potting soil.

Pesticide Any chemical used to kill disease, insects, or mites.

Petal A modified leaf which forms the flower.

Petiole The stem supporting a leaf.

Petiolule A tiny stem attaching a leaflet to a petiole (not all leaves have petiolules).

pH An artificial measurement describing the acidity or alkalinity of soil. The pH scale runs from 1 (totally acidic) to 14 (totally alkaline), with 7 being neutral. It is logarithmic in nature. A soil with a pH of 7 is 10 times as alkaline as a soil with a pH of 6. It is 100 times as alkaline as a soil with a pH of 5. The ideal pH for roses is 6.5. Most organic matter breaks down into the neutral range.

Pheromone Sexual scent used to lure insects into traps. Not recommended in rose gardens, with the exception of Stirrup M, a product mixed into miticides.

Phloem tubes These are like plant arteries. They carry food to all parts of the plant. Their main job is to carry food down to the roots for storage.

Photosynthesis The process used by roses to create food (mostly carbohydrates) from water and carbon dioxide, using light and chlorophyll as aids. Most of this work is done in leaves.

Phyllotaxy Term describing how rose leaves naturally spiral and alternate on a cane, maximizing exposure of leaflets to sunlight for optimal photosynthesis (food production).

Phytotoxic Harmful to plants.

Pillar A rose with stiff wood often wound around a support in warmer climates. The wood dies out after 2 years and is replaced by growth from the base of the plant (unlike a Climber). The wood also is too stiff to act as a ground cover (unlike a Rambler). The names have been confused in the rose world and are almost impossible for the amateur to sort out.

Pinch To remove any growth with pressure between the thumb and forefinger (with fingernails). More common with perennials than roses, although many growers pinch off buds or old blossoms as they fade.

Pistil The female organ in a flower. It consists of the portion that catches the pollen (stigma) and a slender tube (style) running down to the lower opening (ovary with ovules within) that, if fertilized, develop into seeds (see **Flower**).

Pith The spongy material in the center of the cane. If cane is alive, the pith is usually light green or whitish colored. If dead, it turns black or brown.

Plethora An overabundance of small, poor-quality buds found on plants that have been poorly maintained or pruned.

Plunge To submerge an entire potted plant in water.

Pollen Yellow, dustlike material found in sacs (anthers) of each flower. It is the plant equivalent to sperm and must be united with the female part of the flower in order for fertilization to take place (see **Flower**).

Pollination The act of applying pollen to the pistil of a flower. Flowers can self-fertilize. Or, pollen can be transferred from one plant to the next. This is done naturally by insects, especially bees. It is done deliberately by hybridizers to create new varieties (cultivars).

Pot-bound Describes a plant whose roots have filled up a container and have begun to curl around each other. Always unbind or tease long roots out before planting.

Potpourri Rose petals dried and preserved for fragrance. May have oil and herbs added for extended life or varied smells.

Prickles Thorns on rose cane. Will break off easily when cane is ripe and ready for budding (see p. 217).

Procumbent Trailing along the ground.

Prolification The unusual growth of a stem through a flower to form another bud farther out on the stem.

Propagation Creating new plants in a variety of methods, including budding, cuttings, grafting, layering, and seed.

Pruning The removal of any portion of the plant for a specific purpose, such as health, better looks, or more bloom.

Pruning saw A saw with a curved blade. It normally cuts on the pull stroke. Some fold up to protect the sharp teeth and point. Many growers prefer loppers.

Pruning shears A tool with sharp blades to cut branches.

Pup Same as a **Shoot** or **Sucker**. A little plantlet that grows off to the side of a mother plant and can be divided to create a new bush.

Quartered A many-petaled flower that appears to be divided into four separate sections (may actually be divided into fewer or more).

Rabbit pellets Alfalfa meal solidified into small pellets as food for rabbits. Commonly sold in feed stores. Pellets may be dissolved in water to form alfalfa meal tea, an excellent natural nutrient for roses.

Rachis The portion of the petiole (leaf stem) running from the first set of leaves to the terminal leaflet.

Rambler A type of rose that grows long, pliable canes usually trained up a trellis or pillar. Forms clusters of small flowers which often appear once in the season. Left on its own it would "ramble" across the ground, taking root and forming a ground cover of great beauty. Canes form the first year, bloom in the second, and should be removed by the third. Rarely grown in cold climates, where they are almost impossible to winter protect.

Recurrent (see **Repeat blooming**)

Remontant (see **Repeat blooming**)

Repeat blooming The ability of a rose to bloom more than once in a season. Ideal repeat bloom is a plant in continuous flower. More common is intermittent bloom, in which plants bloom off and on. Some roses bloom once in spring and again in fall. The term *repeat bloom,* therefore, has a number of interpretations. It is, however, an important characteristic since many roses bloom only once—in spring or early summer.

Replacement The new cane emerging from the base of the plant after old cane has been removed.

Resistance The ability of a plant to ward off disease or insects. May be natural or bred into a rose.

Respiration The ability of cells to produce energy using chemicals.

Reverse The side of a petal facing away from the center of the flower.

Reversion Can mean several things. A variety may change back to a growth pattern similar to a parent. Or, a plant may be improperly pruned, such as a Climber, and become a bush rose.

Rogue A plant that is not what a grower claims it to be. Also may refer to a plant that grows in an odd way. Some Miniatures shoot up one long cane which can be converted into a Tree rose.

Root ball Roots and surrounding material, usually soil.

Root-bound (see **Pot-bound**)

Rooting hormone Chemical that helps cuttings produce roots more quickly. May be a powder or liquid.

Rooting medium Any material used to grow cuttings.

Roots The portion of the plant that extends underground from the crown. Roots often spread out as far as the rose is tall. Most roses are shallow-rooted and must be cultivated carefully to avoid root damage.

Rootstock The plant that acts as the host for a bud or graft from a superior variety (cultivar). The plant can be grown from seed or cuttings of Species (wild) roses.

Rosa The Latin term for the rose genus, which contains nearly 150 species (naturally occurring, or wild, roses).

Rose water Water saturated with the oils from rose petals. Fragrant and often used in cooking.

Rugose Wrinkled or rough. Refers to the texture of leaves on certain types of wild roses.

Runner (see **Stolon**) May also be a cane that runs along the ground and takes root at a node.

Salt Sodium chloride, which is found in sea water and in sand used to melt ice in cold climates during the winter. In low doses it acts as plant food, but in higher concentrations is toxic. Salt is also used in a general way to refer to chemical compounds that build up in the soil from the use of inorganic fertilizers. These can be toxic to plants in high concentrations.

Sand Coarse particles making up a portion of good soil.

Sap The plant equivalent to blood.

Scion A bud or portion of stem placed on rootstock to create a new plant.

Seed An embryonic plant protected by a thin cover. Best kept cool, dry, and dark until planted.

Seedling A plant raised from seed. Note that in rose circles the term refers to both young and mature plants.

Selective budding Choosing the best bud on a portion of cane to propagate a new plant. If you were to cut off the tip of a cane, the bud closest to the tip would form leggy growth, the bud in the middle would be a "select bud," while the one closest to the base would grow vigorously but form few flowers. Select buds are the ones exceptional growers use for their stock. This explains why roses of the exact same name can be so different. It also suggests that some growers are much more reliable than others.

Semiclimber A rose that sends out longer canes than a bush rose, but not to the same degree as a true Climbing rose.

Semidouble A rose with 9 to 16 petals (or close to that).

Sepal One of the five portions that make up the protective, usually green cover over a rosebud. These petals form the calyx (see **Flower**).

Shank (see **Neck**) The portion of stem between the roots and the bud union. A long shank indicates that the grower started the rootstock from cuttings. A short shank indicates that the grower started the rootstock from seed. Short shanks are preferred, but starting plants from seed takes a long time. Plants with shorter shanks should be more expensive.

Shears Same as **Pruning shears** or **Clippers**. Tool with sharp blades to cut cane.

Shoot Same as **Stem** or **Cane**. Technically, a shoot refers to younger cane, and a stem is a more mature cane.

Shrub A relatively low woody plant with a number of stems or canes.

Side shoot Same as **Lateral**. A branch off a cane.

Silt One of the inorganic components of loam. Made up of particles larger than clay, smaller than sand.

Single A rose with 4 to 8 petals (or close to that).

Single bud Generally one growth bud where a five-leaflet leaf joins the stem. Occasionally, there will be two or even three buds. All but one should be removed.

Snag Any portion of cane left above an improper pruning cut. It often dies back and can cause disease or invite insect infestation.

Softwood cutting A piece of cane taken during the summer to create a new plant.

Soil A mixture of chemicals, particles, water, air, and millions of living plants and animals. Think of it as a living creature, not an inanimate object.

Soil amendment Anything added to the soil to improve its texture (structure). The ideal soil is loose and airy (you can almost push your hand into it easily).

Soil test A chemical analysis of the soil indicating pH and the availability of major nutrients. Often suggested, but almost useless unless plants will simply not grow in the soil selected.

Species A wild rose that when pollinated produces seed that will replicate the parent.

Sphagnum moss Stringy plant material used to line baskets. Used to keep upper cane moist in Tree roses (see p. 170). Wear gloves and a mask when working with this material which has recently been linked to a rare fungal infection (*Cutaneous sporotrichosis*).

Split center An unusual blossom form in which there seems to be an abnormal gap down the middle. Very undesirable. It may be caused by rootstock that grows too vigorously for the budded or grafted upper portion of the rose. If a breeder changes rootstock, the problem often disappears.

Sport Mutations occur in roses as they do in all other living creatures. When a cane produces flowers or shows a growth pattern different from its parent, it's called a sport. The sport, although genetically different from the parent, can be propagated. The new plants will have the new genetic code. Some of the most beautiful roses in the rose world are sports. Sports may mutate again, producing yet another new rose (sport). Sports may also revert to the parent plant.

Spotting Blemishes on blossoms caused by wet weather, improper watering, or improper spraying. May also be a sign of insect infestation.

Spray A stem with many flowers. A liquid applied to a plant as a pesticide.

Spreading A habit of growth in which the plant tends to bend outward.

Staking Can mean two things. Staking a plant for support. Or, bending a long cane over and attaching it to a stake in the ground to induce new cane growth from the base of a one-time blooming rose. This results in a much bushier plant (as with 'Harison's Yellow,' listed under "Shrub Roses").

Stamen The male part of a flower, consisting of a slender, stemlike growth (filament) with a pollen sac (anther) at the tip (see **Flower**).

Standard The name often used for a Tree rose.

Starting medium or **mix** (see **Growing medium**)

Stem An old cane. May also refer to the base of the plant.

Sterile Descriptive of a plant that produces no seed. Or, a seed that will not germinate. Also refers to soil that has been steamed to kill off all disease-causing organisms or to tools disinfected properly to kill disease-causing organisms.

Stigma The part of the female organ (pistil) that gets sticky and traps pollen so that the flower can be fertilized (see **Flower**).

Stipule A small growth at the base of a leafstalk.

Stolon An underground stem. It often shoots off from a mature plant to form another plant to the side. The new plant is called a sucker.

Stomata The breathing pores on a leaf.

Stratification A moist-chilling process that breaks the dormancy of seeds. Seeds planted outdoors go through this naturally. Collected seeds may be placed in moist peat moss in a plastic bag in the crisper of the refrigerator to duplicate nature (never with fruit, however, which gives off ethylene gas which may destroy the seeds).

Striations Lines of tissue forming streaks on older cane as its blooming capacity diminishes. Old cane is often removed to generate new growth.

Strike A grower's term for getting a cutting to root successfully.

Stub The portion of cane left above the crown. Cane should be cut even with the crown or bud union to avoid stubs, which may die back, become infected, or host insects.

Style The slender tube in the female portion of a flower that bears the sticky stigma which traps pollen (see **Flower**).

Sublateral A little branch off a larger branch, off a main cane. Sublaterals are often cut back on Climbing roses to initiate a new round of bloom.

Subshrub A woody plant that regenerates each year, after losing the tips of its canes in winter.

Subsoil The soil underneath the area normally cultivated for planting.

Substitute Many catalogs offer customers substitutes if the plant ordered is not available. Write "No substitutes" in bold print on your order if you don't want to receive any.

Sucker Can mean two things. A shoot coming up from the rootstock of a budded or grafted plant. Remove these immediately. Or, a plant produced to the side of a mother plant identical to the parent (most common with Shrub or Species roses).

Sunscald The effect of sun on a plant that is moved from an indoor location into bright light too quickly. Leaves often turn pale and may even drop off.

Superphosphate An inorganic material added to planting holes to provide phosphorus. Made by treating rock phosphate with sulfuric or phosphoric acid.

Synthetic Man-made, as opposed to organic (occurring naturally).

Syringe (see **Mist**)

Systemic A type of chemical absorbed directly by the plant and distributed throughout the tissue. There are systemic herbicides (to kill perennial weeds) and systemic pesticides (to kill bacteria, fungi, insects, and mites).

Take A grower's term for getting a bud or graft to grow successfully on rootstock.

Tamp To firm soil with your hands (not your feet).

Tender Lack of hardiness in cold climates.

Terminal knot (see **Cane head**)

Thinning The removal of weak or spindly cane at the base of the plant to create more vigorous growth in the remaining canes and to open the plant so that light can reach all areas of the bush to prevent disease.

Thorn A prickle found on most rose canes. It also occurs on the leaves and hips of some species. Roses vary widely in their thorniness.

Tissue culture New method of producing young plants from cells of the parent plant. Also known as micropropagation and in vitro propagation.

Topsoil The uppermost layer of soil containing lots of organic matter. It ideally consists of loam, commonly referred to as *black dirt*.

Trace elements Chemicals needed in extremely small amounts for the health of a rose. Often referred to as microelements.

Transpiration The exhalation of water from a rose. Heat and drying winds cause high water loss and may result in damage to roses.

Transplanting Moving a plant from one location to the next. Should be done as early in the season as possible. Keep soil around the roots, treat as a bare root plant, and water well. Not recommended for roses.

Tree rose A rose grown in the form of a tree. Generally, Tree roses are budded or grafted plants. Remove all suckers that appear on the "trunk" or main cane of the tree. Also known as a **Standard**.

Triacontonal A substance derived from alfalfa that stimulates roses into more active and vigorous growth. Discovered by a grower who interplanted tomatoes with alfalfa. The substance is known in liquid form as "alfalfa meal tea." Usually, it is sprayed onto the foliage for best results.

True Climber (see **Climbing rose**) A rose that naturally produces long canes which can be tied up to a support. A true Climber will not have the word "Climbing" in front of its name; for example, 'New Dawn.'

Truss Many flowers connected to a single stem by their pedicels.

Understock (see **Rootstock**)

Union (see **Bud union**)

Urea Potent nitrogen fertilizer (42-0-0) made from concentrated organic material or, now more commonly, through a synthetic process. Considered nonorganic. Available in solid and liquid form and best left in the hands of experienced or commercial growers. Requires a certain amount of nickel and bacteria in the soil not to be detrimental to plants. Leave it alone.

Variety Technically, any plant that occurs naturally in the wild as a variation from the original parent plant (species). However, almost everyone uses the term to mean any plant distinctly different from another plant whether bred or occurring naturally in the wild. Plants that are bred should be called "cultivars."

Vermiculite Mica heated until it pops. Used for starting plants from seed. Completely sterile and weed free. Holds moisture well.

Very double A flower with 26 to 40 petals (now called *full*). An extremely double flower has more than 41 petals (now called *very full*).

Viable Refers to seed that will grow. Seed viability varies by plant and by how seed is stored (best kept dry, cool, and dark).

Virus A disease-causing agent for which there is no cure. So small it can be seen only with an electron microscope.

Weed Any plant growing where you don't want it to. Weeds steal valuable nutrients and water from roses. They also act as hosts for destructive insects, many of which carry disease. Annual weeds produce seed and die within one year. Perennial weeds last for years, reproducing by both seed and tiny portions of root left in the soil. Kill perennial weeds with a systemic herbicide such as Roundup® before planting your rose garden. Kill annual weeds with an herbicide or simply pull them up by hand.

Weeping standard A Rambler or Polyantha budded or grafted onto tall rootstock. The branches hang down, giving the plant its name. Must be buried using the Minnesota Tip Method of winter protection.

Wetting agent Any substance added to spray to help it adhere to leaves. Should be added only if label indicates that this is okay. Some chemicals already contain a wetting agent.

Wild rose Same as a **Species** or naturally occurring rose. Not a rose bred by a person.

Wilting Leaves and cane hang limply. Usually, a sign of water stress. Water immediately. May also indicate disease or insect problem if watering does not bring the plant back to life. Never allow roses to wilt.

Windbreak Anything that provides protection from wind. Helpful for larger roses.

Winterkill The death of all or part of a bush because of severe winter conditions.

Witch's broom Lots of spindly branches off one bud or node on a cane.

Woody Referring to plants, those that form woody stems rather than shoots that die back at the end of the season (as do perennials).

Xeriscaping A fancy term for using as little water as possible in growing plants.

Xylem tubes Structures like plant arteries carrying water and food to all parts of the rose. Xylem makes up the woody portion of a plant.

INDEX

Key

Cl=Climber

F=Floribunda

Gr=Grandiflora

HT=Hybrid Tea

HMsk=Hybrid Musk

Min=Miniature

MinFl=Miniflora

OGR=Old Garden Rose

Pol=Polyantha

S=Shrub Rose

Sp=Species or Wild Rose

Tree=Tree

ABOUT THE AUTHORS

Richard Hass has grown more than 500 hundred roses in the last 40 years. His passion grew from his first-time experience growing three roses, one of which was 'Peace.' In its second year, that rose bloomed so profusely with such large blossoms that Hass was hooked. Each year he grows approximately 250 roses, adding new roses to his collection on a regular basis. Of these roses approximately half are winter protected using the Minnesota Tip Method. He is the volunteer coordinator for the Minnesota Rose Society in its work lifting, pruning, deadheading, and tipping hundreds of roses each year in the Wilson Rose Garden at the Minnesota Landscape Arboretum. In 1998 he received a bronze medal from the American Rose Society. Then, in 2009 he was honored as the outstanding consulting rosarian from the North Central District division of that society. He has been a rose judge since 2001. Rich Hass shares Jerry Olson's passion for roses. In this book both men share their vast experience to help you select the best roses on the market and tell you exactly how to grow them in a simple, easy-to-understand manner so that you can grow roses as beautiful as theirs.

Jerry Olson grew roses for nearly 60 years and judged them for the American Rose Society for over three decades. His knowledge comes from firsthand experience growing hundreds of varieties of roses in cold climates and from information passed on to him by the late Albert I. Nelson, who helped develop the Minnesota Tip Method of winter protecting roses.

He has won so many ribbons and trophies for exhibiting roses and so many awards for his service to the rose world that naming them all would offend this extremely modest man. However, a few stand out. He won the silver honor medal from the American Rose Society's North Central District for his work in promoting roses, the title of Outstanding Consulting Rosarian both from the American Rose Society and its North Central District, the title of outstanding judge in the North Central District, and virtually every award possible for rose exhibition.

He is presently a lifetime judge and lifetime consulting rosarian for the American Rose Society, which once honored him with the the title of Outstanding Consulting Rosarian in the United States. He is an honorary life member of the Minnesota Rose Society, Northstar Rose Society, and the Twin Cities Rose Club. He is also an honorary life member of the Minnesota State Horticultural Society and the Men's Garden Club of Minneapolis (Gardeners of America) as well as being a member of the Canadian Rose Society and Royal National Rose Society. He has written numerous articles on growing roses for the American Rose Society and Minnesota State Horticultural Society magazines and helped others do the same.

John Whitman, an award winning writer and photographer, has been writing non-fiction books for 45 years. He is an avid gardener with over 50 years of gardening experience. He was a grower at Bachman's, the largest retail florist and nursery in the United States. As with Jerry Olson, all of his gardening knowledge comes from hands-on experience. His book, *Starting from Scratch: A Guide to Indoor Gardening*, was published by Quadrangle: The New York Times Book Company and was chosen as a main selection of the Organic Gardening Book Club and an alternate selection of the Book-of-the-Month Club. He was one of seven contributing writers to the *Better Homes and Gardens New Garden Book* and the sole writer of the *Better Homes and Gardens New Houseplants Book*. He is co-author of *Growing Perennials in Cold Climates* and *Growing Shrubs and Small Trees in Cold Climates*. He is the creator of the cold-climate series of books, which have received national recognition as Landmark books in their field.